Gel Electrophoresis of Proteins:

a practical approach

Edited by:

B. D. Hames

Department of Biochemistry,
University of Leeds, Leeds, England

D. Rickwood

Department of Biology, University of Essex,
Colchester, Essex, England

ISBN 0 904147 22 3

Published by IRL Press Limited, London and Washington DC

IRL Press Limited,
1 Falconberg Court,
London W1V 5FG,
England.

British Library Cataloguing in Publication Data
Gel electrophoresis of proteins.
  1. Proteins - Laboratory manuals
  2. Electrophoresis - Laboratory manuals
  I. Hames,B.D. II. Rickwood,D.
  574. 19'245 QD431.5

ISBN 0-904147-22-3

Cover photograph (SDS-polyacrylamide 10-15% linear gradient slab gel analysis of soybean seed proteins) by courtesy of Susan Marcus and Dianna Bowles.

Printed in England by Information Printing Limited, Eynsham.

Preface

The development of separation methods has played a significant role in the elucidation of biological systems. Of the various techniques in common use, one of the most important is gel electrophoresis. This book and its companion volume (see back cover for contents) are designed to provide details of gel electrophoretic procedures for the separation of macromolecules. The main emphasis of each book is on the practical aspects of the electrophoretic techniques in current use. Several revisions of some chapters were necessary in order to prevent undue repetition whilst including important practical topics and we thank the authors concerned and particularly the publishers for their patience and understanding during this exercise. Thanks are also due to Irene Hames for her skillful proofreading of the text at all stages.

B.D.Hames and D.Rickwood

Contributors

Birgit An der Lan
Bureau of Biologics, Division of Bacterial Products, Food and Drug Administration, Bethesda, MD 20205, USA.

Andreas Chrambach
Endocrinology and Reproduction Research Branch, National Institute of Child Health and Human Development, National Institutes of Health, Bethesda, MD 20205, USA.

B. David Hames
Department of Biochemistry, University of Leeds, Leeds LS2 9JT, England.

Richard D. Jurd
Department of Biology, University of Essex, Wivenhoe Park, Colchester, CO4 3SQ, Essex, England.

Nga Y. Nguyen
Endocrinology and Reproduction Research Branch, National Institute of Child Health and Human Development, National Institutes of Health, Bethesda, MD 20205, USA.

David Rickwood
Department of Biology, University of Essex, Wivenhoe Park, Colchester, CO4 3SQ, Essex, England.

David Rodbard
Endocrinology and Reproduction Research Branch, National Institute of Child Health and Human Development, National Institutes of Health, Bethesda, MD 20205, USA.

John Sinclair
Department of Genetics, School of Biological Sciences, University of Sussex, Falmer, Brighton BN1 9QG, England.

Contents

IMPORTANT NOTICE

HEALTH HAZARDS OF GEL ELECTROPHORESIS

(i) A number of chemicals commonly used for gel electrophoresis are toxic whilst the status of others remains unknown. It is very important that experimenters acquaint themselves with the precautions required for handling all chemicals mentioned in this text. Particular care should be taken when handling acrylamide since this is a known potent neurotoxin. Polyacrylamide gel is not toxic unless it contains unpolymerised monomer.

(ii) Care should be taken when using gel electrophoresis apparatus that no electrical safety hazard exists. Particular care should be taken when using apparatus not obtained from commercial sources since this may not meet the usual required safety standards. It is recommended that all apparatus is checked by a competent electrician before use.

Abbreviations

A	amps
ACES	N-2-acetamido-2-aminoethanesulphonic acid
AEPD	2-amino,2-ethyl,1,3-propanediol ($\equiv$ Ammediol)
AMP	2-amino-2-methyl-propanol
ANS	1-anilino-8-naphthalene sulphonate
BAC	N,N'-bisacrylylcystamine
BES	N,N-bis(2-hydroxyethyl)-2-aminoethanesulphonic acid
Bicine	N,N-bis(2-hydroxyethyl)glycine
Bisacrylamide	N,N'-methylene bisacrylamide
Bistris	[bis(2-hydroxyethyl)-amino]tris(hydroxymethyl)methane
Bistrispropan	1,3-bis[tris(hydroxymethyl)methylamino]propane
% C	percentage crosslinker (as a percentage of the total monomer)
% C_{Bis}	percentage bisacrylamide crosslinker
% C_{DATD}	percentage DATD crosslinker
CTAB	cetyltrimethylammonium bromide
CZE	continuous zone electrophoresis (zone electrophoresis using a continuous buffer system)
DATD	N,N'-diallyltartardiamide
D.C.	direct current
DMAPN	3-dimethylamino-propionitrile
DMSO	dimethylsulphoxide
EDTA	ethylenediaminetetra-acetate
EF	electrofocusing
EPPS	N-2-hydroxyethylpiperazine-N'-3-propanesulphonic acid
g	gram(me)
xg	centrifugal force (x unit gravitational field)
GABA	γ-aminobutyric acid
GACA	γ-aminocaproic acid
HEPES	N-2-hydroxyethylpiperazine-N'-2-ethanesulphonic acid
I.D.	internal diameter
ITP	isotachophoresis
K_R	retardation coefficient (slope of the Ferguson plot); a measure of molecular size
M	molarity
M	mobility
M_o	free electrophoretic mobility ($cm^2/s/V$)
mA	milliamps
MDPF	2-methoxy-2,4-diphenyl-3(2H)-furanone
MES	2-(N-morpholino)ethanesulphonic acid
MOPS	3-(N-morpholino)propanesulphonic acid
MTT	methyl thiazolyl tetrazolium
MW	molecular weight
MZE	multiphasic zone electrophoresis (zone electrophoresis using a multiphasic buffer system)

NBT	nitroblue tetrazolium
O.D.	outside (external) diameter
PAGE	polyacrylamide gel electrophoresis
PCA	perchloric acid
pI	isoelectric point
pI$'$	apparent isoelectric point
PITC	phenylisothiocyanate
pK	$-$ log dissociation constant (the pH at half dissociation)
PMS	phenazine methosulphate
PMSF	phenylmethylsulphonyl fluoride
POPOP	1,4-bis[2-(5-phenyloxazolyl)]benzene
PPO	2,5-diphenyloxazole
$\bar{R}$	geometric mean radius
R_f	relative electrophoretic mobility (e.g. relative to a dye front or to a moving boundary 'front')
SDS	sodium dodecyl sulphate (sodium lauryl sulphate)
SDS-PAGE	polyacrylamide gel electrophoresis in the presence of SDS
SSS	steady-state stacking
%T	polyacrylamide gel concentration defined as percentage total monomers (i.e. acrylamide + crosslinking agent, g/100 ml)
T_{max}	gel concentration for maximum separation between two proteins
T_{opt}	gel concentration for maximum resolution between two proteins
TAPS	3-{[tris(hydroxymethyl)methyl]amino}propanesulphonic acid
TCA	trichloroacetic acid
TEMED	N,N,N',N'-tetramethylethylenediamine
TES	2-{[tris(hydroxymethyl)methyl]amino}ethanesulphonic acid
Tricine	N-[tris(hydroxymethyl)methyl]glycine
U.V.	ultraviolet
V	volts
V	molecular valence (net protons/molecule)
W	watt
Y_o	y intercept on the Ferguson plot; a measure of molecular net charge

CHAPTER 1

An Introduction to Polyacrylamide Gel Electrophoresis

B.DAVID HAMES

INTRODUCTION

Any charged ion or group will migrate when placed in an electric field. Since proteins carry a net charge at any pH other than their isoelectric point, they too will migrate and their rate of migration will depend upon the charge density (the ratio of charge to mass) of the proteins concerned; the higher the ratio of charge to mass the faster the molecule will migrate. The application of an electric field to a protein mixture in solution will therefore result in different proteins migrating at different rates towards one of the electrodes. However, since all proteins were originally present throughout the whole solution, the separation achieved is minimal. Zone electrophoresis is a modification of this procedure whereby the mixture of molecules to be separated is placed as a narrow zone or band at a suitable distance from the electrodes such that, during electrophoresis, proteins of different mobilities travel as discrete zones which gradually separate from each other as electrophoresis proceeds. In theory, separation of different proteins as discrete zones is therefore readily achieved provided their relative mobilities are sufficiently different and the distance allowed for migration is sufficiently large. However, in practice there are disadvantages to zone electro-phoresis in free solution. Firstly, any heating effects caused by electrophoresis can result in convective disturbance of the liquid column and disruption of the separating protein zones. Secondly, the effect of diffusion is to constantly broaden the protein zones and this continues after electrophoresis has been terminated. To minimise these effects, zone electrophoresis of proteins is rarely carried out in free solution but in-stead is performed in a solution stabilised within a supporting medium. As well as reducing the deleterious effects of convection and diffusion during electrophoresis, the supporting medium allows the investigator to fix the separated proteins at their final positions immediately after electrophoresis and thus avoid the loss of resolution which results from post-electrophoretic diffusion. The fixation process employed varies with the supporting medium chosen.

Many supporting media are in current use, the most popular being sheets of paper or cellulose acetate, materials such as silica gel, alumina, or cellulose which are spread as a thin layer on glass or plastic plates, and gels of agarose, starch, or polyacryl-amide. These media fall into two main classes. Paper, cellulose acetate, and thin-layer materials are relatively inert and serve mainly for support and to minimise con-vection. Hence separation of proteins using these materials is based largely upon the charge density of the proteins at the pH selected, as with electrophoresis in free solution. In contrast, the various gels not only prevent convection and minimise diffusion but in some cases they also actively participate in the separation process by interacting with the migrating particles. These gels can be considered as porous media in which the pore size is the same order as the size of the protein molecules such that a molecular sieving effect occurs and the separation is dependent on both charge density and size. Thus two proteins of different sizes but identical charge densities would probably not be well separated by paper electrophoresis, whereas, provided the size difference is large enough, they could be separated by polyacrylamide gel electrophoresis since the molecular sieving effect would slow down the migration rate of the larger protein relative to that of the smaller protein.

The extent of molecular sieving depends on how close the gel pore size approximates the size of the migrating particle. The pore size of agarose gels is sufficiently large that molecular sieving of most protein molecules is minimal and separation is based mainly on charge density. In contrast, starch and polyacrylamide gels have pores of the same order of size as protein molecules and so these do contribute a molecular sieving effect. However, the success of starch gel electrophoresis is highly dependent on the quality of the starch gel itself, which, being prepared from a biological product, is not reproducibly good and may contain contaminants which can adversely affect the quality of the results obtained. On the other hand, polyacrylamide gel, as a synthetic polymer of acrylamide monomer, can always be prepared from highly purified reagents in a reproducible manner provided that the polymerisation conditions are standardised. The basic components for the polymerisation reaction are commercially available at reasonable cost and high purity although for some purposes extra purification may be required. In addition, polyacrylamide gel has the advantages of being chemically inert, stable over a wide range of pH, temperature, and ionic strength, and is transparent. Finally, polyacrylamide is better suited to a size fractionation of proteins since gels with a wide range of pore sizes can be readily made whereas the range of pore sizes obtainable with starch gels is strictly limited. For these and other reasons, polyacrylamide gels have become the medium of choice for zone electrophoresis of most proteins although starch gels have been widely used for the analysis of isoenzymes. Starch gel electrophoresis has been reviewed by Gordon (1) and Smith (2). Agarose gels are used for the fractionation of molecules or complexes larger than can be handled by polyacrylamide gels, especially certain nucleic acids and nucleoproteins. In addition, agarose is widely used in immuno-electrophoresis where zone electrophoresis of proteins is coupled to immunological detection and quantitation (Chapter 7).

This chapter is concerned with the basic techniques of analytical zone electrophoresis of proteins in polyacrylamide gels plus modifications which allow small-scale preparations of proteins of interest. A more advanced text dealing with detailed quantitative approaches to analytical zone electrophoresis (including the determination of optimum gel pore size for maximum separation of two proteins) plus special techniques for large-scale preparation of proteins by zone electrophoresis is given in Chapter 2.

PROPERTIES OF POLYACRYLAMIDE GEL

Chemical Structure

Polyacrylamide gel results from the polymerisation of acrylamide monomer into long chains and the crosslinking of these by bifunctional compounds such as N,N'-methylene bisacrylamide (usually abbreviated to bisacrylamide) reacting with free functional groups at chain termini. Other crosslinking reagents have also been used to impart particular solubilisation characteristics to the gel for special purposes (p. 58). The structure of the monomers and the final gel structure are shown in *Figure 1*.

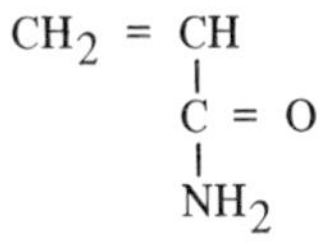

```
    CH2 = CH                          CH2 = CH
         |                                 |
         C = O                             C = O
         |                                 |
         NH2                               NH
                                           |
                                           CH2
                                           |
    Acrylamide                             NH
                                           |
                                           C = O
                                           |
                                      CH2 = CH

                               N,N'—methylene bisacrylamide
```

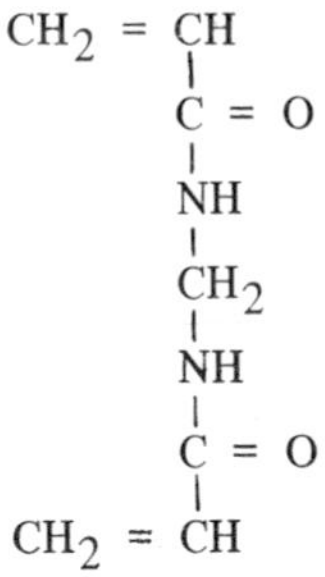

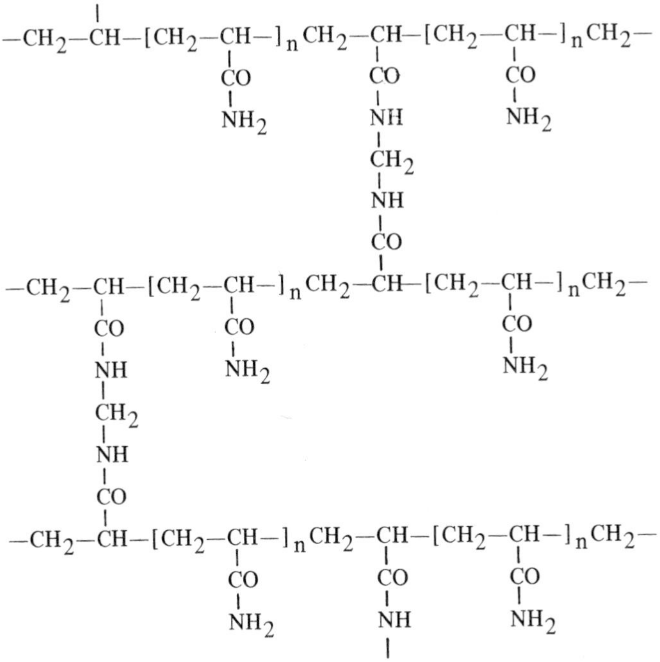

```
        |
 —CH2—CH—[CH2—CH—]n CH2—CH—[CH2—CH—]n CH2—
          |             |          |
          CO            CO         CO
          |             |          |
          NH2           NH         NH2
                        |
                        CH2
                        |
                        NH
                        |
                        CO
                        |
 —CH2—CH—[CH2—CH—]n CH2—CH—[CH2—CH—]n CH2—
    |         |                     |
    CO        CO                    CO
    |         |                     |
    NH        NH2                   NH2
    |
    CH2
    |
    NH
    |
    CO
    |
 —CH2—CH—[CH2—CH—]n CH2—CH—[CH2—CH—]n CH2—
              |         |         |
              CO        CO        CO
              |         |         |
              NH2       NH        NH2
                        |

                   Polyacrylamide gel
```

Figure 1. The chemical structure of acrylamide, N,N'-methylene bisacrylamide, and polyacrylamide gel.

Polymerisation Catalysts

Polymerisation of acrylamide is initiated by the addition of either ammonium per-sulphate or riboflavin. In addition, *N,N,N',N'*-tetramethylethylenediamine (TEM-ED) or, less commonly, 3-dimethylamino-propionitrile (DMAPN) are added as accelerators of the polymerisation process.

In the ammonium persulphate-TEMED system, TEMED catalyses the formation of free radicals from persulphate and these in turn initiate polymerisation. Since the free base of TEMED is required, polymerisation may be delayed or even prevented at low pH. Increases in either the TEMED or ammonium persulphate concentration increase the rate of polymerisation.

In contrast to chemical polymerisation with persulphate, the use of the riboflavin -TEMED system requires light to initiate polymerisation. This causes photo-decomposition of riboflavin and production of the necessary free radicals. Although gelation occurs when solutions containing only acrylamide and riboflavin are irradiated, TEMED is usually also included since under certain conditions polymerisation occurs more reliably in its presence.

Oxygen inhibits polymerisation and so gel mixtures are usually degassed prior to use.

Effective Pore Size

The effective pore size of polyacrylamide gels is greatly influenced by the total acrylamide concentration in the polymerisation mixture, effective pore size decreasing as acrylamide concentration increases. Gels with concentrations of acrylamide less than about 2.5%, which are necessary for the molecular sieving of molecules above a molecular weight of 10^6, are almost fluid but this can be remedied by the inclusion of 0.5% agarose (p.79). At the other extreme, polyacrylamide gels will form at over 30% acrylamide at which concentration polypeptides with a molecular weight as low as 2,000 experience considerable molecular sieving. As one might expect, the choice of acrylamide concentration is critical for optimal separation of protein components by zone electrophoresis and will be considered in more detail later (p.12).

For any given total monomer concentration, the effective pore size, stiffness, brittleness, light scattering, and swelling properties of the polyacrylamide gel vary with the proportion of crosslinker used. Polymerisation in the absence of crosslinker leads to the formation of random polymer chains resulting only in a viscous solution. When bisacrylamide is included in the polymerisation mixture, gelation occurs with random polymer chains crosslinked at intervals to form a covalent meshwork. As the proportion of crosslinker is increased, the pore size decreases, reaching a minimum when the bisacrylamide represents about 5% of the total monomer (i.e. $C_{Bis} = 5\%$). With higher proportions of bisacrylamide the polymer chains become crosslinked into increasingly large bundles with large spaces between them and so the effective pore size increases again. This dependence of pore size upon proportion of crosslinker occurs irrespective of the total monomer concentration used, and so a gel containing a greater total monomer concentration can have a larger effective pore size than one of lower concentration if the proportion of bisacrylamide is either below or above the optimum value of 5%.

EXPERIMENTAL APPROACH

Before the researcher concerns himself with the detailed methodology of zone electrophoresis in polyacrylamide gels, it is worthwhile considering the general strategy for the separation to be attempted. In practice, he must decide upon the physical form the gel should best take (gel rods or slabs), whether to use a dissociating or non-dissociating buffer system (and if this should be a continuous or discontinuous system), what pH and buffer ionic strength to use for the separation, and finally what gel concentration would be most appropriate for the sample to be fractionated. The answers to some of these questions will be immediately obvious once one considers the information desired and that obtainable by the methods available. Other questions will require preliminary experimentation in order that they be answered satisfactorily.

Rod or Slab Gels

Originally analytical zone electrophoresis in polyacrylamide made use of cylindrical rod gels in glass tubes but now flat slab gels, 0.75-1.5 mm thick, are usually preferred instead. One of the most important advantages of slab gels is that many samples, including molecular weight marker proteins, can be electrophoresed under identical conditions in a single gel such that the band patterns produced are directly comparable (*Figure 2*). In contrast, due to minor differences in polymerisation efficiency, gel length and diameter, etc., rod gels even of the same sample are rarely identical. Additional advantages of slab gels are:

(i) any heat produced during electrophoresis is more easily dissipated by the standard slab gel than the thicker rod gels usually used, thus reducing distortion of protein bands due to heating effects;

(ii) their rectangular cross-section allows densitometry and photography with less risk of optical artifacts and they can be easily dried for storage or autoradiography;

(iii) less time is required for the preparation of gels for a large number of samples to be electrophoresed under identical conditions. Up to 25 samples can be easily accommodated on a standard-size slab gel.

Given so many advantages one might ask whether there is still a requirement for rod gels for analytical separations. There are several situations where rod gels will continue to be used. Firstly there are those instances where the investigator wishes to slice the gel after electrophoresis and either determine the radioactivity present in each slice (for radioactive proteins) or elute and assay proteins of interest by their biological activity. Although both rod and slab gels can be sliced, the normal dimensions of the gels used mean that the former have greater sample capacity and may be preferred for this reason. Furthermore, if the investigator wishes to use an automatic fractionation device he may choose rod gels since many of these devices were designed for rod gels. Secondly, rod gels are often the preferred format when determining the optimum pH or gel concentration for separation of protein components since a large number of conditions can be tested with minimum effort, especially using a rod

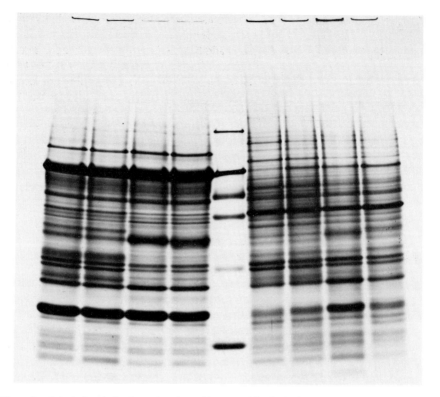

Figure 2. A typical analysis of sample polypeptide composition by SDS-PAGE using the slab gel format. Soluble proteins from the lumen of different regions of rat epididymis were fractionated using the SDS-discontinuous buffer system in a 10-15% linear gradient gel run at 20mA for 5h at room temperature. The track in the centre contained the following molecular weight marker polypeptides; β-galactosidase, bovine serum albumin, γ-globulin heavy chain, ovalbumin, γ-globulin light chain, and cytochrome *c*. Polypeptides were visualised after electrophoresis by staining with Coomassie blue R250. (Photograph by courtesy of Dr D.Brooks).

gel apparatus modified for this purpose (see Chapter 2). Finally, increasing use is being made of two-dimensional electrophoretic techniques whereby the protein mixture is separated in the first dimension in a rod gel and then this is attached along one edge of a slab gel for electrophoretic separation of the components in the second dimension. This high-resolution method is described in detail in Chapter 5.

Dissociating or Non-dissociating Buffer System

The vast majority of studies employing zone electrophoresis of proteins in polyacrylamide gel use a buffer system designed to dissociate all proteins into their individual polypeptide subunits. The most common dissociating agent used is the ionic detergent, sodium dodecyl sulphate (SDS). The protein mixture is denatured by heating at 100°C in the presence of excess SDS and a thiol reagent (to cleave disulphide bonds). Under these conditions, most polypeptides bind SDS in a constant weight ratio (1.4g SDS per gram of polypeptide). The intrinsic charges of the polypeptide are insignificant compared to the negative charges provided by the bound detergent, so that the SDS-polypeptide complexes have essentially identical charge densities and migrate in polyacrylamide gels of the correct porosity strictly according to polypeptide size (3,4). Thus, in addition to analysing the polypeptide composition of the sample, the investigator can determine the molecular weight of the sample polypeptides by reference to the mobility of polypeptides of known molecular weight under the same electrophoretic conditions (p. 15). The simplicity and speed of the method, plus the fact that only microgram amounts of sample proteins are required, have made SDS-polyacrylamide gel electrophoresis (SDS-PAGE) the most widely used method for determination of the complexity and molecular weights of constituent polypeptides in a protein sample. Proteins from almost any source are readily solubilised by SDS so that the method is generally applicable.

Urea has also been used as a dissociating agent and works by disrupting hydrogen bonds. High urea concentrations (~ 8 M) are necessary, a thiol reagent is also required for complete denaturation of proteins containing disulphide bonds, and urea must be present during electrophoresis to maintain the denatured state. The advantage of urea for some applications is that it does not affect the intrinsic charge of proteins and so separation of the constituent polypeptides will be on the basis of both size *and* charge, in contrast to the use of SDS. One disadvantage is that this combined size and charge fractionation prevents accurate molecular weight determinations. Furthermore, urea is not as good as SDS in dissociating proteins; up to 50% of a complex protein mixture may fail to enter the gel whereas at least 90% of even crude cell lysates will enter the gel if SDS is the dissociating agent used. However some proteins require the presence of both urea and SDS if most of the material is to enter the gel (pp. 66-70).

In contrast to the above systems, zone electrophoresis of native proteins under non-dissociating buffer conditions is designed to fractionate a protein mixture in such a way that subunit interaction, native protein conformation, and biological activity are preserved. Separation of the native proteins occurs on the basis of both size *and* charge. Further details of the types of non-dissociating buffer systems available are given in the next section and later in this chapter.

Continuous or Discontinuous (Multiphasic) Buffer System

Zone electrophoretic systems in which the same buffer ions are present throughout the sample, gel, and electrode vessel reservoirs (albeit possibly at different concentrations in each) at constant pH, are referred to as continuous buffer systems. In these systems the protein sample is loaded directly onto the gel in which the separation will occur, the resolving gel (*Figure 3a,b*) which has pores sufficiently small to cause a size fractionation of the sample components during electrophoresis. In contrast, discontinuous (or multiphasic) buffer systems employ different buffer ions in the gel compared to those in the electrode reservoirs. Most discontinuous buffer systems have discontinuities of both buffer composition and pH. In these, the sample is loaded onto a large-pore 'stacking' gel polymerised on top of the small-pore resolving gel (*Figure 3c,d*).

The major advantage of these discontinuous buffer systems over continuous buffer systems is that relatively large volumes of dilute protein samples can be applied to

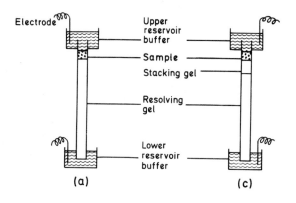

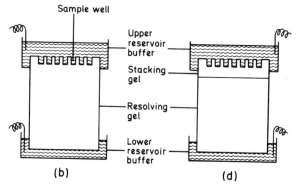

Figure 3. Use of continuous and discontinuous buffer systems with rod and slab gels. (a) continuous buffer system used in conjunction with a rod gel; the sample is loaded directly onto the resolving gel. (b) continuous buffer system used in conjunction with a slab gel; samples are loaded into wells formed directly in the resolving gel. (c) and (d), discontinuous buffer system used in conjunction with rod and slab gels, respectively; samples are loaded directly onto the stacking gel or into wells formed in the stacking gel, respectively.

the gels but good resolution of sample components can still be obtained. The reason for this is that the proteins are concentrated into extremely narrow zones (or stacks) during migration through the large-pore stacking gel prior to their separation during electrophoresis in the small-pore resolving gel. The production of thin protein starting zones by discontinuous systems can be understood by consideration of the original discontinuous system of Ornstein (5) and Davis (6) as an example.

Consider a weak acid, such as glycine, at a pH near its pK_a. Only part of the population of molecules will be negatively charged at any one time. If x is the proportion of the total molecules which is dissociated, and hence present as negatively charged ions, then each molecule can be regarded as being charged for x proportion of the time and uncharged the rest of the time. Hence, if the mobility of the charged species is M, the effective mobility = Mx and the velocity of migration = V. Mx (where V is the voltage gradient). Therefore, an ion of lower mobility can migrate as fast as one with higher mobility if the products of voltage and effective mobility are equal. Now, in the Ornstein-Davis system, the sample itself plus the stacking gel con-

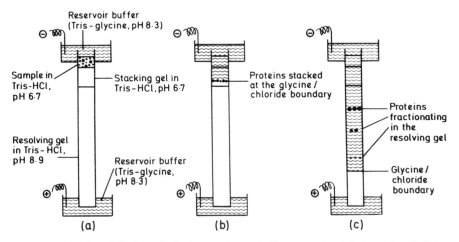

Figure 4. Operation of the Ornstein-Davis discontinuous buffer system. (a) at the beginning of electrophoresis, (b) during stacking, (c) during separation in the resolving gel. For detailed explanation, see the text.

tain Tris-HCl buffer whilst the upper electrode reservoir contains Tris-glycine buffer (*Figure 4a*). At the pH of the sample and stacking gel (pH 6.7), glycine is very poorly dissociated so that its effective mobility is low. Chloride ions have a much higher mobility at this pH whilst the mobilities of proteins are intermediate between that of chloride and glycine. The moment the voltage is applied, the chloride ions (the leading ions) migrate away from the glycine (the trailing ions) leaving behind a zone of lower conductivity. Since conductivity is inversely proportional to field strength, this zone attains a higher voltage gradient which now accelerates the glycine so that it keeps up with the chloride ions. A steady-state is established where the products of mobility and voltage gradient for glycine and chloride are equal, these charged species now moving at the same velocity with a sharp boundary between them. As this glycine/chloride boundary moves through the sample and then the stacking gel, a low-voltage gradient moves before the moving boundary and a high-voltage gradient after it. Any proteins in front of the moving boundary are rapidly overtaken since they have a lower velocity than the chloride ions. Behind the moving boundary, in the higher voltage gradient, the proteins have a higher velocity than glycine. Thus the moving boundary sweeps up the proteins so that they become concentrated into very thin zones or 'stacks', one stacked upon the other in order of decreasing mobility, with the last protein followed immediately by glycine (*Figure 4b*). The concentration of protein in the moving boundary depends only on the concentration of Tris-HCl in the sample and the stacking gel and not on the initial concentration of proteins in the sample. The thickness of the steady-state protein stack is independent of the initial concentration of protein in the sample and dependent only on the total amount of protein loaded onto the gel. At the protein loads typical of analytical polyacrylamide gel electrophoresis, the protein stacks are only a few microns thick. Since the stacking gel is a large-pore gel, no molecular sieving occurs at this stage.

When the moving boundary reaches the interface of the stacking and resolving gels, the pH value of the gel increases markedly and this leads to a large increase in the degree of dissociation of the glycine. Therefore the effective mobility of the glycine increases so that glycine overtakes the proteins and now migrates directly behind the chloride ions. At the same time, the gel pore size decreases markedly, retarding the migration of the proteins by molecular sieving. These two effects cause the proteins to be unstacked. The proteins now move in a zone of uniform voltage gradient and pH value (now Tris-glycine, pH 9.5, instead of the original resolving gel buffer of Tris-HCl, pH 8.9) and are separated according to their intrinsic charge and size, the latter depending on the molecular sieving effect of the small-pore resolving gel (*Figure 4c*).

Because of the high resolution obtainable with discontinuous buffer systems, the SDS-discontinuous system (a discontinuous buffer system with SDS added to all buffers) is usually the system of choice for high-resolution fractionation of protein mixtures under dissociating conditions. The most commonly used SDS-discontinuous system is that originally described by Laemmli (7) consisting of the discontinuous Ornstein-Davis buffer system with SDS present (see p. 26 for details). Indeed slab gels are now used almost exclusively with this Laemmli SDS-discontinuous system. Neville (8) has described an alternative SDS-discontinuous system based on Tris-borate buffer. Rod gels are often used with either the Laemmli SDS-discontinuous system or a continuous buffer system, usually sodium phosphate buffer containing SDS. The merits of the SDS-phosphate system are its simplicity and the fact that it is less susceptible than the Laemmli system to artifacts caused by contaminants in some commercial sources of SDS (p. 24). Nevertheless the superior resolving power of the Laemmli SDS-discontinuous buffer system means that this will undoubtedly continue to be the main dissociating buffer system used, especially for complex protein mixtures.

Unfortunately there is no universal buffer system ideal for electrophoretic separations of native proteins. The choice must be based upon the conditions required to maintain activity of the proteins of interest whilst achieving sufficient resolution of the protein components for the problem under investigation. Until recently only three discontinuous buffer systems had been described for the separation of native proteins, the Ornstein-Davis high pH system (see above) resolving proteins at pH 9.5 at 25°C, a 'neutral' pH system resolving proteins at pH 8.0 (9), and a low pH system resolving proteins at pH 3.6 (10). Details are given later (p.28). However, based on a knowledge of pK and ionic mobility data of buffer constituents and the theory of discontinuous (multiphasic) zone electrophoresis, several thousand discontinuous buffer systems have now been designed for use at any pH in the useful range of pH 2.5-11.0 and are available as a computer output. A description of the computer output and its use is given in Chapter 2.

Some native proteins aggregate and may precipitate at the very high protein concentrations reached in the sharply stacked zones of discontinuous buffer systems and then either fail to enter the resolving gel or cause 'streaking'. This phenomenon occurs when aggregated protein accumulates at the gel surface and then slowly dissolves during electrophoresis, causing protein streaks running parallel to the direction of migration. The problem can sometimes be overcome by using a continuous buffer system since this avoids the concentration of sample proteins to such an extent that they precipitate. Fortunately the use of a continuous buffer system can still give good resolution provided that certain conditions are met. Firstly, the sample must be applied in as small a volume as possible to give a thin starting zone. Depending on the method used to detect the separated protein zones after electrophoresis this usually requires that a concentrated solution of sample proteins is available (about 1mg/ml). Further zone-sharpening occurs as the proteins enter the single resolving gel in which the mobilities of the proteins in question are considerably less than in free solution. Secondly, additional zone-sharpening can be obtained by loading the protein sample in a buffer which has a lower ionic strength (say $^1/5$th to $^1/10$th) than that of the gel and electrode buffer. The proteins will initially be in a zone of lower ionic strength (lower conductivity) and hence higher voltage and so move still faster in free solution, slowing down as they move into the gel as a result of the sieving effect produced by the gel and the drop in voltage gradient as they enter the more concentrated gel buffer. Virtually any buffer can be used for electrophoresis of native proteins in a continuous buffer format and so for most proteins it is a matter of experimentation to determine which buffer is most appropriate. Some common non-dissociating buffers are given later (p. 26). However, certain classes of proteins such as histones, nuclear non-histone proteins, ribosomal proteins, and membrane proteins are not soluble in the usual non-dissociating buffers so that to analyse these on a charge and size basis requires additional agents such as urea, chloral hydrate, or non-ionic detergents. These are discussed on pages 65-71.

Choice of pH

Polyacrylamide gel electrophoresis can be carried out at a pH anywhere between 2.5 and 11, but in practice the limits are pH 3 and 10 since some hydrolytic reactions (such as deamidation) occur at the extremes of pH.

In SDS-PAGE the SDS-polypeptide complexes are negatively charged over a wide range of pH such that the pH of the SDS-phosphate continuous buffer system is not critical. The pH of the SDS-discontinuous buffer system is important in such separations only in that it permits sample concentration via the stacking phenomenon to occur. In contrast, pH *is* critical in polyacrylamide gel electrophoresis of proteins in non-dissociating buffers where native proteins are separating on the basis of size and charge density. Here changes in the pH alter the net charge of the protein components and hence the separation which can be achieved.

In choosing the pH of a buffer to be used in a continuous buffer system for native proteins, the initial consideration must be the pH range over which the proteins of interest are stable. This pH range may be narrower if one wishes the proteins to retain biological activity (either for detection purposes or because the polyacrylamide gel electrophoretic step is being used preparatively) than if detection is on the basis of standard protein stains where all that is required is to prevent dissociation of the native protein into its substituent subunits. Within this pH range, the selection of pH is a compromise between two opposing considerations. The further the pH of the electrophoresis buffer from the isoelectric points of the proteins to be separated, the higher the charge on the proteins. This leads to shorter times required for electrophoretic separation and hence reduced band spreading due to diffusion. On the other

11

hand, the closer the pH to the isoelectric points of the proteins the greater the charge differences between proteins, thus increasing the chance of separation. Many proteins have isoelectric points in the range pH 4-7 so a common compromise is to use buffers in the pH 8.0-9.5 region. Ideally a systematic study should be carried out in which the pH is progressively adjusted nearer the isoelectric points of the proteins until a pH is found which yields optimal resolution and separation for the protein mixture. A similar study can be made with basic proteins which need to be separated at acid pH.

Choice of Polymerisation Catalyst

For most situations the choice between the ammonium persulphate-TEMED and riboflavin-TEMED systems is a matter of personal preference, although ammonium persulphate is the usual initiator for resolving gels. Stacking gels of discontinuous buffer systems are frequently polymerised with either ammonium persulphate or riboflavin. If native proteins are being separated which are particularly sensitive to persulphate ions and yet these have been used to polymerise the resolving gel in a continuous buffer system, excess persulphate ions may need to be removed by a period of pre-electrophoresis prior to loading the sample. This is not possible for discontinuous buffer systems and so in these cases it would be wise to polymerise at least the large-pore stacking gel with riboflavin.

The other advantage of using riboflavin as catalyst is that polymerisation will not begin until the gel mixture is illuminated. This is useful if there is likely to be some delay in overlayering the mixture with buffer, or adjusting the sample comb in slab gel electrophoresis, or if multiple gels are being made. If riboflavin is used as catalyst in the resolving gel, the time period of illumination should be standardised since this affects the gel porosity and hence protein mobility.

Choice of Gel Concentration

Separation of protein bands via zone electrophoresis in polyacrylamide gels refers to the distance between bands irrespective of band width, whereas resolution refers to separation relative to band width. Resolution between two components is influenced by all the factors that affect band sharpness. Thus the use of small amounts of sample proteins to prevent overloading, the use of a discontinuous buffer system with dilute protein samples, and removal of high concentrations of ions from samples will all maximise resolution. The major factors affecting separation are the pH of the gel (in polyacrylamide gel electrophoresis of native proteins but not in SDS-PAGE; p. 11) and the gel concentration. At the extremes, choice of an incorrect gel concentration can lead to total exclusion of the proteins which would be unable to penetrate the gel, or conversely, lack of fractionation with the proteins running with the buffer front. Between these two extremes the proteins will be separated to variable extents depending on the gel concentration. Although there is a single gel concentration which is optimal for the resolution of any two proteins (see below) there can be no gel concentration which will give maximum separation of all the components from each other in a complex protein mixture. Therefore the gel concentration chosen for fractionating a heterogeneous mixture is usually one which gives an adequate display of all the components of interest. A reasonable approach for initial analysis of a protein mixture using a non-dissociating buffer system would be to start with a 7.5% acrylamide gel and then attempt a number of gel concentrations between 5% and 15% and choose the most desirable concentration for further studies. A similar strategy can be attempted for use with SDS-PAGE unless the molecular weight range of the polypeptide mixture is known, when a suitable gel concentration can be selected from existing calibration curves of polypeptide molecular weight versus mobility (see p. 16, *Figure 6*). An alternative approach, which is strongly recommended for initial analysis of protein mixtures by SDS-PAGE, is to use a concentration gradient polyacrylamide gel in which the concentration of acrylamide increases (and hence pore size decreases) in the direction of protein migration. Gradient gels have two considerable advantages over uniform concentration gels in the analysis of complex protein mixtures. Firstly, they are able to fractionate proteins over a wider range of molecular weights than any uniform concentration gel. Secondly, the gradient in pore size causes significant sharpening of protein bands during migration. The result is that gradient gels are unsurpassed in the resolution of protein mixtures covering a wide range of molecular weights where a display of all the components on one gel is desired. A useful gradient gel for initial SDS-PAGE analysis is a 5-20% or 6-18% linear gradient slab gel. Details of gradient gel preparation and use are given later (p. 71).

If the aim of polyacrylamide gel electrophoresis is to obtain optimal resolution of any two proteins, rather than to simply display the protein components of interest, this will only be achieved using the optimal concentration of acrylamide in a uniform concentration polyacrylamide gel. The gel concentration required will depend on the size and charge of the proteins under study and can be determined by measuring the mobility of each protein in a series of gels of different acrylamide concentration and then constructing a Ferguson plot (11) for each protein of interest, that is a plot of $\log_{10}$ relative mobility (R_f) versus gel concentration, %T, (percentage total monomer i.e. grams acrylamide plus bisacrylamide/100ml). The relative mobility, R_f, refers to the mobility of the protein of interest measured with reference to a marker protein or to a tracking dye where:

$$R_f = \frac{\text{distance migrated by protein}}{\text{distance migrated by dye}}.$$

Each Ferguson plot can be characterised by its slope K_R and its ordinate intercept Y_0. Since the Ferguson plot relates to mobility during electrophoresis when only the gel pore size (as determined by %T) is varying, then the slope of the Ferguson plot, K_R, is a measure of the retardation of the protein by the gel, that is, K_R is a retardation coefficient which can be related to molecular size. The ordinate intercept, Y_0 (when %T = 0), is a measure of the mobility of the protein in free solution. Rodbard *et al.* (12) have classified four sets of separation problems which are highlighted by such an analysis (*Figure 5*).

Case A: is the situation when the proteins have identical charge densities and show identical mobilities in free solution. In polyacrylamide gels these proteins migrate

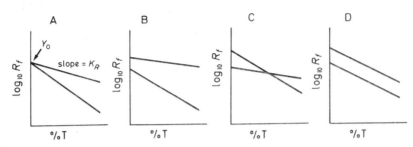

Figure 5. Classification of separation problems as determined by Ferguson plot analysis. See the text for details.

strictly according to size. This is the situation approximated to by most SDS-denatured proteins (p. 17).

Case B: the protein with the greater mobility in free solution (greater charge density) also has the smaller size. Thus the effects of fractionation on the basis of size and charge are synergistic and increased gel concentration leads to increased separation.

Case C: the larger protein has the higher free mobility such that size and charge fractionation are antagonistic. This case is commonly found in non-dissociating systems.

Case D: the proteins have the same size but different free mobilities (e.g. isozymes such as lactate dehydrogenase, haemoglobins). In this case, increasing acrylamide concentration has no effect upon the relative separation between the two proteins and one should consider using a charge separation method such as electrofocusing or isotachophoresis.

The use of Ferguson plots to determine conditions for the optimal separation and resolution of proteins by polyacrylamide gel electrophoresis is detailed in Chapter 2.

Molecular Weight Estimation

Native Proteins

During the electrophoresis of native proteins in polyacrylamide gel, separation takes place according to both size and charge differences of the molecules. By constructing Ferguson plots (*Figure 5*) the charge aspect is eliminated in that the slope, K_R (retardation coefficient), is a measure only of molecular size (see p. 13). Indeed, Hedrick and Smith (13) found that there is a linear relationship between K_R and the molecular weight of native proteins so that, by first using a series of standard native proteins of known molecular weight to construct a plot of K_R against molecular weight, one can determine the molecular weight of any sample protein by simply determining its K_R and then referring to the standard curve. Rodbard and Chrambach (14) have argued that a better relationship exists between $(K_R)^{1/2}$ and the molecular radius for globular proteins (see Chapter 2). Nevertheless, the problem of molecular weight determination with native proteins is that it is only valid if the standard proteins used to generate calibration curves have the same shape with the same degree of hydration and partial specific volume. For this reason, molecular weight determination of proteins using polyacrylamide gel electrophoresis is now usually performed in the presence of SDS after reduction of protein disulphide bonds with a thiol reagent. The ionic detergent SDS virtually eliminates conformational and charge density differences amongst proteins and reduces the effect of variability in partial specific volume and hydration. However, it should be noted that the molecular weight obtained using SDS-PAGE is the polypeptide subunit molecular weight and not that of the native protein if it is oligomeric.

Denatured Proteins (SDS-PAGE)

Basic method. When denatured by heating in the presence of excess SDS and a thiol reagent (usually 2-mercaptoethanol or dithiothreitol), most polypeptides bind SDS in a constant weight ratio such that they have essentially identical charge densities and migrate in polyacrylamide gels of the correct porosity according to polypeptide size (p. 6). Under these conditions, a plot of log_{10} polypeptide molecular weight versus relative mobility (R_f) reveals a straight-line relationship (3,4). The approach is therefore to electrophorese a set of marker polypeptides of known molecular weight and use the distance migrated by each to construct a standard curve from which the molecular weight of the sample polypeptides can be calculated based on their mobility under the same electrophoretic conditions. When using rod gels the marker polypeptides and sample polypeptides are necessarily run on separate gels. Since rod gels experience considerable variability from gel to gel in terms of distance migrated by any given polypeptide, it is always wise to express all distances migrated as R_f values relative to the dye front. The position of the dye front is marked by insertion of a piece of thin wire into the gel or injection of India ink before staining to locate the polypeptide bands. However, a better method is to use slab gels for molecular weight estimation since sample and marker polypeptides can be electrophoresed on a single gel and therefore under identical conditions. In this situation it is sufficient simply to measure the distance migrated by all the polypeptides after staining (measured from the top of the resolving gel), and to construct a standard curve by plotting log_{10} molecular weight against the distance migrated by the marker polypeptides (*Figure 6*). Practical details are given later (p. 38).

It must be emphasised that for any given gel concentration the relationship between log_{10} molecular weight and relative mobility is linear over only a limited range of molecular weight. As a general guide for the SDS-phosphate buffer system, the linear relationship holds true over the following ranges: 15% acrylamide, mol. wt. 12,000 - 45,000; 10% acrylamide, mol. wt. 15,000 - 70,000; 5% acrylamide, mol. wt. 25,000 - 200,000. Improved resolution of dilute protein samples can be achieved using the SDS-discontinuous buffer system instead, with the linear relationship holding over a similar range for 15% gels. However, with 10% and 5% gels, polypeptides with molecular weights less than about 16,000 and 60,000, respectively, migrate with the buffer front (*Figure 6*).

Anomalous behaviour of polypeptides. It is implicit in the linear relationship between log_{10} molecular weight and mobility for SDS-PAGE that all polypeptides bind a constant weight ratio of SDS. Therefore it is important that an excess of SDS to polypeptide of at least 3:1 is present during protein dissociation; otherwise polypeptides

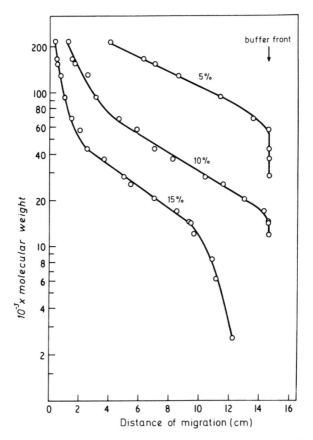

Figure 6. Calibration curves of $\log_{10}$ polypeptide molecular weight versus distance of migration during SDS-PAGE in slab gels using the SDS-discontinuous buffer system. The polyacrylamide gels used were uniform concentration 5%, 10%, or 15%. The polypeptide markers, in order of decreasing molecular weight, were myosin (mol. wt. 212,000), RNA polymerase β' (165,000) and β (155,000) subunits, β-galactosidase (130,000), phosphorylase a (92,500), bovine serum albumin (68,000), catalase (57,500), ovalbumin (43,000), glyceraldehyde-3-phosphate dehydrogenase (36,000), carbonic anhydrase (29,000), chymotrypsinogen A (25,700), soybean trypsin inhibitor (20,100), horse heart myoglobin (16,950), horse heart myoglobin cyanogen bromide cleavage fragment I + II (14,404), lysozyme (14,300), cytochrome c (11,700), horse heart myoglobin cyanogen bromide cleavage fragments I (8,159), II (6,214), and III (2,512). The molecular weights of horse heart myoglobin and its cyanogen bromide cleavage fragments are calculated from the primary sequence given in ref. 129. References to the molecular weights of the other polypeptides are given in *Table 6*.

which do obey the linear relationship under the correct conditions may fail to do so. An excess of thiol reagent is also essential to ensure breakage of disulphide bridges which otherwise oppose denaturation and prevent saturation of the polypeptide with SDS. Anomalies can be detected by carrying out the electrophoresis of both sample proteins and molecular weight markers at a number of different gel concentrations and compiling a Ferguson plot for each polypeptide. Ideally, all SDS-polypeptide complexes have identical charge densities and so have an identical Y_0 value (*Figure 5, case A*). In practice this ideal is often not achieved, but most polypeptides which

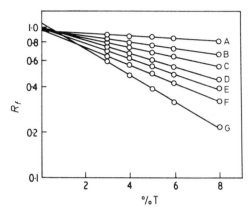

Figure 7. A typical Ferguson plot analysis of standard polypeptides analysed by SDS-PAGE. A, myoglobin; B, chymotrypsinogen; C, lactate dehydrogenase; D, ovalbumin; E, glutamate dehydrogenase; F, bovine serum albumin; G, phosphorylase a. Redrawn with permission from ref. 127 (see also refs. 8, 128).

behave 'normally' do approximate to this situation and have an R_f value similar to that of the standard polypeptides at or near %T = 0 (e.g. *Figure 7*) whereas anomalous polypeptides show up as having markedly different Y_0 values.

Many glycoproteins behave anomalously even when SDS and thiol reagent are in excess, probably because they bind SDS only to the protein part of the molecule. The reduced net charge resulting from reduced SDS binding lowers the polypeptide mobility during electrophoresis, yielding artifactually high molecular weight estimates. However, with increasing polyacrylamide gel concentration, molecular sieving predominates over the charge effect and the apparent molecular weights of glycoproteins decrease and approach their real molecular weights. Based on this, Segrest and Jackson (15) have presented a method for molecular weight estimation of a glycoprotein which involves determining the apparent molecular weight, relative to standard proteins, at a number of polyacrylamide gel concentrations to yield an asymptotic minimal molecular weight (*Figure 8*) which approximates to the real molecular weight of the glycoprotein. An alternative approach may be to use concentration gradient gels (p. 71).

Very basic proteins, for example histones, also behave aberrantly in SDS-PAGE, migrating slower than would be expected on the basis of their known molecular weights (p. 67).

Finally, polypeptides with a molecular weight below about 10,000 are not well resolved on uniform concentration polyacrylamide gels, even 15% polyacrylamide, when SDS is the sole dissociating agent (*Figure 6*). Separation of these oligopeptides can be improved using 8M urea as well as SDS in 12.5% polyacrylamide gels containing a high percentage of crosslinker (p. 64).

SDS-PAGE in concentration gradient gels. Although, in the past, polypeptide molecular weight estimation using SDS-PAGE has usually been carried out in gels of uniform acrylamide concentration, increased use has recently been made of concentration gradient gels for this purpose (see p. 71).

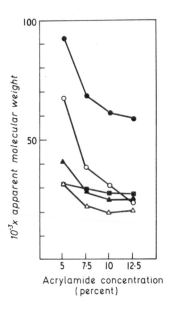

Figure 8. Molecular weight estimation of glycoproteins by electrophoresis in polyacrylamide gels of different acrylamide concentration (15). Human erythrocyte glycoprotein (●); human erythrocyte membrane tryptic glycoprotein (○); fragmented human erythrocyte membrane glycopeptide (■); porcine ribonuclease, higher molecular weight species (▲); porcine ribonuclease, lower molecular weight species (△) (reproduced from ref. 15 with permission).

The glycoproteins were electrophoresed in uniform concentration polyacrylamide gels at the gel concentrations shown on the abscissa, using the SDS-phosphate buffer system. The apparent molecular weight of each glycoprotein was determined for any given gel concentration by comparison of the electrophoretic mobility of the glycoprotein with that of standard polypeptides of known molecular weight. The anomalous behaviour of glycoproteins in SDS-PAGE is shown by the decreasing glycoprotein apparent molecular weight with increasing acrylamide concentration; non-glycosylated polypeptides give consistent molecular weights at all acrylamide concentrations. Although the asymptotic minimal molecular weight for each glycoprotein shown here approximates to the real molecular weight, sialic acid-containing and desialylated glycoproteins behave quite differently in SDS-PAGE even by this protocol (15) so that the molecular weight obtained for any unknown glycoprotein is still very much only an approximation.

APPARATUS

Gel Holders and Electrophoresis Tanks

During electrophoresis, heat is generated by the passage of electric current through the gel. Since the mobility of migrating ions is increased as the temperature rises, it is apparent that the temperature must be controlled if electrophoretic separations are to be reproducible. Furthermore, if the heating effect is significant, there will be a temperature gradient from the centre of the gel to the gel surface which will cause proteins to migrate faster at the gel centre than the surface, resulting in distorted bands. The increased temperature may also inactivate labile proteins.

The actual temperature will depend on both the rate of heat production and its rate of dissipation. The rate of heat production is proportional to the current passed. Hence heat production is minimised by using low-conductivity buffers or, if high-conductivity buffers must be used, by extending the duration of the run using a lower power input. Suitable buffers and electrophoresis conditions are described later. Heat dissipation is facilitated by using rod and slab gels as thin as possible to allow the efficient loss of heat whilst being thick enough to allow a reasonable loading capacity. In practice, for one-dimensional zone electrophoresis of proteins on an analytical scale, gels should not be thicker than about 0.6 cm and preferably less. In addition, electrophoretic separations of labile native proteins in non-dissociating buffer systems a.e best carried out using an apparatus which has some means of cooling the gel. Cooling is often not necessary for simple comparative analyses of protein mixtures by SDS-PAGE using a slab gel apparatus since minor changes in the mobility of sample polypeptides due to small temperature variations are compensated for by changes in the mobility of the molecular weight marker polypeptides. However, with SDS-PAGE as with all other buffer systems, accurate temperature regulation is essential for quantitative studies of protein mobility (see Chapter 2).

Rod Gel Apparatus

The cylindrical tubes in which the gels are formed should be made of glass, usually with an internal diameter of 0.4-0.6 cm and a 0.1 cm wall thickness for one-dimensional electrophoretic separations, although gels are often much narrower for some two-dimensional separations (Chapter 5). Perspex tubes should be avoided for analytical zone electrophoresis of proteins since they do not allow heat dissipation as efficiently as glass tubes. The exact length of the tubing employed varies from one laboratory to another and will depend on the experience of the user. In general, rod gel tubes should be about 2-3 cm longer than the final gel. A convenient length is 12 cm which allows for the use of continuous buffer systems with a single resolving gel 9-10 cm long or discontinuous buffer systems with a resolving gel of similar length but with an additional stacking gel (about 1 cm). Tubing should be of uniform diameter throughout its length and fire polished at the ends.

During gel preparation the rod gel holders must be kept exactly vertical to ensure a horizontal, sharp meniscus. Racks for holding the gel tubes in this way are available commercially or can be made easily in the laboratory from wooden battens and suitably sized clips. For electrophoresis, the rod gels are transferred to an apparatus such as that shown in *Figure 9*. The apparatus is made mainly of Perspex. The gel tubes are suspended by means of silicone-rubber grommets held in holes drilled in the base of the upper buffer reservoir; any holes not required are blocked with rubber stoppers. In the apparatus shown, most of the gel length is immersed in buffer held in the lower buffer reservoir which can be cooled by means of a water jacket. It is important that the platinum electrodes should be the same distance from each gel tube and hence the electrodes are circular in shape and located centrally. Simpler apparatuses may lack the water jacket and rely solely on the large heat capacity of the lower buffer reservoir for cooling. More sophisticated apparatuses have interchangeable upper buffer reservoirs capable of holding rod gels of different diameters, plus a number of other refinements (Chapter 2, p. 93).

Figure 9. A typical rod gel apparatus. The model shown (Bio-Rad model 150A) is constructed of Perspex, incorporates a water-jacketed lower buffer reservoir, and holds 12 tubes, 0.7 to 0.85 cm O.D. (optional 0.5 cm O.D.) x 15 cm long. A more recent model (Bio-Rad 155) has core cooling, rather than a water-jacket, and interchangeable upper buffer reservoirs to hold a range of tube sizes from 0.4-1.7 cm O.D.

Slab Gel Apparatus

Apparatus has been designed for protein and nucleic acid fractionation either by vertical or horizontal slab gel electrophoresis. Whilst considerable use is still made of horizontal slab electrophoresis for protein electrofocusing and immunoelectrophoresis, zone electrophoresis of proteins in polyacrylamide slab gels is almost always now carried out in the vertical format; indeed discontinuous buffer separations can *only* be done using vertical gels. One of the most popular designs has been that of Studier (16). A modification of this apparatus which can be easily constructed in a laboratory workshop is shown in *Figure 10*. Essentially the apparatus consists of two buffer reservoirs, the upper of which is notched, supported on an integral Perspex stand. The gel is formed between two glass plates each about 0.3 cm thick. One plate is rectangular in shape, 17.0 cm x 19.5 cm, and the second is the same size but with a notch 2.0 cm deep and 14.0 cm long cut in one of the 17.0 cm edges. The two plates are placed together to form the gel holder with a Perspex spacer (22.0 cm long, 1.0 cm wide and 0.15 cm thick) running down each vertical side of the sandwich. Sample wells are formed in the gel during polymerisation using a Perspex sample comb; sample well number and dimensions can then easily be altered by changing the number and dimensions of the comb teeth. Some useful combs are shown in *Figure 10*. The detailed pouring of such slab gels is described later. The polymerised gel, held between its glass plates, is attached to the electrophoresis apparatus by means of strong metal spring clips in such a way that the notched glass plate is aligned with and adjacent to the notch in the upper buffer reservoir to allow contact with the top of the gel and the upper reservoir buffer. The bottom of the gel is immersed in the buffer in the lower reservoir. Since many samples can be analysed on one slab gel, the platinum electrodes are placed so as to be equidistant from each sample which means that they are positioned along the length of both upper and lower buffer reservoirs. Similar apparatus is also commercially available from a number of sources (e.g. Raven Scientific Ltd., Bethesda Research Labs Inc.).

It is important to note that much of the gel is uncooled using this apparatus, the investigator relying on the minimal thickness (0.75-1.5 mm) of the slab gel to allow efficient heat dissipation. The apparatus works well for SDS-PAGE run at room temperature. However, care must be taken not to exceed the heat dissipative ability of the slab gel by using a too-high power input; otherwise the centre section of the gel increases in temperature relative to the rest of the gel so that polypeptides migrate faster if loaded at the gel centre than if loaded at the gel edge, resulting in a curved display of polypeptide components. This effect can be prevented, whilst allowing separation in SDS-PAGE at higher voltage in less time, by using one of the many cooled vertical slab gel electrophoresis systems which are available commercially, (see, for example, *Figure 11*.) Nevertheless, the cheap, easily-constructed Studier-type of apparatus works sufficiently well for most comparative analyses of sample polypeptide composition and molecular weight estimation by SDS-PAGE. Electrophoretic separations of labile native proteins in non-dissociating buffer systems are best carried out using a cooled slab gel electrophoresis apparatus at 4-8°C.

Additional Items of Equipment Required for Electrophoresis

Additional items of equipment which may be required apart from standard laboratory glassware and magnetic stirrers are given below.

(i) A 15 W daylight fluorescent lamp; for photopolymerisation of riboflavin-catalysed polyacrylamide gels.

(ii) Microlitre syringes or micropipettes; for loading samples onto the gel. A number of alternatives are available including various pipettors with plastic disposable pipette tips, microcaps (calibrated glass capillary tubes of known volume) and calibrated plunger-type microsyringes. The choice depends upon personal preference when loading rod gels but the control needed for careful sample loading plus the narrow sample wells in slab gels is best coped with by using a 50 μl or 100 μl microsyringe.

(b)

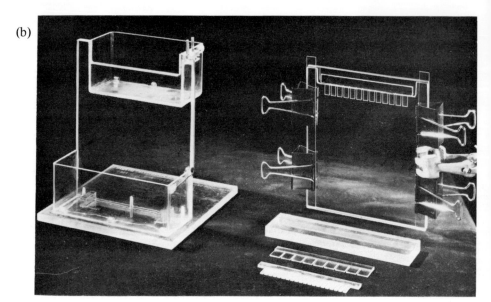

(a)

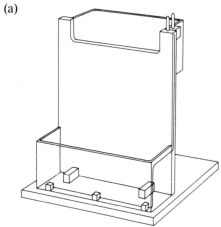

Figure 10. A Studier-type slab gel apparatus. (a) a line drawing of the electrophoresis apparatus, (b) a photograph of the electrophoresis apparatus together with the notched glass-plate assembly used to form the slab gel. The apparatus itself is constructed of Perspex and consists of a base (20.5 cm x 20.5 cm x 0.9 cm) supporting a vertical section (21.3 cm x 17.8 cm x 0.9 cm) which has a notch 3.0 cm deep and 14.0 cm long cut in one of the 17.8 cm edges. The other 17.8 cm long edge is cemented and screwed into the base of the apparatus 12.5 cm from one of the base edges. The base of the upper buffer reservoir and three sides are formed from 0.3 cm thick Perspex to give a chamber 14.6 cm long, 6.0 cm wide, and 7.0 cm deep (outside edge measurements). This is cemented to the vertical section of the apparatus, which forms the final side of the upper reservoir, flush with the top edge. The lower buffer reservoir is 17.8 cm long, 6.0 cm wide, and 6.5 cm deep (outside edge measurements) and consists of a sheet of 0.3 cm thick Perspex, 29.8 cm long and 6.5 cm wide, moulded to form the three sides of the reservoir and then cemented to the lower part of the vertical section (which forms its final side) and to the base of the apparatus (which forms the reservoir base). Two Perspex blocks, 2.5 cm x 0.9 cm x 0.9 cm, are also cemented, 12 cm apart (from the centre of each block), to the base of the lower reservoir and abutting the vertical section so as to form supports for the slab gel glass-plate sandwich. Platinum wire electrodes run along the front and back edge of the lower and upper buffer reservoirs, respectively, where they are held in position with small blocks of cemented Perspex. The elec-

22

(iii) A peristaltic pump; required for the preparation of acrylamide concentration gradient gels (see p. 73) and also useful for the careful overlayering of all slab gels to obtain a flat gel surface.

(iv) A power pack; capable of supplying about 500 V and 100 mA. In many cases it is not too important whether constant current or constant voltage conditions should be used, but for versatility it is worthwhile obtaining a power pack which has these as alternative modes of operation. A number of suitable power packs are available commercially although simple inexpensive power supplies can be built which usually prove to be entirely satisfactory. If electrophoretic destaining of gels (p. 45) is envisaged, it is worth having a second power supply with a high current output (1-2 A) and low voltage; in practice a car battery charger is adequate.

PREPARATION AND ELECTROPHORESIS OF POLYACRYLAMIDE GELS

Reagents

Acrylamide and bisacrylamide. Both acrylamide and bisacrylamide are neurotoxins. Therefore, disposable plastic gloves should be worn when handling all solutions containing these reagents and mouth-pipetting must be avoided. Polyacrylamide itself is not toxic but usually contains some unpolymerised monomer which is.

Several manufacturers now offer highly purified grades of acrylamide and bisacrylamide which can be used quite successfully without further purification in most investigations. However, where necessary, acrylamide and bisacrylamide can be recrystallised from a number of solvents. One method is to dissolve 70 g acrylamide in 1 litre of chloroform at 50°C, filter the solution hot without suction, and store at −20°C to allow recrystallisation. The crystals are collected by filtration, washed briefly with cold chloroform and then dried exhaustively under vacuum. Bisacrylamide may be purified by dissolving 10 g in 1 litre of acetone at 50°C, filtering hot, and storing at −20°C. The crystals are collected by filtration, washed briefly with cold acetone and dried.

---

trodes connect to male terminals screwed into a Perspex block (3.0 cm x 1.5 cm x 1.5 cm) cemented alongside the upper reservoir. For safety reasons, a lid can be made of 0.3 cm thick Perspex and incorporating female sockets which interconnect with the male terminals on the apparatus and lead to the power supply. The electrical circuit is then complete only with the lid in position. However, it is often useful to be able to cool the slab gel with a fan during electrophoresis. In this situation the apparatus terminals are connected to the power pack without a lid but it must be realised that this then constitutes a potential safety hazard. The dimensions of the slab gel glass plates and side spacers are given in the text. Each comb for sample-well formation is constructed from a 0.15 cm x 2.5 cm x 13.0 cm Perspex strip with teeth cut 1.25 cm into one 13.0 cm edge. One face of the uncut half of the strip is then cemented to a second Perspex strip (0.15 cm x 1.25 cm x 13.0 cm) which serves to hold the comb horizontal during use. A comb we find useful (shown in place in the glass-plate sandwich of *Figure 10*) is for 13 wells (0.7 cm wide teeth with 0.25 cm spaces between teeth) since it is a good compromise between adequate track width, sample-well volume, and the number of samples which can be analysed simultaneously. However, the number and dimensions of the sample wells are entirely dependent on the preferences of the investigator. The Perspex trough shown immediately in front of the glass-plate sandwich is used in one method of sealing the base of the glass-plate sandwich during gel polymerisation and is described in the legend to *Figure 12*.

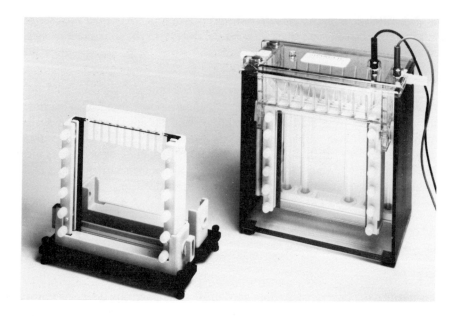

Figure 11. A cooled vertical slab gel apparatus. The model shown is the Protean Cell available from Bio-Rad Laboratories Ltd. (manufactured by Hoefer Scientific Instruments). Up to two slab gels can be electrophoresed simultaneously. Each slab gel is formed between two glass-plates 16.0 cm x 18.0 cm x 0.3 cm, neither of which are notched, which are held apart by PVC spacers, 0.075, 0.15, 0.3, or 0.6 cm thick, placed down each of the two vertical sides of the sandwich. The spacers are held in position by one-piece plastic clamps. The bottom of the sandwich is sealed using polycarbonate cams which press the base against a silicone rubber gasket in a casting stand, as shown on the left-hand side of the above photograph. After gel pouring, polymerisation, sample-well formation, and sample loading, as described in the text, the slab gel sandwich is released from the casting stand and locked in position against another silicone rubber gasket in the upper buffer reservoir using the same cam system. The upper and lower reservoirs are filled with reservoir buffer and that in the lower reservoir is cooled by coolant passing through a glass tube heat exchanger, placed between the two slab gels, and stirred by a magnetic stir bar. A very similar cooled slab gel apparatus, also manufactured by Hoefer Scientific Instruments, is marketed by LKB Instruments Ltd. Several other types of cooled vertical slab gel apparatus are also commercially available. In many of these the slab gels are held in place by rubber grommets in the upper buffer reservoir instead of the cam-operated sealing system.

SDS. Several grades of SDS are commercially available but only highly purified grades should be used. Many problems in SDS-PAGE can be traced to the purity of the SDS used; Swaney *et al.* (17) showed that the entire migration pattern (and apparent molecular weights) of polypeptide mixtures were changed with certain sources of SDS. The problem was only observed using the SDS-discontinuous buffer system and not with the continuous SDS-phosphate system. Fortunately SDS in a form specially purified for electrophoresis is now available from a number of sources (e.g. BDH Chemicals Ltd., Bio-Rad Laboratories Ltd.) and appears to be satisfactory. However, one would still be advised to select and use SDS from one source only. If necessary, SDS can be recrystallised from ethanol. Approximately 200 g of SDS is dissolved in 3 litres of boiling ethanol, filtered hot without suction, and stored at 4°C to allow the SDS to recrystallise. The crystals are recovered by filtration and dried.

Urea. When urea is used, the main problem is the accumulation of cyanate ions in stock solutions as a result of chemical isomerisation. The cyanate reacts with amino groups to form stable carbamylated derivatives thus altering the charge of the proteins. If this reaction does not go to completion, several artifactual species of proteins with differing charges will result. The simplest remedy is to use fresh urea solutions and, where appropriate, to buffer the solutions with Tris, the free amino groups of which neutralise the cyanate ions. Since cyanate ion formation is accelerated with increasing temperature, heating of urea-containing solutions should be avoided if possible. For exacting situations, urea solutions should be deionised, just before use, by passage through a mixed-bed ion-exchange resin.

TEMED. The TEMED used should be a colourless liquid. Most commercial sources are at least 99% pure.

All other reagents should be of the highest purity available.

Stock Solutions

Details of methodology vary with the type of gel and buffer system selected but certain chemicals and solutions are common to many methods and are described here. Once made, most of these are stable for several months under the conditions described, so that to attempt different gel systems the investigator need only prepare fresh buffer solutions.

(i)   Acrylamide - bisacrylamide (30:0.8); is prepared by dissolving 30 g of acrylamide and 0.8 g bisacrylamide in a total volume of 100 ml water. The solution is filtered through Whatman No. 1 filter paper, and stored at 4°C in a dark bottle. Hydrolysis of acrylamide monomer to yield acrylic acid and ammonia will occur upon prolonged storage. To ensure reproducibility of data, only enough acrylamide-bisacrylamide stock solution to last for about 1-2 months should be prepared.

(ii)   TEMED; used as supplied. It is stable in undiluted solution at 4°C in a dark bottle.

(iii)   Ammonium persulphate (1.5%, w/v); 0.15 g of ammonium persulphate is dissolved in 10 ml water. This solution is unstable and should be made fresh just before use.

(iv)   Riboflavin (0.004%, w/v); 4 mg riboflavin are dissolved in 100 ml water. The solution is stable when stored at 4°C in a dark bottle.

(v)   SDS (10%, w/v); prepared by dissolving 10 g SDS in water to 100 ml. The solution should be both clear and colourless. If the water temperature is too low, not all the SDS will dissolve and heating is necessary. The solution is stable at room temperature for several weeks but precipitates in the cold.

(vi)   Electrophoresis buffers; the exact buffer required will depend on the buffer system chosen (see below).

Table 1. Recipe for Gel Preparation using the SDS-phosphate (Continuous) Buffer System.

Stock solution	Final acrylamide concentration (%)[a]							Reservoir buffer[b]
	20.0	17.5	15.0	12.5	10.0	7.5	5.0	
Acrylamide-bisacrylamide (30:0.8)	20.0	17.5	15.0	12.5	10.0	7.5	5.0	-
0.5M sodium phosphate, pH 7.2	6.0	6.0	6.0	6.0	6.0	6.0	6.0	200
10% SDS	0.3	0.3	0.3	0.3	0.3	0.3	0.3	10
1.5% ammonium persulphate	1.5	1.5	1.5	1.5	1.5	1.5	1.5	-
Water	2.2	4.7	7.2	9.7	12.2	14.7	17.2	790
TEMED	0.015	0.015	0.015	0.015	0.015	0.015	0.015	-

[a]The columns represent volumes (ml) of the various reagents required to make 30 ml of gel mixture.
[b]Volumes (ml) of reagents required to make 1 litre of reservoir buffer.

Gel Mixture Preparation

The recipes tabulated below yield a sufficient volume of gel mixture for 12-15 rod gels or one standard-size slab gel. The volume of stacking gel mixture for discontinuous buffer systems is ample for both rinsing the resolving gel surface prior to stacking gel polymerisation and the final stacking gel itself. Higher concentration gels usually polymerise more rapidly than lower concentration gels at any given TEMED concentration. Thus, the volume of TEMED given in the following tables is only a guide and should be adjusted to obtain gel polymerisation within 10-30 min.

Dissociating Buffer Systems (SDS-PAGE)

Details of buffer composition and gel mixture preparation for the SDS-phosphate (continuous) system, essentially as described by Weber and Osborn (4), and the SDS-discontinuous buffer system based on the method of Laemmli (7), are given in *Tables 1* and *2*, respectively. Note that the acrylamide concentration of the stacking gel for the SDS-discontinuous system is constant irrespective of the acrylamide concentration chosen for the resolving gel.

Non-dissociating Buffer Systems

Proteins differ widely in their sensitivity to ionic strength, ionic species, and cofactor requirements. Therefore the buffer chosen for the zone electrophoresis of native proteins will depend entirely on the proteins under study.

With regard to continuous buffer systems, almost any buffer between pH 3 and 10 may be used for electrophoresis. In general, only solutions of relatively low ionic strength (and thus only weakly conductive) are suitable as electrophoresis buffers since these keep heat production to a minimum. On the other hand, if the ionic strength is too low, protein aggregation may occur. Obviously the buffer concentration which satisfies these requirements will depend on the particular buffer ions chosen and the proteins under study, but in general the concentration limits for electrophoresis are from about 0.01 M to 0.1 M. Typical buffer systems which have been used are Tris-glycine (pH range 8.3-9.5); Tris-borate (pH range 8.3-9.3); Tris-acetate (pH range 7.2-8.5); Tris-citrate (pH range 7.0-8.5). Usually Tris concentra-

Table 2. Recipe for Gel Preparation using the SDS-discontinuous Buffer System.

Stock solution	Stacking gel (ammonium persulphate as catalyst)	Stacking gel (riboflavin as catalyst)	Final acrylamide concentration in resolving gel (%)[a]							Reservoir buffer[b]
			20.0	17.5	15.0	12.5	10.0	7.5	5.0	
Acrylamide-bisacrylamide (30:0.8)	2.5	2.5	20.0	17.5	15.0	12.5	10.0	7.5	5.0	-
Stacking gel buffer stock[c]	5.0	5.0	-	-	-	-	-	-	-	-
Resolving gel buffer stock[d]	-	-	3.75	3.75	3.75	3.75	3.75	3.75	3.75	-
Reservoir buffer stock[e]	-	-	-	-	-	-	-	-	-	100
10% SDS	0.2	0.2	0.3	0.3	0.3	0.3	0.3	0.3	0.3	-
1.5% ammonium persulphate	1.0	-	1.5	1.5	1.5	1.5	1.5	1.5	1.5	-
0.004% riboflavin	-	2.5	-	-	-	-	-	-	-	-
Water	11.3	9.8	4.45	6.95	9.45	11.95	14.45	16.95	19.45	900
TEMED	0.015	0.015	0.015	0.015	0.015	0.015	0.015	0.015	0.015	-

Final concentration of buffers: stacking gel; 0.125M Tris-HCl, pH 6.8
resolving gel; 0.375M Tris-HCl, pH 8.8
reservoir buffer; 0.025M Tris, 0.192M glycine, pH 8.3

[a] The columns represent volumes (ml) of the various reagents required to make 30 ml of gel mixture.

[b] Volumes (ml) of reagents required to make 1 litre of reservoir buffer.

[c] Stacking gel buffer stock: 0.5M Tris-HCl (pH 6.8); 6.0 g Tris is dissolved in 40 ml water, titrated to pH 6.8 with 1M HCl (~ 48 ml), and brought to 100 ml final volume with water. The solution is filtered through Whatman No. 1 filter paper and stored at 4°C.

[d] Resolving gel buffer stock: 3.0M Tris-HCl (pH 8.8); 36.3 g Tris and 48.0 ml 1M HCl are mixed and brought to 100 ml final volume with water. This buffer is then filtered through Whatman No. 1 filter paper and stored at 4°C.

[e] Reservoir buffer stock: 0.25M Tris, 1.92M glycine, 1% SDS (pH 8.3); 30.3 g Tris, 144.0 g glycine, and 10.0 g SDS are dissolved in and made to 1 litre with water. The solution is stored at 4°C.

tions are 0.02-0.05 M. For basic proteins one can use β-alanine-acetate (pH range 4.0-5.0) with 0.01-0.05 M β-alanine. The choice of a suitable pH was discussed in an earlier section (p. 11). If reducing agents are found to be essential at all times to retain protein activity, dithiothreitol (1 mM) can be added to the gel mixture. Once a suitable gel concentration (p. 12) and buffer have been selected, the gel mixture is prepared according to the directions in *Table 3* which assumes that the buffer can be prepared as a five-times-concentrated stock.

Details of the three discontinuous buffer systems originally devised for the electrophoresis of native proteins are given in *Table 4*. The Ornstein-Davis high pH system has been the most used whereas the low pH system is useful for basic proteins. The neutral pH system is poorly buffered and little used but may preserve some enzyme activities which are not stable at extremes of pH. The volumes of reagents required for gel mixture preparation using any of these three buffer systems are given in *Table 5*. In addition, Chapter 2 describes the wealth of discontinuous buffer systems now available as a computer output.

Preparation of Rod Gels

Continuous Buffer System

1. Use glass tubes 2-3 cm longer than the required gel length. The most important point is to ensure that gel tubes are absolutely clean to obtain uniform adhesion of the gel to the glass. This is done by soaking the glass tubes in chromic acid overnight, rinsing them first with distilled water and finally with ethanol before drying.

Although low percentage gels (below 12%) are usually easily recovered after electrophoresis by rimming (p. 42), this becomes increasingly difficult with higher percentage gels. One way to facilitate gel removal by this method is to use tubes siliconised with an anhydrous solution of 0.5% dimethyldichlorosilane in carbon tetrachloride (commercially available as 'Repelcote', Hopkin & Williams Ltd.). Instructions on the use of siliconising reagents are supplied by the manufacturer but it is worth emphasising that they should be used with care and in an efficient fume cupboard since they are highly toxic. The reduced adhesion of siliconised tubes means that low percentage glass gels in particular may slip out of the tube during electrophoresis. This can be prevented by covering the bottom of the gel tube, before electrophoresis, with a square piece of nylon mesh (about 100 micron size) held in place with a 2-3 mm section of Tygon tubing.

2. Gel tubes are marked at the required gel length with a fine-tipped water-insoluble marker pen (unsiliconised tubes only), or a glass knife, and sealed at the opposite end with Parafilm.

3. The gel tubes are now placed in racks which should hold the tubes snugly in an exactly vertical position.

4. Having selected the resolving gel concentration to be used, the gel mixture, except for TEMED, is prepared according to the volumes indicated in *Table 1* (SDS-phosphate) or *Table 3* (non-dissociating buffers) in a small thick-walled flask and degassed for 1 min using a water-pump. As well as aiding reproducible polymerisation rates, degassing prevents bubble formation in the gel.

Table 3. Recipe for Gel Preparation using Non-dissociating Continuous Buffer Systems.

Stock solution	Final acrylamide gel concentration (%)[a]							Reservoir buffer[b]
	20.0	17.5	15.0	12.5	10.0	7.5	5.0	
Acrylamide-bisacrylamide (30:0.8)	20.0	17.5	15.0	12.5	10.0	7.5	5.0	-
Continuous buffer (5 x conc.)	6.0	6.0	6.0	6.0	6.0	6.0	6.0	200
1.5% ammonium persulphate[c]	1.5	1.5	1.5	1.5	1.5	1.5	1.5	-
Water	2.5	5.0	7.5	10.0	12.5	15.0	17.5	800
TEMED[d]	0.015	0.015	0.015	0.015	0.015	0.015	0.015	-

[a]The columns represent volumes (ml) of the various reagents required to make 30 ml of gel mixture.

[b]Volumes (ml) of reagents required to make 1 litre of reservoir buffer.

[c]Riboflavin (0.004% w/v), 2.5 ml, may be used in place of ammonium persulphate as polymerisation catalyst if desired. The water volume is adjusted accordingly. Riboflavin is usually more effective than ammonium persulphate/TEMED at low pH whilst the latter is more effective at high pH.

[d]The concentration of TEMED may need to be increased for low pH buffers.

Table 4. Buffers for Non-dissociating Discontinuous Systems.

High pH discontinuous (6)

Stacks at pH 8.3, separates at pH 9.5.

Stacking gel buffer:	Tris-HCl (pH 6.8); 6.0g Tris is dissolved in 40 ml water and is titrated to pH 6.8 with 1M HCl (~ 48 ml). Water is added to 100 ml final volume.
Resolving gel buffer:	Tris-HCl (pH 8.8); 36.3g Tris and 48.0 ml 1M HCl are mixed and brought to 100 ml final volume with water. The solution is titrated to pH 8.8, with HCl, if necessary.
Reservoir buffer:	Tris-glycine (pH 8.3) at the correct concentration for use; 3.0g Tris and 14.4g glycine are dissolved in and made to 1 litre with water.

Neutral pH discontinuous (9)

Stacks at pH 7.0, separates at pH 8.0.

Stacking gel buffer:	Tris-phosphate (pH 5.5); 4.95g Tris is dissolved in 40 ml water and titrated to pH 5.5 using 1M orthophosphoric acid. Water is added to 100 ml final volume.
Resolving gel buffer:	Tris-HCl (pH 7.5); 6.85g Tris is dissolved in 40 ml water and titrated to pH 7.5 with 1M HCl. Water is added to 100 ml final volume.
Reservoir buffer:	Tris-diethylbarbiturate (pH 7.0); 55.2g diethylbarbituric acid and 10.0g Tris are dissolved in and brought to 1 litre with water.

Low pH discontinuous (10)

Stacks at pH 5.0, separates at pH 3.8.

Stacking gel buffer:	acetic acid-KOH (pH 6.8); 48.0 ml 1M KOH and 2.9 ml glacial acetic acid are mixed and then water is added to 100 ml final volume.
Resolving gel buffer:	acetic acid-KOH (pH 4.3); 48.0 ml 1M KOH and 17.2 ml glacial acetic acid are mixed and water added to 100 ml final volume.
Reservoir buffer:	acetic acid-β-alanine (pH 4.5); 31.2g β-alanine and 8.0 ml glacial acetic acid are dissolved in and made to 1 litre with water.

5. The correct volume of TEMED is now added and rapidly but gently mixed in by swirling. Without delay, each tube is now filled with gel solution to the mark, being careful to avoid trapping any air bubbles. This can be achieved using either a Pasteur pipette or a 10 ml hypodermic syringe fitted with a long blunt needle, lowering the tip to the tube bottom before expelling the contents.

6. Once the gel tubes have been filled, the gel solution is overlayered, preferably with buffer of the same composition as in the gel (although water is often used instead) to a height of about 0.5 cm, both to exclude oxygen (which inhibits polymerisation) and to ensure a flat gel meniscus. Great care should be taken to ensure that the overlay is added as gently as possible to prevent mixing with the gel solution; if mixing occurs it will result in a diffuse boundary which will lower the acrylamide concentration at the gel top and reduce zone sharpness. There are a number of ways of adding the overlay but perhaps the most convenient is to use a small-volume syringe with needle attached. The syringe is filled with overlay buffer, the needle wiped to remove any droplets, and then gently touched to the gel-mixture surface, holding the tip firmly against the

Table 5. Recipe for Gel Preparation using Non-dissociating Discontinuous Buffer Systems.

Stock solution	Stacking gel (riboflavin as catalyst)	Final acrylamide concentration in resolving gel (%)[a]							Reservoir buffer[b]
		20.0	17.5	15.0	12.5	10.0	7.5	5.0	
Acrylamide-bisacrylamide (30:0.8)	2.5	20.0	17.5	15.0	12.5	10.0	7.5	5.0	-
Stacking gel buffer stock[c]	5.0	-	-	-	-	-	-	-	-
Resolving gel buffer stock[c]	-	3.75	3.75	3.75	3.75	3.75	3.75	3.75	-
Reservoir buffer stock[c]	-	-	-	-	-	-	-	-	1000 (i.e. undiluted)
1.5% ammonium persulphate[d]	-	1.5	1.5	1.5	1.5	1.5	1.5	1.5	-
0.004% riboflavin	2.5	-	-	-	-	-	-	-	-
Water	10.0	4.75	7.25	9.75	12.25	14.75	17.25	19.75	-
TEMED[d]	0.015	0.015	0.015	0.015	0.015	0.015	0.015	0.015	-

[a]The columns represent volumes (ml) of the various reagents required to make 30 ml of gel mixture.

[b]Volumes (ml) of reagents required to make 1 litre of reservoir buffer.

[c]Stock solution prepared as described in *Table 4.*

[d]When the low pH discontinuous buffer system is used with ammonium persulphate as catalyst, the volume of TEMED should be increased to 0.15 ml for the resolving gel and the water volume adjusted accordingly. Riboflavin is usually more effective than ammonium persulphate/TEMED at high pH whilst the latter is more effective at low pH.

31

tube wall. The tip is then raised about 2 mm to leave a wet track to the gel-mixture surface. Keeping the tip firmly against the wall, the syringe contents are slowly expelled whilst raising the tip to keep it 2 mm above the liquid surface. At this stage, a sharp boundary should be visible between the gel mixture and the overlay. Other workers prefer to use a syringe fitted with a needle bent at 90°, or a Pasteur pipette, for applying the overlay. Whatever the method, it is imperative that the overlay is added with as little mixing as possible.

7. The gels are left undisturbed to polymerise (10-30 min). Any vibration at this stage will lead to an uneven gel surface. During this time the original interface disappears and is replaced with another slightly below; this is the polymerised gel surface.

8. The gels are left another 10-15 min after the new interface forms and then the water or buffer overlay is removed. If the gel is to be used immediately, the gel surface is rinsed with reservoir buffer and the space above the gel is filled with this buffer. The gels may also be stored, provided that dehydration is prevented. In this case, the overlay is removed and the gel surface is rinsed with fresh buffer of the same composition as in the gel. Next, more of this buffer is layered onto the gel surface and the tube is sealed with Parafilm or a rubber bung.

Discontinuous Buffer System

The two gel layers required for use with the discontinuous buffer system are prepared by first polymerising the resolving gel in the tube and then polymerising the stacking gel on top of this. As a general rule, the stacking gel should not be less than twice the height and volume of the sample to be applied. A stacking gel of about 1.0 cm will suffice for most sample volumes used; in the event that an exceptionally large volume of dilute protein need be loaded, the stacking gel dimensions should be adjusted accordingly.

Steps 1-7 inclusive are identical to *steps 1-7* given above for the continuous buffer system except that the gel tubes are marked at two places, once at the desired height of the resolving gel, and then again at a position 1 cm above this to mark the height of the stacking gel. The gel tubes are filled to the first mark with resolving gel mixture prepared according to the details given in *Table 2* (for the SDS-discontinuous buffer system) or *Table 5* (for non-dissociating discontinuous buffer systems).

8. Once the resolving gel has polymerised, the overlay is poured off and the stacking gel mixture is prepared (see *Table 2* for SDS-discontinuous buffers or *Table 5* for non-dissociating buffers). If ammonium persulphate is to be used as polymerisation catalyst, all the components except TEMED are mixed and then the correct volume of this reagent is added just prior to pouring the stacking gel. If riboflavin is used instead of ammonium persulphate, TEMED can be added to the stacking gel mixture any time prior to pouring the gel since polymerisation will not occur until the gel solution is illuminated.

9. The resolving gel surface is rinsed with a small volume of stacking gel mixture which is applied using a Pasteur pipette, removed, and then discarded. Next, each gel tube is filled to the second mark with stacking gel mixture and overlayered with water or stacking gel buffer according to the procedure detailed in *step 6* above. If riboflavin has been used as polymerisation catalyst for the stacking gel, a daylight

fluorescent lamp is placed 2-5 cm away from the gel to initiate polymerisation.
10. After polymerisation of the stacking gel, the overlay is removed and the gel surface rinsed with reservoir buffer. After discarding this, the space above the gel is filled with reservoir buffer.

These gels should be used as soon as possible after preparation since the different buffers of the stacking and resolving gels, and the reservoir buffer, essential for the stacking phenomenon to occur, mix by diffusion upon storage. A convenient alternative is to polymerise the resolving gel in place, rinse this with buffer of the same composition as in the resolving gel, and then store with a fresh overlay of this buffer. The stacking gel is polymerised in place just before the gel is required.

Preparation of Slab Gels

The following method for pouring slab gels is described for the modified Studier apparatus using the notched plate system (*Fig. 10*) which is widely used for SDS-PAGE. The many cooled slab apparatuses commercially available vary in the final gel dimensions (including gel thickness), whether notched or unnotched glass plates are used, and the method of sealing the plates. However, the method for pouring uniform concentration polyacrylamide slab gels and for producing sample wells is essentially the same irrespective of the equipment used. The main difference encountered would be in the method of sealing the plates in which case the manufacturer's instructions should be consulted.

Disposable plastic gloves should be worn during slab gel preparation to prevent contamination of clean glass plates with skin proteins.

Continuous Buffer System

1. As with rod gels, it is most important to ensure that the slab gel plates are perfectly clean to obtain good gel adhesion to the glass. The glass plates are cleaned by soaking them in chromic acid overnight, rinsed in water, and then with ethanol. Then the plates are put down onto clean tissue paper, with the side which is to be in contact with the gel uppermost, and swabbed with an acetone-soaked tissue held in a gloved hand. After a final rinse with ethanol, the plates are allowed to air dry.
2. The glass plates are usually held the correct distance apart by thin Perspex or Teflon spacers which *must* be of uniform thickness both with respect to each other and along their length to ensure good contact with the plates and a gel of uniform thickness. Usually for SDS-PAGE, these are about 1.5 mm thick although gels as thin as 0.75 mm can be used successfully. Several methods of sealing the glass-plate sandwich have been used. For the notched plate system of Studier (16), the easiest method we have tried is to grease the side spacers lightly with Petroleum jelly (Vaseline) and then lay these down the sides of the unnotched plate. The notched glass plate is then placed face down onto this using gloved hands and taking care not to allow the grease to touch anywhere but the plate edges. The plate assembly is clamped together with strong metal clips which are positioned to press on the sandwich just over the spacer positions. Next the assembly is stood vertically in a Perspex trough (*Figure 12a*) and the base sealed by filling the trough with 10 ml acrylamide

mixture prepared as described in *step 3* below. For gradient gels (p. 73) the trough is filled with 10 ml of the most concentrated acrylamide mixture. The polyacrylamide gel is poured after this sealing gel has been polymerised. The advantage of using a gel plug to seal the plate bottom is that after the polyacrylamide gel has been polymerised the assembly is ready for electrophoresis without the further manipulation associated with other methods.

The alternative common method of assembly is to use three spacers, one for each vertical side and one for the base of the glass-plate assembly (*Figure 12b*). These are sealed by coating with grease before assembly, or by dripping molten 2% agarose around the edges after assembly, or using a special adhesive tape (e.g. Tape UFT1/AT; Universal Scientific Ltd.) which is resistant to immersion in SDS-containing buffers. The spacer along the bottom edge is removed after the polyacrylamide gel has polymerised and prior to electrophoresis.

3. The clamped plate assembly is held vertically during pouring of the gel. Having selected an appropriate resolving gel concentration, the gel mixture is prepared by adding the correct volumes of all components (*Table 1* for SDS-phosphate or *Table 3* for non-dissociating buffer systems), except TEMED, to a small thick-walled flask. After degassing for 1 min using a water pump, the TEMED is added, gently mixed in, and the gel solution poured without delay between the glass plates to within 0.5 cm of the top. Immediately, a Perspex comb is inserted between the glass plates and into the gel mixture. The teeth of the comb should fit snugly against the glass plates. Special care is required to ensure that air bubbles are not trapped beneath the comb otherwise irregularly shaped sample wells will be formed.

4. The assembly is left undisturbed for the gel to polymerise (10-30 min), as evidenc-

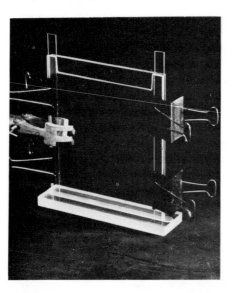 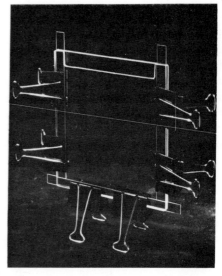

Figure 12. Methods for sealing the base of the glass-plate sandwich. (a) use of a Perspex trough; the trough (0.6 cm deep, 1.3 cm wide, 19.0 cm long) is milled in a Perspex block (1.25 cm thick, 5.0 cm wide, 19.5 cm long) and has a capacity of approximately 10 ml with the glass-plate sandwich in position. (b) use of three spacers. Details of each method are given in the text.

ed by the appearance of a sharp boundary below areas of gel/air interfaces. After a further 10 min, the comb should be carefully and slowly removed.

5. If the gel is to be used immediately, the sample wells formed by removal of the comb should be rinsed out with reservoir buffer using a Pasteur pipette, or a syringe fitted with a needle, and then filled with reservoir buffer. Any of the gel sections bordering the sample wells which have become distorted during removal of the comb can be repositioned vertically using a syringe needle or microspatula. If the gel is to be stored before use, the sample wells should be rinsed out with buffer of the same composition as in the gel and then filled with this buffer.

In some situations when a continuous buffer system is required, the resolving gel is too brittle to allow the ready formation of sample wells. This can be overcome by polymerising a low concentration gel (3.75-5.0%) on top of the resolving gel, and the sample wells formed in this in an analogous manner to that described for a discontinuous buffer system (see below) but with the low concentration gel containing the same buffer as the resolving gel (e.g. *Table 10*).

Discontinuous Buffer System

The use of a stacking gel polymerised on top of the resolving gel, as required by the discontinuous buffer system, means that sample wells are formed in the stacking gel.

1-2. The gel plates are cleaned and assembled as described for *steps 1-2* of slab gel preparation using the continuous buffer system.

3. Having selected an appropriate resolving gel concentration, the gel mixture is prepared in a small, thick-walled flask by mixing the components (except TEMED) in the volumes listed in *Table 2* for SDS-discontinuous or *Table 5* for non-dissociating buffer systems. The mixture is degassed for 1 min using a water pump, the correct volume of TEMED is added, and gently mixed in.

4. The resolving gel mixture is poured into the space between the glass plates leaving sufficient space at the top for a stacking gel to be polymerised later and sample wells formed. The stacking gel needs to be at least twice the height of the sample; 2 cm will be sufficient in most cases. Taking into account the depth of the sample wells, this means a space of about 3.5 cm needs to be left above the resolving gel. The volume of resolving gel required will obviously depend on the gel dimensions but using the apparatus shown in *Figure 10* it is 30 ml. Gel buffer, of the same composition as in the resolving gel, is now gently layered onto the gel surface using a peristaltic pump fitted with fine plastic tubing, or with a hypodermic syringe fitted with fine tubing. The tip of the tubing should be positioned so as to deliver the buffer just above the gel surface and at the centre of the glass plates.

5. After polymerisation (10-30 min), as evidenced by the presence of a sharp interface between the polymerised gel and the overlay, the assembly is tilted to pour off the overlay. Although the stacking gel can be polymerised in place and the gel used immediately, routinely the resolving gel is overlayered with buffer of the same composition as in the resolving gel and left overnight before use.

6. The stacking gel is prepared according to the protocol listed in *Table 2* for the SDS-discontinuous, or *Table 5* for non-dissociating buffers.

7. A small volume of the stacking gel mixture is used to rinse the surface of the

resolving gel. This is poured off and then the remaining space between the gel plates is filled with stacking gel mixture.

8. The comb is inserted immediately into the stacking gel mixture, being careful to avoid trapping any air bubbles beneath it. For riboflavin-catalysed stacking gel polymerisation only, a daylight fluorescent lamp is placed within 2-5 cm of the gel to initiate polymerisation. The assembly is left undisturbed whilst the stacking gel polymerises.

9. After polymerisation, the comb is carefully removed to expose the sample wells which are rinsed with reservoir buffer and then filled with this buffer. Any divisions between wells that have become displaced during comb removal may be straightened with a syringe needle or microspatula provided that care is taken to avoid damage. Once the stacking gel has been polymerised in place, the slab gel should be used immediately.

Sample Preparation

Dissociating Buffer System (SDS-PAGE)

It is important not to overload the gel or bands will be distorted. In the case of slab gels an overloaded sample in one track can also distort the electrophoretic pattern of bands in adjacent tracks. At the other extreme, underloading will result in one not detecting minor components and even major component bands may be too faint after staining for a good photographic record. Therefore it is wise to use standard assays to determine sample protein concentration before electrophoretic analysis is attempted. About 1-10 μg of each polypeptide should be loaded on the the the gel to give optimal results, such that for a complex mixture about 50-100μg is usually sufficient. The volume in which this is loaded is also important. The thickness of the starting zone and hence the volume of the sample applied has a large effect on protein band sharpness using the continuous buffer system (SDS-phosphate) and so the sample volume should be as small as possible. The stacking effect which occurs with the SDS-discontinuous system means that the method is essentially volume independent, but in practice the sample volume is limited in slab gel electrophoresis by the size of the sample wells and the fact that with large volumes large stacking gels are needed and some sideways spreading of polypeptide bands occurs. Best results using 1.5 mm thick slab gels with 7 mm wide sample wells are obtained using sample volumes of 10-30 μl but 60 μl still gives reasonable results. In rod gels, sample volumes should also be kept small for optimal resolution but the amount of sample loaded (in micrograms) and sample volumes can be higher because of the larger cross-sectional area of these gels.

A number of methods can be used for concentrating protein samples too dilute for immediate electrophoretic analysis. These include lyophilisation, ammonium sulphate precipitation, and dialysis against a high concentration of polyethylene glycol (mol.wt. 20,000). Alternatively, solid polyethylene glycol may be placed in contact with the outside of a dialysis bag containing the dilute protein sample. Sephadex G-100 or G-200 can be used in a similar manner. When the sample has been concentrated by any of these methods it should be dialysed against 0.01 M sodium phosphate buffer (pH 7.2) for the SDS-phosphate buffer system, or 0.0625 M Tris-HCl (pH 6.8) for the SDS-discontinuous system, to remove salts or low molecular weight polyethylene glycol impurities which may interfere with electrophoresis. Potassium ions in particular must be removed since they precipitate SDS. An appropriate volume of sample protein in 0.01 M phosphate buffer (pH 7.2) or 0.0625 M Tris-HCl (pH 6.8) is then brought to 2% SDS, 5% 2-mercaptoethanol, 10% sucrose (or glycerol) and 0.002% bromophenol blue using concentrated stock solutions. Since molecular weight estimations by SDS-PAGE depend on all polypeptides having the same charge density, and hence the same amount of SDS bound per unit weight of polypeptide (see p. 15), it is important that the SDS is present in excess. Knowing the protein content of the sample, one should calculate whether the SDS is indeed in excess; a ratio of at least 3:1 is required. At least 5% SDS can be added to the sample buffer of the SDS-discontinuous system without deleterious effects. Dithiothreitol can be used instead of 2-mercaptoethanol to disrupt polypeptide disulphide bonds and has the advantages that it is odourless and does not tend to auto-oxidise. The sucrose (or glycerol) is present to increase the density of the sample so that, when applied to the gel, the sample remains as a well-defined overlay and does not undergo convective mixing with the reservoir buffer during the early stages of electrophoresis.

A common method of sample protein concentration, especially for multiple samples, is to precipitate the protein with 10% trichloroacetic acid (TCA) (incubate in ice for 30 min), followed by centrifugation (10,000xg for 5 min), followed by repeated washing with ethanol-ether (1:1 vol/vol) to remove the TCA. Apart from speed in concentrating large numbers of samples, the method simultaneously removes interfering salts as well as some non-dialysable contaminants. Nevertheless, TCA precipitation should be used with caution since some glycoproteins and histones are soluble in low concentrations of TCA whilst in other cases the protein will precipitate but prove extremely difficult to redissolve completely in the sample buffer. Therefore, before using TCA precipitation routinely, it is wise to check on the efficiency of the procedure for the particular sample under study. After washing with ethanol-ether, TCA precipitates are dissolved directly in 0.01 M sodium phosphate buffer (pH 7.2), 2% SDS, 5% 2-mercaptoethanol, 10% sucrose or glycerol, 0.002% bromophenol blue for the SDS-phosphate buffer system or 0.0625 M Tris-HCl (pH 6.8), 2% SDS, 5% 2-mercaptoethanol, 10% sucrose or glycerol, 0.002% bromophenol blue for the SDS-discontinuous buffer system. The sample should be blue in colour. If it is yellow then there is still sufficient TCA present to interfere with electrophoresis and the pH must be adjusted by addition of microlitre volumes of concentrated Na_2HPO_4 (for the SDS-phosphate system) or Tris (for the SDS-discontinuous system). However, excessive addition of concentrated buffers can lead to problems during stacking and separation of proteins during electrophoresis and so a better approach is to remove the TCA efficiently initially. If the protein pellet does not completely dissolve with mixing in the neutralised solution it is possible that insufficient SDS is present.

An even simpler method of sample protein concentration which avoids the problems of acid neutralisation is acetone precipitation. Five volumes of cold acetone are added to the sample, mixed, and incubated at $-20°C$ for 10 min. The precipitated protein is collected by centrifugation (10,000xg for 5 min) and may be washed by repeated precipitation. After drying, the protein pellet is dissolved directly in the appropriate sample buffer and is then ready for analysis by polyacrylamide gel electrophoresis. As with TCA precipitation, it is important to check that the proteins under study are quantitatively precipitated before using this concentration method routinely.

Prior to electrophoresis, samples for SDS-PAGE are heated in a boiling waterbath for 3 min. This ensures denaturation of the protein. Use of lower temperatures may not dissociate certain 'metastable' protein complexes, especially certain proteases which may then proceed to degrade other sample proteins. After heating, the sample is allowed to cool to room temperature. Finally, and of crucial importance, any insoluble material should now be removed by centrifugation (10,000xg for 5 min) or this will cause protein 'streaking' during gel electrophoresis. The sample may be used immediately or stored in the freezer at $-20°C$. When cooled, SDS crystallises out of solution and so stored samples must be warmed before use.

Some proteins, particularly nuclear non-histone proteins, require the presence of 8M urea in the SDS-sample buffer if most of the protein is to enter the gel (p. 69). Similarly, immunoprecipitates and membrane proteins are often dissolved in SDS-sample buffer with urea added to aid solubilisation. When urea is present there is no need to add sucrose or glycerol to the sample buffer to increase its density. If the sample must be heated prior to loading, the sample should contain Tris as the buffer to minimise cyanate modification of proteins (p. 25).

Molecular weight standards. Whenever analytical SDS-PAGE is used, it is wise to include a mixture of polypeptides of known molecular weight. Whilst this is essential for determining the molecular weight of sample polypeptides (see p. 15), it is also worthwhile if only undertaking a qualitative assessment of the protein composition of samples since it provides a measure of reproducibility between different gel runs. A number of protein standards that are commercially available are listed in *Table 6*. Each of the standard proteins should be dissolved in the appropriate sample buffer at a concentration of 1 mg/ml, heated at 100°C for 3 min, and stored frozen in small aliquots. When a sample is to be analysed, an aliquot of molecular weight markers can be thawed out, warmed to dissolve any precipitated SDS, and run on a parallel rod gel or in parallel tracks of a slab gel. After electrophoresis, the gel is stained and destained and the distances migrated by the sample and marker polypeptides measured. The molecular weight of radioactive sample proteins can be determined by including radioactive marker proteins, either prepared in the laboratory (20) or purchased from commercial sources, followed by autoradiography or fluorography. Alternatively one can use unlabelled marker proteins and then indicate the position of each in the stained gel with a small spot of radioactive ink (p. 51) prior to autoradiography. A calibration curve of $\log_{10}$ molecular weight versus distance migrated is constructed using the data from the standard polypeptides (p. 15) and then the molecular weight of the sample polypeptides determined from this, knowing their migration distances. To rule out potential anomalies (see p. 16) the analysis should be repeated with at least one different gel concentration and shown to give no

Table 6. Molecular Weights of Polypeptide Standards.[a,b,c]

Polypeptide	Molecular weight
myosin (rabbit muscle) heavy chain	212,000
RNA polymerase *(E. coli)* β'-subunit	165,000
β -subunit	155,000
β-galactosidase *(E. coli)*	130,000
phosphorylase a (rabbit muscle)	92,500
bovine serum albumin	68,000
catalase (bovine liver)	57,500
pyruvate kinase (rabbit muscle)	57,200
glutamate dehydrogenase (bovine liver)	53,000
fumarase (pig liver)	48,500
ovalbumin	43,000
enolase (rabbit muscle)	42,000
alcohol dehydrogenase (horse liver)	41,000
aldolase (rabbit muscle)	40,000
RNA polymerase *(E. coli)* α-subunit	39,000
glyceraldehyde-3-phosphate dehydrogenase (rabbit muscle)	36,000
lactate dehydrogenase (pig heart)	36,000
carbonic anhydrase	29,000
chymotrypsinogen A	25,700
trypsin inhibitor (soybean)	20,100
myoglobin (horse heart)	16,950[d]
α-lactalbumin (bovine milk)	14,400
lysozyme (egg white)	14,300
cytochrome *c*	11,700

[a]Several proteases, for example trypsin, chymotrypsin, and papain, have been used as molecular weight standards by various workers but these may sometimes cause proteolysis of other polypeptide standards and so are omitted here.

[b]The polypeptide molecular weights given here are mainly from refs. 4, 89, and the references given in Appendix III, and are the molecular weights in the presence of excess thiol reagent. A more comprehensive list of suitable polypeptides is available from the original sources.

[c]The molecular weight range ~ 12,000 to 68,000 is reasonably well covered but there are few suitable proteins with subunit molecular weights above this range. This can be overcome by using polypeptides which are crosslinked to form an oligomeric series (18,19). Kits of these are commercially available.

[d]Calculated from the sequence data given in ref. 129.

significant change in the estimated molecular weight. Ideally, if sufficient material is available, several gel concentrations are used and a Ferguson plot constructed.

Non-dissociating Buffer Systems

The sample proteins should be present at 1-10 μg per protein in the volume which is to be loaded onto a slab gel, or about 100 μg for a complex protein mixture. To increase zone-sharpening using a continuous buffer system, the sample proteins should be dissolved in buffer diluted 5-10 fold over the concentration used in the gel (see p. 11) and containing 10% sucrose (or glycerol) plus a tracking dye (0.002%) the choice of which will depend on the pH at which the electrophoresis is to be performed. Bromophenol blue is a widely used tracking dye for separations at alkaline pH and methyl green can be used for acid pH. Whether diluted sample buffer can be used or not will depend upon the concentration of buffer used in the gel, since if the ionic strength of the sample buffer falls too low, protein aggregation may occur. Samples for analysis using one of the discontinuous buffer systems should be present in 1/4-1/8 diluted stacking gel buffer stock, 10% sucrose (or glycerol), plus a suitable tracking dye (final concentration 0.002%).

Depending on the protein under study, concentration of dilute samples can be achieved by protein precipitation (with acetone, alcohol, or ammonium sulphate), ion-exchange chromatography, use of Diaflo membrane filters, vacuum ultra-filtration, lyophilisation, or use of hydrophilic polymers such as polyethylene glycol or Sephadex (p. 36). Obviously the trichloroacetic acid precipitation method used for concentrating samples for SDS-PAGE cannot be used for the concentration of native protein samples. If necessary, samples should be dialysed against the sample buffer to be used in the electrophoretic analysis (but lacking sucrose, glycerol, or tracking dye) to remove salts or other low molecular weight contaminants which could interfere with electrophoretic analysis. All samples are then centrifuged (10,000xg for 15 min at 4°C) to remove any insoluble material which would interfere with electrophoresis. Sucrose or glycerol may then be added to 10% final concentration and finally a tracking dye is added.

Usually all steps involved in the preparation of native protein samples are performed at low temperature (0-4°C) both to reduce loss of protein activity through denaturation and to minimise attack by any proteases in the sample. The presence of protease inhibitors, such as phenylmethylsulphonyl fluoride (PMSF) for serine active site proteases, is also helpful in this respect. The problem with all these reagents is that they can only be used if they do not also inactivate the protein of interest. Once ready for electrophoresis the samples may be used immediately, or stored in sample buffer containing glycerol at 4°C, or frozen if the proteins are stable under these conditions.

Sample Loading and Electrophoresis

The method of loading liquid samples is essentially the same for both rod and slab gels using any buffer system. The method given here for slab gels is for the notched glass-plate system of Studier (16) where the slab gel is attached to the electrophoresis apparatus before sample loading. Slab gels formed between the unnotched glass plates of some cooled commercial apparatuses are attached to the apparatus only after sample loading. The method of attachment varies with the commercial apparatus being used (see *Figure 11* legend).

1. Any gel formers, rubber stoppers, Parafilm, etc. are removed from the base of the rod or slab gels and the gels are mounted in place on the electrophoresis apparatus. Rod gels are placed in rubber grommets located in the upper buffer reservoir of the electrophoresis apparatus (*Figure 9*). Slab gels prepared in notched glass-plate sandwiches are clamped in position using metal clips with the notch of the glass plate aligned with the notch in the upper reservoir (*Figure 13*). A seal between the glass plate and the electrophoresis apparatus can be made by clamping a silicone-rubber sheet (cut to the dimensions of the notched glass plate) between the glass-plate sandwich and the apparatus. The best seal is obtained if the silicone rubber is slightly

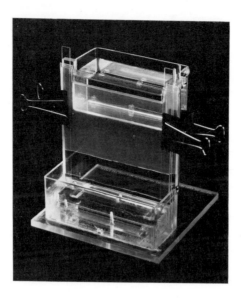

Figure 13. Attachment of the slab gel to the Studier-type electrophoresis apparatus. In the example shown, a seal has been made by clamping a silicone-rubber sheet between the glass-plate sandwich and the apparatus (see the text).

prewarmed just before use. An alternative method is simply to coat the area around the apparatus notch with Vaseline so that this forms a seal when the glass-plate sandwich is clamped in position.

2. Reservoir buffer is added to the lower reservoir of the electrophoresis apparatus. Next, it is essential to remove any air bubbles from the bottom of the gel or these will prevent uniform electrical contact between the gel and reservoir buffer. This can be achieved using a Pasteur pipette with a bent tip. Alternatively for rod gels, the bubbles are easily removed by flicking the top of the gel tube with a gloved finger.

3. Any unused holders in the rod or slab gel apparatus are plugged and electrode buffer is added to the upper reservoir.

4. After checking for leaks the gel surface is washed by directing a gentle stream of electrode buffer into each sample well of a slab gel or onto the gel surface of a rod gel using a Pasteur pipette. This also serves to fill the space above the gel with reservoir buffer, hence removing any air bubbles present.

5. The sample is now carefully loaded onto the gel surface using a microsyringe or micropipette. The tip of the sample applicator should be held only 1-2 mm above the gel surface to minimise sample mixing with the reservoir buffer during loading. The dense sample solution will flow onto the gel surface and form a sharply-defined layer. Best results with slab gels are achieved if unused wells are filled with an equivalent volume of blank sample buffer.

6. The electrophoresis apparatus is connected to the power pack (switched off) with the anode (+) connected to the bottom reservoir for SDS-PAGE and electrophoresis of negatively charged proteins in other buffer systems, and the cathode (−) connected to the upper reservoir. For the electrophoresis of positively charged proteins, the polarity of the buffer reservoirs is reversed.

7. The power pack is now connected to the mains, switched on, and adjusted to deliver the necessary current and voltage. Electrophoresis conditions are quoted as either constant current or constant voltage. Either mode of operation is permissible although only constant voltage conditions give constant protein mobility during electrophoresis. The exact electrophoresis conditions will depend on the buffer and gel conditions employed and usually need to be determined empirically. In general, too-high current has the danger of excessive heating whilst too-low voltage increases electrophoresis time and may decrease resolution as a result of band diffusion. For the SDS-phosphate buffer system, 4 mA constant current is applied per rod gel until the sample has entered the gel and then 6 mA constant current per 9 cm rod gel for about 3h (5% gels), 4h (10% gels) or 8h (15% gels). For the lower conductivity SDS-discontinuous system, 2-3 mA constant current per rod gel or 100 V constant voltage allows electrophoresis to be completed in similar time periods. Longer rod gels take proportionately longer to run at constant voltage. Similarly, the electrophoresis conditions for slab gels with the SDS-discontinuous system depend on the gel dimensions and gel concentration. For the slab gel format described earlier (p. 20) one can electrophorese 10% acrylamide gels at room temperature at about 50 V overnight. Higher voltages are necessary for overnight runs with higher percentage slab gels. Alternatively, slab gels can be electrophoresed during the day at 25-30 mA constant current or by stacking at 120 V followed by electrophoresis at 200 V constant voltage.

Electrophoresis conditions with non-dissociating buffers vary enormously depending on the conductivity of the buffer system used. In general one should not exceed about 2 mA per rod and 100-200 V to avoid excessive heating.

ANALYSIS OF GELS FOLLOWING ELECTROPHORESIS

It is very important that disposable plastic gloves are worn when handling gels to prevent proteins from the skin being transferred to the gels. In addition the gels must not be allowed to come into contact with paper surfaces since polyacrylamide readily sticks to such surfaces.

Recovery of Gels

Slab gels are easily recovered by removing the side spacers and gently levering the glass plates apart at the end away from the notch to avoid damage to the fragile notched end. Leave the gel resting on the unnotched plate. Higher percentage gels (above about 10% acrylamide) may be carefully lifted between thumb and forefinger, using both hands, at the gel base and transferred to a suitable rectangular solvent-resistant plastic tray (ideally at least 5 cm longer and wider than the gel) for soaking in buffers or stains involved in analysis. Some workers find a broad plastic spatula useful in handling lower percentage slab gels.

Rod gels are more difficult to recover. Basically one of two methods may be used; either the gel is removed from the intact tube by rimming or the gel tube is broken and the gel washed free of tube fragments. The aim of the rimming procedure is to gently pry the gel away from the tube wall, without damaging it, using water

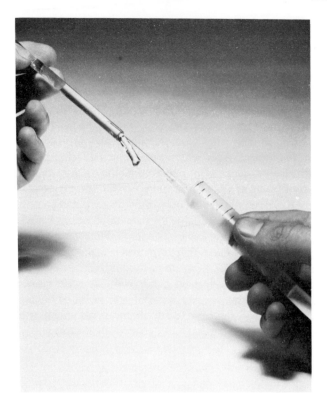

Figure 14.   Recovery of a rod gel from its glass tube by rimming.

pressure. The rod gel is held in one hand whilst a syringe fitted with a long, fine, blunt needle (e.g. 2.5 inch 23 gauge) is used to squirt water in between the gel surface and the tube wall whilst slowly rotating the gel *(Figure 14)*. With siliconised glass tubes the gel usually slides out fairly easily, so easily that the operation should be carried out over a tray (not the sink!). If the gel does not detach itself easily then begin the whole procedure again at the other end of the tube. If required, pressure can be exerted in the final stages of removing the loosened gel by applying a Pasteur pipette bulb to one end of the gel tube. This rimming procedure works well with low concentration polyacrylamide gels (below about 12%) but at gel concentrations above this, especially if the gel is in an unsiliconised glass tube, it may be necessary to use the second method which is to break the tube itself. There are several ways of doing this. Perhaps the simplest is to wrap the gel tube in tissue and place it between the jaws of a workshop vice. The jaws are tightened until a crack is heard and then released. The gel tube fractures into a number of pieces along its length and the gel can be washed free of these without damage. Alternatively one can use a heavy hammer to carefully break the tube, starting from one end, without damaging the gel.

   Once the rod or slab gel has been recovered it is essential to mark which end is which. In addition it is sometimes necessary to mark the position of the tracking dye for later R_f determinations. Both of these requirements are met by inserting a small piece of fine-gauge wire (about 0.1 mm in diameter) into the gel at the dye front. Alternatively one can inject a small amount of India ink at this point.

Protein Staining and Quantitation

Protein bands of sufficiently high concentration may be localised by direct photometric scanning of unstained gels at 280 nm. The gel is placed in a suitable quartz trough (or actually run in quartz tubes) and scanned using any suitable modern spectrophotometer fitted with a gel scanning attachment. Various U.V. absorbing impurities are present in many monomer stocks and these give high background absorbance, as will any unreacted acrylamide monomer. Although much of this problem can be eliminated by using pure reagents and correct polymerisation protocols, the method is still unsuitable for most applications because the extinction coefficient for proteins is rather low at 280 nm and light scattering can be a serious problem. These limitations mean that only major components can be quantitated with precision. Because of this, the usual method is to increase the sensitivity of protein detection by reacting the protein bands with an easily visualised reagent. A number of reagents will react with any protein irrespective of biological activity and these are discussed below. Of these, easily the most commonly used is Coomassie blue staining because of its simplicity and general applicability.

Coomassie Blue Staining

Most of the early work with polyacrylamide gel electrophoresis of proteins used amido black as the stain but this has been almost completely superceded by Coomassie blue which is far more sensitive. Usually 0.2-0.5 μg of any protein in a sharp band can be detected using this dye and staining is quantitative to 15 μg for at least some proteins.

In early procedures it was common to fix proteins in the gel first and then stain them but this has been superceded by methodology in which protein fixation and staining occur simultaneously. Since Coomassie blue is predominantly non-polar, it is usually used in methanolic solution and excess dye removed from the gel later by destaining. Stacking gels are usually discarded before staining the resolving gels unless the investigator wishes to test for proteins unable to enter the resolving gel.

Various recipes have been used for the staining solution containing dye, methanol, and acetic acid with the exact proportions of each varying. A common protocol is as follows.

Dissolve Coomassie blue R250 (0.1%) in water:methanol:glacial acetic acid (5:5:2 by volume) and filter through a Whatman No. 1 filter paper to remove any insoluble material before use. Each rod gel is placed in a test-tube filled with stain and left for at least 4h at room temperature. The minimum time required for staining depends on the gel thickness and the gel concentration, increasing as the gel concentration increases. The rate of staining is increased at higher temperatures, for example 40-50°C, but in practice it is often more convenient simply to allow the gels to stand in stain overnight at room temperature and destain the following day. Gel slabs are placed in a plastic tray containing stain solution and are stained fully within about

4-6h at room temperature if only 1.5 mm thick. Some workers find it useful to suspend the gel slab on a nylon mesh screen held on a Perspex frame during staining and destaining. It has been reported that high concentrations of SDS interfere with Coomassie blue staining but this is avoided if at least 10 volumes of stain solution are used with each gel, for example 500 ml stain solution per slab gel.

After staining is complete, excess stain must be removed to allow protein bands to be seen clearly. Two different methods of destaining may be employed; diffusion destaining or electrophoretic destaining. In diffusion destaining the gel is transferred to 12.5% isopropanol, 10% acetic acid and the destain solution is simply renewed as stain leaches out of the gel over a period of about 48h at room temperature (faster at higher temperatures). Although most background staining has been removed by this time, longer destaining may be necessary to obtain completely clear backgrounds. Faster destaining is possible using 30% methanol, 10% acetic acid but destained gels should not be stored in this since protein bands are eventually destained also. An alternative to the replacement of destaining solution is to include a few grams of an anion-exchange resin to absorb the stain as it leaches from the gel. Commercial diffusion destainers are available for both rods and slabs (e.g. Bio-Rad Laboratories Ltd.) and these usually contain a cartridge of a suitable resin or charcoal for absorbing the dye.

Electrophoretic destaining is possible because Coomassie blue is an anionic dye. Although electrophoretic destaining can be carried out transversely or longitudinally, the former is preferred because of the shorter gel electrophoretic distance and hence shorter time period involved. Transverse electrophoretic destaining apparatuses for both rod and slab gels are available commercially but these can be easily manufactured in the average departmental workshop. The gels are held between porous plastic sheets with horizontal electrodes placed either side. The destaining apparatus is filled with 7% acetic acid and connected to a power pack capable of delivering a high current (0.25-1.0 A); either a commercially available model or even a car battery charger (delivering $\sim$ 12 V, 3 A). Because of the high power input the electrode buffer reservoirs should be fitted with efficient cooling coils. Destaining occurs within 15-30 min. It is important to note that if destaining is continued too long some destaining of the protein bands may occur and weakly stained bands may be destained completely.

Staining methods have also been described where destaining is not necessary in order to observe the protein banding pattern. In these methods Coomassie blue is dissolved in TCA (21, 22) or perchloric acid (23) in which it is relatively insoluble and so forms dye-protein complexes preferentially. Therefore, staining is rapid; dense protein bands appear within seconds and almost full intensity is reached after about 45 min although the staining method is reported by some workers to not be quite as sensitive as those involving methanol-based stains followed by destaining. The advantage of these methods is their speed. Furthermore, TCA is a much better fixative than acetic acid and so may be preferred if one suspects loss of protein from the gel during staining. This is discussed more fully in Chapter 2 (p.117). However, most workers have continued to use the methanol-based Coomassie blue stain for routine detection and quantitation of protein bands.

Quantitation of Coomassie blue stained proteins. A number of instruments specifically designed for scanning are available as well as gel scanning attachments for standard spectrophotometers. Using the latter instrumentation, each destained rod gel is placed in a shallow glass trough containing 7% acetic acid. Individual tracks may also be cut from slab gels and placed against one trough wall for scanning. Most of these apparatuses move the gel at a constant speed perpendicular to a narrow, fixed, parallel light beam (550 nm), the transmission of which is detected by a photomultiplier. The output of the spectrophotometer is recorded as a series of tracings on chart paper. Unfortunately the optics of many of these apparatuses is such that sharp bands clearly separated by the naked eye often do not give clearly separated peaks by densitometry (*Figure 15*). Some densitometers incorporate automatic integration of the densitometric record which allows automatic integration of the areas under each peak. By this means the amounts of a single component in different gels may be estimated. For machines lacking automatic integration one can cut out the chart paper and weigh the peaks. Alternatively, for well-resolved stained bands, these may be cut out, macerated, and the Coomassie blue eluted overnight by shaking at room temperature with 25% (v/v) pyridine in water (24). The pyridine shifts the dye absorption maximum from 550 nm to 605 nm. Quantitation of dye eluted by absorbance at 605 nm allows quantitation of proteins in bands in the range 1-100 μg. This method may be particularly useful for quantitating proteins in spots after two-dimensional gel electrophoresis (Chapter 5).

Different proteins bind Coomassie blue to different extents and so quantitative determination of the amount of a particular protein by staining requires a standard curve for that particular protein. This applies whether quantitation is by scanning or elution. However, if another protein is used as a standard, this allows the *relative* amounts of the specific protein to be determined in multiple samples although the linearity of dye absorbance with mass still needs to be determined for the proteins under study.

Photography of stained protein bands. For storage purposes, rod gels can be placed in sealed tubes and slab gels in sealed plastic bags each containing 7% acetic acid. Stained bands remain visible almost indefinitely under these conditions. However, gels stored in this wet state take up large amounts of space and are easily mislaid. The best procedure with slab gels is to photograph them and then to dry them down (photography of wet gels gives better results than using dried gels). Both photographs and dried gels are easily labelled and stored in notebooks. Rod gels are also best photographed for record purposes. They can be dried down after longitudinal slicing into thin strips (25) but this is not commonly practised since the major applications of dried gels (autoradiography and fluorography; see pp. 50-55) are best served using the slab rather than the rod gel format.

Wet rod or slab gels are photographed by being laid directly onto an illuminator with an opal white screen (avoiding trapped air bubbles) and kept wet during photography by addition of 7% acetic acid. A fine-grain, panchromatic film should be used, for example Ilford Pan F, and a medium-red filter will increase band contrast. It is worth including a gel title in the photograph to identify the samples later. Also, ensure that the entire gel is photographed to enable R_f values of proteins to be

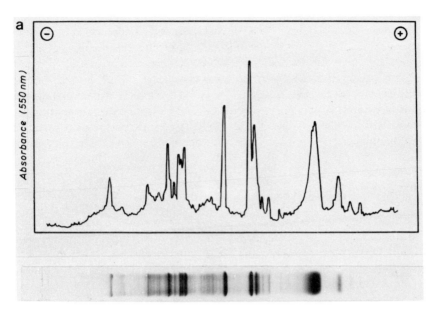

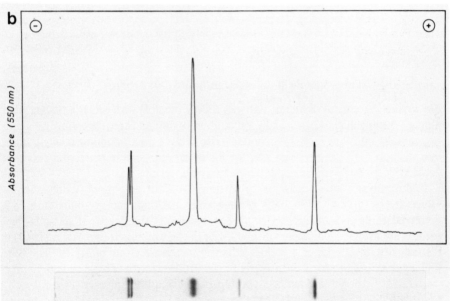

Figure 15. Densitometric analysis of Coomassie blue stained protein bands. (a) A human placental membrane subfraction obtained by wheat germ agglutinin precipitation (kindly donated by Dr. A.G.Booth) was analysed on a 6-18% polyacrylamide linear gradient slab gel, and then stained with Coomassie blue R250. A photograph of the stained gel track is shown and above this is the densitometric scan of this track at 550 nm. (b) A track of several molecular weight marker polypeptides (RNA polymerase α, β, β' subunits, bovine serum albumin, and soybean trypsin inhibitor) analysed on a 6-18% polyacrylamide linear gradient slab gel, stained and scanned as in (a).

Figure 16. Bio-Rad gel drier, model 224. The silicone-rubber sheet attached to the apparatus has been rolled back to show the stainless steel screen used to support the gels during drying. The porous plastic sheet, also required for gel drying, is not shown here.

calculated from photographs if necessary in future data analysis.

Gel drying. Successful drying of slab gels is only possible with gels not thicker than about 1.5 mm; thicker gels usually crack upon drying and must therefore be sliced horizontally into thin sheets. Apparatus to do this is available commercially but the best method is to use thin slab gels for the initial electrophoresis so that slicing is avoided.

If the polyacrylamide gels are dried down without any support they shrivel. However, by drying down onto a filter-paper backing, the gel dimensions are preserved as the gel attaches itself to the paper support. Many types of gel drying apparatus are available commercially and differ in the size of gels they will accept and the presence or absence of a built-in heater to increase the rate of drying. The Bio-Rad gel drier (*Figure 16*) is particularly recommended since it is large enough to take two standard-size slab gels and has a built-in heater coupled to a timer. Two slab gels, 1.5 mm thick, will usually dry in less than one hour using this apparatus. Best results are obtained if the gel is soaked overnight in a solution containing 3% glycerol prior to drying. The solution should contain isopropanol or methanol in sufficiently high concentration to prevent swelling of the gel to greater than its original size, or cracking may result. Some concentration gradient gels (e.g. 6-18% acrylamide) can be dried successfully if they are first soaked in 30% methanol, 3% glycerol, but difficulties may be experienced at higher gel concentrations. Two sheets of 3 MM filter paper are placed on the stainless-steel support screen and wetted with water. The slab gel or rod gel slices are then aligned on this, being careful not to trap air bubbles under the gel since this can result in cracking during drying. This is overlaid with a sheet of pre-wetted cellophane, then a porous plastic sheet, and finally the silicone sheet (attached to the apparatus) forming a leak-proof seal. Vacuum is supplied by a good water aspirator or vacuum pump fitted with a cold-finger water-trap and the heating block turned on. The exact time required for drying depends on the size and concentration of the gels being used. If air is allowed to enter the assembly before drying is complete the gel will crack irreparably. The resulting dried gel is sandwiched between its filter paper backing and a protective cellophane sheet. When radioactive gels are dried, the cellophane is replaced with Saran Wrap (cling film) to prevent contamination of the porous plastic sheet.

A number of gel driers have also been described which can be made in the laboratory, the simplest of these being that of Maizel (26).

Silver Stain

Recently a highly sensitive silver stain has been reported for proteins in polyacrylamide gel (27) which is claimed to be up to 100 times more sensitive than Coomassie blue, apparently being able to detect 0.38 ng/mm^2 of bovine serum albumin. The relative density of silver stain is approximately linear with protein concentration for some proteins (e.g. phosphorylase) but markedly non-linear for others (e.g. bovine serum albumin) over the same range. The general usefulness and reliability of this method remain to be proven. A simplified protocol which still retains the sensitivity of the original method has recently been reported (28).

Fluorescent Protein Labels

Methods have also been devised which rely on covalent labelling of proteins with fluorescent molecules prior to electrophoresis and detection of the fluorescent bands after electrophoresis by scanning. Although dansylation using dansyl chloride was the first method to be described for labelling proteins in this way, recent analytical studies have preferred 'fluorescamine' {4-phenylspiro-[furan-2(3H),1-phthalan]-3,3'-dione}, trade name 'Fluram', since, unlike dansyl chloride, neither free fluorescamine nor its hydrolysis products are fluorescent, leaving the labelled protein as the only fluorescent component. The major advantages of these reagents are the increased sensitivity of detection which they afford (1 ng of protein can be detected once it is labelled with fluorescamine) and the ability to monitor the progress of electrophoretic separation simply by exposing the gel to U.V. light in a darkened room. The disadvantage of fluorescamine is that it converts an amino group to a carboxyl group when it reacts with proteins and so will alter the mobility of proteins in non-SDS containing buffers. In fact, fluorescamine has also been reported to alter the mobility of some proteins in SDS-PAGE (29), although other workers using the same proteins found that these still obeyed the usual relationship between R_f and $\log_{10}$ molecular weight (30). Barger *et al.* (31) have introduced another fluorescent label called MDPF [2-methoxy-2,4-diphenyl-3(2H)-furanone] which has the advantage that MDPF-labelled proteins in polyacrylamide gels retain their fluorescence for several months whereas that due to fluorescamine is reported to decay over a period of 24 h. Using MDPF the fluorescence signal is linear with protein from 1-500 ng and labelled proteins obey the linear relationship between $\log_{10}$ molecular weight and R_f when analysed by SDS-PAGE. Details on the use of fluorescamine and MDPF and the equipment required for scanning are given elsewhere (32).

Possibly because of the requirement for special scanning equipment when quantitating fluorescent proteins and the applicability of Coomassie blue staining to most situations, the technique of pre-labelling proteins with fluorescent molecules prior to electrophoresis has not found widespread use in analytical polyacrylamide gel electrophoresis.

Protein bands can also be detected *after* electrophoresis by labelling with fluorescent molecules (33,34), but again these methods have been little used in analytical zone electrophoresis.

Detection of Radioactive Proteins

Proteins can be radiolabelled either during synthesis, using labelled amino acids, or post-synthetically by iodination, reductive methylation, etc. (Appendix II). A major advantage of radiolabelling is that detection methods for radiolabelled proteins following gel electrophoresis are far more sensitive than staining methods for unlabelled proteins. Two basic approaches are possible. The first is to place the gel next to X-ray film whereupon the radioactive emissions cause the production of silver atoms within silver halide crystals and these are visualised after developing the film (*Figure 17a*). Important variations of this method exist which serve to increase the sensitivity of detection whilst, in some cases, maintaining the linear relationship between film image absorbance and sample radioactivity (see later). The film image may be recorded photographically or densitometrically by using a spectrophotometer fitted with a scanning attachment. The latter method also enables quantitation by measurement of the peak area, although, as with scanning stained bands, the resolution of closely migrating bands is poor. The second approach involves slicing the gel after electrophoresis and determining the radioactive protein content of each slice by scintillation counting (*Figure 17b*). This is the approach most commonly used for quantitation, especially in dual isotope experiments. However, the resolution of the method is dependent upon the size of the fractions produced which for gel slicing techniques is not usually less than 0.5 mm. Higher resolution is obtained using X-ray film detection methods on intact gels and so these should be used initially, especially with more complex protein mixtures, to identify the number of polypeptides present.

Methods using X-ray Film

The optimum exposure time has to be determined empirically for each gel since it will depend on the isotope used, the amount of radioactivity present in each protein band, and whether the researcher wishes to detect only major components or also more minor ones. Several exposures of different durations will probably be required to detect all or most components in a protein mixture since, whilst a long exposure will detect minor bands, band spreading of the photographic image occurs for major components during long exposures and so bands become indistinct and may overlap.

Direct autoradiography. ^{32}P- and ^{125}I- labelled proteins can be easily detected in wet

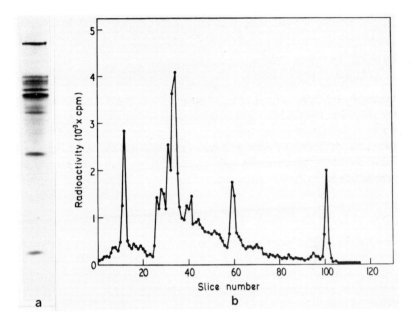

Figure 17. Detection of radioactive protein bands by (a) autoradiography or (b) gel slicing followed by scintillation counting of the slice contents. Detailed protocols are given in the text. The sample shown here is a subfraction of [^{35}S]-methionine-labelled spore proteins of *Dictyostelium discoideum*.

gels simply by sealing the slab gel or rod gel slices with cling-film (to prevent dehydration of the gel and wetting of the film) and placing this against an X-ray film. However, resolution is much improved if the gel is first dried down onto a sheet of Whatman 3 MM filter paper (p. 48). Usually the gel is stained with Coomassie blue and destained before drying so that the photographic image obtained after autoradiography can be compared with the stained protein banding pattern. To facilitate alignment of the stained, dried gel and the developed X-ray film, it is wise to place a few spots of radioactive ink (~ 1 μCi^{14}C per ml ink) on the filter paper at the gel perimeter prior to autoradiography. This ink is also useful for labelling the gel (and hence the X-ray film) to avoid errors later. The stained dried gel with radioactive ink marker spots is placed in direct contact with the X-ray film, clamped in a radiographic cassette or between glass or hardboard plates using metal clips, and then wrapped in a black plastic bag or placed in a light-tight box. After exposure (in a location away from external sources of ionising radiation), the film is developed according to the manufacturer's instructions.

Direct autoradiography of dried polyacrylamide gel using Kodirex X-ray film will give a film image absorbance of 0.02 A_{540} units (just visible above background) in 24h exposure with about 6,000 dpm/cm^2 of ^{14}C or ^{35}S, 1,600 dpm/cm^2 of ^{125}I, or about 500 dpm/cm^2 of ^{32}P (35,36). The film image absorbance is proportional to sample radioactivity. Unfortunately, ^{3}H is not detected since the low-energy β-particles fail to penetrate the gel matrix. However, fluorography (see below) can detect ^{3}H and also increases the sensitivity of ^{14}C and ^{35}S detection over that possible using direct autoradiography (35,37). In addition, Laskey and Mills (36) have developed a variation of this method, called 'indirect autoradiography' which increases the detection efficiency of ^{125}I and ^{32}P considerably compared to direct autoradiography. The methodologies of fluorography and indirect autoradiography have been recently reviewed by Laskey (38).

Fluorography (for ^3H, ^{14}C, ^{35}S). In fluorography the gel is impregnated with a scintillator (2,5 diphenyloxazole; PPO), dried down and exposed to the X-ray film at $-70°C$. The method is less sensitive when exposure is carried out at higher temperatures. Although low-energy β-particles produced by ^3H-labelled proteins are unable to penetrate the gel matrix and expose the X-ray film directly, they are able to interact with the PPO molecules in the gel to convert the energy of the β-particle to visible light which then forms an image on blue-sensitive X-ray film. Thus absorption of β-particles by the sample and gel matrix is overcome by the increased penetration of light.

The basic fluorographic procedure is given in *Table 7*. It may be used on either un-

Table 7. The Basic Fluorographic Technique.[a]

Caution. Rubber gloves should be worn throughout the procedure to avoid skin contact with DMSO.

1. Directly after electrophoresis, or after staining/destaining, the gel is soaked in about 20 times its volume of DMSO for 30 min followed by a second 30 min immersion in fresh DMSO.[b] These separate batches of DMSO can be stored for use (in the same sequence) with other gels. Prolonged use is to be avoided since all water must be removed or the PPO will not enter the gel.

2. The gel is immersed in 4 volumes of 20% (w/w) PPO in DMSO for 3 h.[c]

3. The gel is then immersed in 20 volumes of water for 1 h. The PPO is soluble in DMSO but insoluble in water. Therefore, PPO precipitates within the gel matrix turning the gel opaque.

4. The gel is dried under vacuum (p.48).[d]

5. X-ray film is placed in contact with the gel and exposed at $-70°C$. The most sensitive film is Kodak X-Omat R, XR1, but where this is not available the slightly less sensitive Fuji RX film should be used instead. The relative sensitivities of films for fluorography are listed in ref. 36. Kodirex and Kodak No-screen X-ray films, much used in autoradiography, are inefficient at recording the visible light produced by the fluorographic technique and so are not used for fluorography.

6. After exposure, the film is unwrapped before it warms up (to avoid physical fogging) and developed according to the manufacturer's instructions.

---

[a]Adapted from refs 35, 37.

[b]Methanol is used instead of DMSO for gels which contain less than 2% acrylamide plus 0.5% agarose. After electrophoresis the gel is soaked in 20 volumes of methanol for 30 min followed by a second 30 min immersion in fresh methanol. Then the gel is immersed in 10% (w/w) PPO in methanol for 3 h and finally dried under vacuum without heat.

[c]Excess PPO may be recovered from the DMSO by adding 1 volume PPO in DMSO to 3 volumes 10% (v/v) ethanol. After 10 min, the suspension is filtered and the PPO precipitate is washed with 20 volumes water, then air dried.

[d]If gel cracking during drying is a problem, the gel can be successfully dried by soaking in methanolic solutions (up to 30% methanol) containing 3% glycerol (see p.48) after step 3 and prior to gel drying.

fixed or stained gels. Exposure to dimethylsulphoxide (DMSO) removes some Coomassie blue from protein bands such that faintly stained bands may not be visible at the end of the procedure. Furthermore the precipitate of PPO which is formed in the gel obscures the staining pattern. For these reasons a photographic record should be made of stained gels before attempting fluorography. Alignment of stained bands in the gel with the final fluorographic image can be made using radioactive ink spots marked on the gel perimeter after drying and prior to clamping the dried gel and X-ray film together (p. 51).

Unfortunately, using untreated X-ray film, the absorbance of the fluorographic image is not proportional to the amount of radioactivity in the sample, small amounts of radioactivity producing disproportionately faint images. This can be overcome by exposing the film to an instantaneous flash of light (≤ 1 msec) prior to using film for the fluorographic detection of radioactive bands (35). The procedure for controlled pre-exposure of X-ray film is detailed in *Table 8*. When the film is pre-exposed to between 0.1 and 0.2 A_{540} units, the absorbance of the fluorographic image becomes proportional to sample radioactivity and the sensitivity of the method increases still further compared to direct autoradiography such that 400 dpm/cm^2 of ^{14}C or ^{35}S, or 8,000 dpm/cm^2 of ^{3}H are detectable (*Table 9*). It is essential to note that only under these conditions of film pre-exposure does the absorbance profile of the fluorographic images represent the true distribution of radioactivity in the sample. Therefore, a pre-exposure of 0.15 A_{540} units is used routinely for all quantitative fluorography. The absorbance of any film image produced by a radioactive sample protein is quantitated using a scanning spectrophotometer at 540 nm. The linearity of film image absorbance to radioactivity should be checked with each batch of film, and over the range required in the experiment, by measuring the actual amounts of radioactivity in selected gel slices using liquid scintillation counting (p. 56).

Increasing the pre-exposure of X-ray film above 0.2 A_{540} units increases the sen-

Table 8. Pre-exposure of X-ray Film for Quantitative Fluorography.[a]

Pre-exposure is performed using a single flash from an electronic flash unit e.g. Vivitar 283. It is essential that the duration of the light flash is short (≤ 1 msec) since flashes of longer duration only increase the background 'fog' absorbance without hypersensitising the film. To overcome slight variations in charging the capacitor, the unit is used only after the charging lamp has been illuminated for at least 30 sec. Three filters are taped to the window of the flash unit to reduce and diffuse the light output:

    (i) an infrared-absorbing filter; placed nearest the flash unit to protect the other filters from the heat generated.

    (ii) a coloured filter; to reduce light output. A 'Deep Orange' Kodak Wratten No. 22 is suitable for the above flash unit whereas an 'Orange' Kodak Wratten No. 21 is used for weaker flash units.

    (iii) porous paper (e.g. Whatman No. 1 filter paper); to diffuse the image of the bulb so that the film is evenly exposed.

Minor adjustments to illumination intensity are made by varying either the distance between the film and the light source (usually 60-70 cm) or the diameter of an aperture in an opaque mask. The film is backed by yellow paper during pre-exposure, and the surface which has been facing the light source is applied to the gel. The degree of pre-exposure is determined by using any conventional spectrophotometer to measure the increase in background 'fog' absorbance at 540 nm of pre-exposed film compared to unexposed film. Storage of films after pre-exposure is not recommended.

---

[a]Method derived from ref. 35.

Table 9. Sensitivities of Methods for Radioisotope Detection in Polyacrylamide Gel.[a]

Isotope	Type of method	dpm/cm² required for detectable image (A_{540} = 0.02) in 24h	Enhancement over direct autoradiography
^{125}I	indirect autoradiography	100	16
^{32}P	indirect autoradiography	50	10.5
^{14}C	fluorography	400	15
^{35}S	fluorography	400	15
^{3}H	fluorography	8,000	>1000

[a]From ref. 38 with permission. The data are for exposure at $-70°C$ using X-ray film pre-exposed to A_{540} = 0.15 above the background absorbance of unexposed film. Direct autoradiography for comparison was performed on Kodirex film.

sitivity of ^3H, ^{14}C, and ^{35}S detection by fluorography still further, but the absorbance of the film image ceases to be proportional to the radioactivity present (35).

Recently, Chamberlain (39) has introduced the use of sodium salicylate as the fluor in fluorography, instead of PPO. One advantage of salicylate over PPO is that it is considerably less expensive. The other main advantage is that salicylate is water-soluble, so that fluorography using this fluor avoids lengthy equilibration of the gel in DMSO followed by washing with water and typically takes 0.5-1h, instead of about 5h required with the DMSO-PPO system. According to the data published so far, fluorography using salicylate with pre-exposed X-ray film appears to be as sensitive as that using DMSO-PPO, with similar linearity of fluorographic image absorbance to sample radioactivity (39). The only disadvantage noted is a slight diffusiveness of the film image using salicylate compared to PPO.

Indirect autoradiography (for ^{32}P or ^{125}I). The method involves placing a pre-exposed X-ray film against the gel (which is either wet or dry, stained or unstained, but otherwise untreated) and then against the other side of the film is placed a calcium tungstate X-ray intensifying screen, the entire sandwich then being placed at $-70°C$ for film exposure (36). Emissions from the sample pass to the film producing the usual direct autoradiographic image, but emissions which pass completely through the film are absorbed more efficiently by the screen where they produce multiple photons of light which return to the film and superimpose a photographic image over the autoradiographic image. The resolution obtained is only slightly less than using direct autoradiography but sensitivity is much increased. Using Kodak X-Omat R film pre-exposed to 0.15 A_{540} units and a Fuji Mach II or Dupont Cronex Lightning Plus intensifying screen, the detection efficiency for ^{32}P is increased 10.5-fold and for ^{125}I 16-fold when compared to direct autoradiography, such that about 100 dpm $^{125}I/cm^2$ or 50 dpm $^{32}P/cm^2$ blackens the film detectably (*Table 9*). The density of the film image is still proportional to the sample radioactivity allowing quantitative work. As with fluorography, pre-exposure of the film above 0.2 A_{540} units increases the detection efficiency still further but the relationship between film image and radioactivity ceases to be linear.

The suitability of X-ray films for indirect autoradiography is the same as for photon detection during fluorography (*Table 7*). The source of the calcium tungstate intensifying screen (which can be used repeatedly) is also important, the Fuji Mach II or Dupont Lightning Plus screens being the most efficient for indirect autoradiography. The performance of other intensifying screens is listed in reference 36.

Dual isotope detection using X-ray film. McConkey (40) has discussed the approaches to using X-ray film for double-labelling studies and described a method for this purpose. The gel is prepared for fluorography and exposed to X-ray film which is sensitive to the light flashes produced and so detects both ^{14}C and ^{3}H (see above). The gel is then placed in contact with Kodak No-Screen X-ray film which is insensitive to the light produced by fluorography and detects only the ^{14}C by direct autoradiography. Comparison of the images produced on both films allows one to identify proteins labelled with either or both isotopes.

Walton *et al.* (41) have published a similar method. The gel is first prepared for fluorography and exposed to sensitive X-ray film to detect both ^{14}C and ^{3}H. Then the gel is painted black using marking ink and a second film exposed (direct autoradiography). Provided that the $^{3}H/^{14}C$ ratio exceeds about 40:1, ^{3}H is detected almost exclusively on the fluorograph whilst ^{14}C only appears on the direct autoradiograph because photons emitted by the ^{3}H are prevented from reaching the film by the black ink. Finally, another method being developed is the use of colour negative film (42). This consists of three photographic emulsions each sensitive to one of the three additive primary colours (red, green, and blue). Since the layers are exposed through different thicknesses of gelatin emulsion, emissions of different energies penetrate to different depths and so produce different ratios of exposure in the three layers and so different final colours.

Methods Based upon Gel Fractionation and Counting

Fractionation of the gel and scintillation counting of the solubilised fraction is still the most widespread method of quantitation of labelled protein components following polyacrylamide gel electrophoresis. One reason for this is the ability of modern scintillation spectrometers to distinguish between emissions of different energies thus allowing the quantitation (rather than just detection) of samples differentially labelled.

Fractionation methods. Slab gels and rod gels which have been dried down onto filter paper (p. 48) may be sectioned with scissors but this is inaccurate since it is difficult to obtain slices of even width and so gels are usually sliced whilst still wet. However, one situation where cutting dried gels may be advantageous is if one wishes to quantitate individual bands or measure the isotopic ratio of doubly-labelled bands detected by autoradiography or fluorography. If the gel used for autoradiography or fluorography was unstained, the band may be located by photographing the autoradiograph (to provide a permanent record) and then cutting out the band image of interest from the film. The film is now laid over the unstained gel where it serves as a template to mark the position of the radioactive band which is then cut out from the gel. Gel slices containing ^{32}P or ^{125}I are counted efficiently by simply adding conventional scintillation fluid to the dry gel slice. Gel slices containing ^{14}C, ^{35}S, or ^{3}H, cut from gels prepared for fluorography using PPO, can be counted in the same way without gel hydrolysis. In fact, Laskey (38) has noted that in the latter case there is no need to add scintillant prior to counting since the gel slice already contains PPO. If necessary, dried gel slices can be hydrolysed directly by the methods given below, or radioactive proteins can be leached out of the gel for counting (p. 59) after rehydration of each gel slice with water.

Numerous sectioning devices exist for wet polyacrylamide gels (see ref. 43) only some of which can be applied to both gel rods and slabs. Using a Mickle gel slicer (Joyce Loebl) the gel rod or gel slab track is held on a carriage which moves past a guillotine knife in calibrated small steps. The smallest practical section size with many polyacrylamide gels is about 0.5 mm. Slices are removed manually during the operation. However, one of the simplest and most widely used fractionators consists of a number of razor blades separated by aluminium, stainless-steel, or plastic spacers, bolted together. A suitable device is easily built in a laboratory workshop but certain commercial models do tend to give more reproducible slices (e.g. Bio-Rad Laboratories Ltd.; *Figure 18*).

Rod gels need to be frozen for easy slicing since low gel concentrations are viscous whilst high gel concentrations are rubbery, both being difficult to cut. The temperature of the frozen gel is important since if the gel is too cold it is extremely difficult to slice and tends to crack during the attempt. The easiest method is to place the gels at $-70°C$ for about 1h. Alternatively the gel is placed in an aluminium-foil trough which is then placed on dry ice until the gel is just frozen. The frozen gel, kept straight during freezing, is placed on a plastic sheet, or in a trough cut from plastic tubing (*Figure 18*), and the slicer lined up along the length of the gel. The gel is then sliced by exerting firm downward pressure whilst rocking the slicer transversely across the gel to help the razor blades penetrate. A syringe needle, a scalpel, or a pair of fine-tipped forceps is used to transfer gel slices from between the blades into counting vials. Next the slicer is washed under the tap, shaken free of excess water, and drained on a pad of filter paper ready for re-use. If gel cracking is a problem during slicing, the gel should be allowed to thaw slightly before attempting to slice. Cracking can also be prevented by incorporating 10% glycerol in the gel mixture prior to polymerisation.

Slab gels may be fractionated by first cutting the gel into individual tracks. Attempts to do this using a scalpel usually result in the gel tearing. A better method is to use a long knife with a sharp, flat cutting edge and slice downwards into the gel along the length of the track in a single movement. The slab gel track can then often be sliced transversely using a rod gel slicer (see above) but freezing is not usually necessary, presumably because of the thickness of the gels used (0.75-1.5 mm).

Counting methods. Liquid scintillation counting is the preferred method of isotope quantitation. For maximum counting efficiency the proteins need to be eluted from the gel slices. Two approaches are possible. Either the gel is solubilised or the gel is made to swell and protein leaches out.

Bisacrylamide crosslinked polyacrylamide gels can be solubilised by heating with hydrogen peroxide.

(i) the slice (1 mm) is allowed to dry in a vial at room temperature overnight or at 50°C for about 2h.

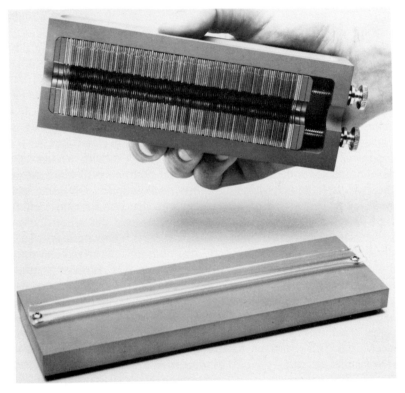

Figure 18. A manual gel slicer (Bio-Rad model 190). (a) The gel slicer consists of a series of razor blades separated by 0.1 cm thick metal spacers; here, four razor blades and three spacers have been removed and are displayed in front of the gel slicer. (b) The gel to be sectioned can be held in a trough cut from plastic tubing, as shown here.

(ii) 0.25 ml 30% (w/v) H_2O_2 is added to each vial which is then tightly capped, tilted so that the slice is immersed in liquid, and incubated at 50°C to allow solubilisation.

(iii) the vials are cooled to room temperature before opening. Then 5 ml (for vial inserts) or 10 ml of a water-miscible scintillation cocktail, consisting of NCS (Amersham-Searle) and toluene scintillation fluid (5.0 g PPO, 0.5 g POPOP per litre of toluene) mixed in the ratio of 1:5 vol/vol, is added per vial. A considerably cheaper alternative is to use a Triton X-100:toluene scintillation cocktail (1 vol Triton X-100:2 vol toluene scintillation fluid). In this case the gel slice is dissolved in 0.5 ml of 30% (w/v) H_2O_2 and then 4.5 ml of the Triton X-100:toluene scintillation cocktail is added. Either method yields a completely clear counting mixture.

Although the loss of ^{14}C as $^{14}CO_2$ does not exceed about 5% with this protocol, this is avoided in the modification of Goodman and Matzura (44) by using alkaline peroxide to trap any $^{14}CO_2$. The procedure is as described above except that the reagent used is 1 volume of concentrated NH_4OH (specific gravity 0.88) added to 99 volumes of 30% (w/v) H_2O_2 that has been previously cooled to 4°C. The mixture is kept on ice and used immediately. The temperature of incubation was originally quoted as 37°C but 50°C is often used. Although the NCS-toluene scintillation cocktail still yields a clear counting mixture, the Triton X-100:toluene cocktail gives a slightly cloudy emulsion. Chemiluminescence is sometimes a problem. In this case vials should be counted twice, 24h apart, or until constant values are obtained. Usually counts due to chemiluminescence decay to background within 48h.

Another approach to gel solubilisation has been to replace bisacrylamide with a crosslinker more susceptible to chemical hydrolysis. A number of such labile crosslinkers are available. Ethylene diacrylate has an ester linkage in place of the amide linkage of bisacrylamide so that ethylene diacrylate crosslinked gels can be solubilised with alkali (45). Unfortunately this alkali lability renders ethylene diacrylate unsuitable for the many types of protein electrophoresis which are carried out at high pH. N,N'-bisacrylylcystamine (BAC) is a disulphide-containing analogue of bisacrylamide such that use of this crosslinker renders polyacrylamide gel soluble in 2-mercaptoethanol (see ref. 46 for details). Obviously this sensitivity to thiol reagents restricts the use of BAC to those situations where thiols are absent. The most generally useful, labile, crosslinker is N,N'-diallyltartardiamide (DATD) which contains a 1,2 diol structure readily oxidised by periodic acid. However, substituting DATD for bisacrylamide on a mole per mole basis produces gels exhibiting reduced physical strength. This can be corrected by increasing the proportion of crosslinker. Some workers have used acrylamide: DATD (30:3) instead of acrylamide: bisacrylamide (30:0.8) whilst Spath and Koblet (47) find that DATD needs to be increased to 27% to give gels with similar physical properties to those made with 5% bisacrylamide (crosslinker percentages quoted as proportion of total monomer). In addition DATD crosslinked gels exhibit greater porosity than gels crosslinked with 2.6% or 5% bisacrylamide and so proteins migrate further at any given acrylamide concentration. This can be compensated for by increasing the concentration of acrylamide gel used.

Solubilisation of DATD crosslinked gels is effected by adding 0.5 ml of 2% (w/v) periodic acid followed by incubation at room temperature for 1-2h. After incubation, vigorous mixing ensures dispersal of the gel and then a water-miscible scintillation cocktail is added and the sample counted. Anderson and McClure (48) found that addition of 5 ml of 0.3% (w/v) PPO, 25% (w/v) Triton X-114 in xylene to be especially useful for this, giving high counting efficiencies for both ^{3}H (47%) and ^{14}C (93%). Other water-miscible scintillants may also be used.

The alternative approach, of leaching the protein from the gel in order to increase the efficiency of scintillation counting, can be carried out according to several protocols. In the original procedure, each gel slice was swollen by soaking in 0.5 ml NCS for 2h at 65°C, followed by addition of 10 ml toluene scintillation fluid (49). High percentage polyacrylamide gels may resist swelling unless the NCS is diluted 9 parts NCS:1 part water. A variation on this is to add 10 ml scintillant (14.0 g PPO, 0.21 g POPOP, 143 ml NCS, 3.75 l toluene) to each gel slice and incubate at 37°C overnight before counting (50). Most recently Aloyo (51) suggests adding 10 ml scintillation cocktail (6.0 g PPO, 10 ml NCS, 10 ml hyamine hydroxide, per litre toluene) directly to each wet 1 mm gel slice, vortexing briefly and measuring the radioactivity after 48h at room temperature. The yields and counting efficiencies by this method compare favourably with the alkaline peroxide solubilisation method. Finally, New England Nuclear (NEN) markets ready-mixed scintillation cocktails for recovery of labelled proteins from gel slices by leaching.

Determination of Specific Radioactivity of Proteins

Martin *et al.* (52) have described a method for determining the specific radioactivity of leucine in proteins isolated by polyacrylamide gel electrophoresis. The protein is hydrolysed in HCl followed by radioactivity determination and estimation of leucine content using amino-acyl synthetase. A more recent method based upon the use of radiolabelled dansyl chloride (53) allows one to determine the specific radioactivity or amino acid content for several amino acids in any protein separated by polyacrylamide gel electrophoresis.

Detection of Glycoproteins and Phosphoproteins

Glycoproteins can be readily located in polyacrylamide gels and distinguished from other non-glycosylated proteins. In the past, the most usual detection method has been the periodic acid-Schiff (PAS) stain but recently a number of other methods have become available, including selective radiolabelling of the carbohydrate moiety prior to electrophoresis (either by *in vivo* labelling with sugar precursors or *in vitro* labelling) then glycoprotein detection by fluorography/autoradiography after electrophoresis, and the use of radiolabelled, fluorescent, or peroxidase-conjugated lectins. These and other glycoprotein detection methods are referenced in Appendix I.

Phosphoproteins can be detected according to any of several protocols (Appendix I) but the most sensitive method is to label *in vivo* with ^{32}P then detect phosphoproteins after gel electrophoresis using autoradiography. Lipid and nucleic acids will also be labelled and should be removed from the sample prior to electrophoresis. Alternatively, labelled nucleic acids can be removed by acid hydrolysis after gel electrophoresis (54); if desired, the proteins are first stained using Coomassie blue (p. 44) then the gel is equilibrated in 7% TCA overnight followed by hydrolysis in 7% TCA at 90°C for 30 min. The labelled nucleotides are removed by soaking the gel in several changes of 7% TCA at room temperature over a period of 24h. The gel is shaken gently on a reciprocating shaker during this time. If the gel has been stained prior to acid hydrolysis, it is now destained, dried, and autoradiographed as usual (p. 50). Labelled lipid migrates with the buffer front and most is extracted by the staining and destaining steps.

Detection of Proteins using Immunological Methods

Particular proteins can often be detected after electrophoresis, even if this was carried out in the presence of SDS, if monospecific antibodies are available. Several methods have been devised for this purpose. Thus antigens may be visualised by incubating the washed gel with specific antibody followed by incubation with anti-IgG coupled to horse-radish peroxidase (55). After further incubation with 3,3′-diaminobenzidine and hydrogen peroxide, the antigen-antibody-antibody-peroxidase complexes become visible as stained bands. Alternatively, antigens can be detected by exposing the gel to radioactive antibody followed by autoradiography. A double-antibody method is preferred with the radiolabel on a second antibody directed against the first, or using ^{125}I-labelled *Staphylococcus aureus* protein A (which binds to the Fc portion of the IgG molecule) instead of a second antibody (56). Careful washing is needed to obtain good ratios of signal to background. A promising modification is to first transfer the proteins to diazobenzyloxymethyl paper where they are coupled covalently (57). Specific antigens are then detected by autoradiography after sequential incubation with unlabelled antibody and ^{125}I-labelled *S. aureus* protein A. Antibody and protein A can be removed with urea and 2-mercaptoethanol and the paper challenged again with antibody of different specificity.

Another approach is to cut out the gel strip and identify specific antigens by crossed immunoelectrophoresis during which the antigens migrate electrophoretically from the polyacrylamide gel into an agarose gel containing the antibody, yielding characteristic antibody-antigen precipitation patterns (58). A simpler method has also been described whereby agarose containing antibody is cast directly onto the polyacrylamide slab gel surface following electrophoresis. The protein antigen diffuses out of the polyacrylamide gel and reacts with the antibody in the agarose gel to produce an immune precipitation pattern called an immunoreplica, which can be visualised by staining or by using ^{125}I-labelled protein A (59,60). Other references to several of these methods are given in Appendix I.

Detection of Enzymes

Enzymes can be assayed after gel slicing followed by diffusion elution into a suitable buffer (p. 63). However, this is laborious for large numbers of samples and, furthermore, slight differences in the migration of two enzymes may be missed. The alternative method is to stain for specific enzyme activity *in situ*. The major limitations of the method are that detection methods exist for only a relatively small number of enzymes and secondly the results are not easily quantifiable. It is therefore mainly useful in analytical comparisons of a particular enzyme in multiple samples run on slab gels or possibly to locate enzyme for preparative purposes, although other proteins are, of course, not visualised and may contaminate the preparation.

Dehydrogenases can be localised by incubating the gel in a solution of a tetra-zolium salt which then acts as the terminal electron acceptor and becomes reduced to yield a coloured formazan. Nitroblue tetrazolium (NBT) and methyl thiazolyl tetra-zolium (MTT) have been widely used for this purpose although other suitable tetra-zolium salts are also available. Phenazine methosulphate (PMS) is sometimes includ-ed to serve as hydride-ion carrier between the reduced coenzyme or enzyme prosthetic group and the tetrazolium salt. Since tetrazolium salts are light-sensitive, the incuba-tion has to be carried out in the dark. Staining in the absence of dehydrogenase is sometimes observed (61) and so controls should be included to ensure that only the enzyme reaction is being studied. Hydrolases are most readily detected using chromogenic or fluorogenic substrates which yield coloured or fluorescent products after enzyme action. Transferases and isomerases are not amenable to this method and usually require coupling enzymes to be included to the detection mixture which convert the enzyme product into one which can be acted upon by a dehydrogenase also included in the mixture. The dehydrogenase product, and hence the original en-zyme, is located as described above. Since the coupling enzymes penetrate polyacryl-amide gel only poorly, the coloured product is formed only on the gel surface. Zone spreading and the volume of reagents can be minimised by incorporating the detec-tion mixture, including the coupling enzymes, in an indicator gel, often agarose, which is poured onto the surface of the polyacrylamide gel or preformed and then the gels placed in contact. If this strategy is adopted then a drop of sample should be ap-plied to the indicator gel as a positive control to ensure that the coupling enzymes have survived the agarose treatment.

Detection methods for individual enzymes are referenced in Appendix I.

RECOVERY OF SEPARATED PROTEINS

The preparation of large quantities of a single protein is best carried out using specialist polyacrylamide gel apparatus which allows the collection of protein zones as they elute from the base of a large-scale gel column and is described in Chapter 2. However, the isolation of several proteins in sub-milligram amounts can be achieved by excision of individual bands from rod gels or modified slab gels. In the latter case, instead of forming individual sample wells the gel mixture is either overlayered with buffer to form a continuous flat meniscus or a sample comb with only a single 'tooth' spanning the width of the slab gel may be used to form a single sample well. After polymerisation, the sample is loaded onto the gel and forms a continuous layer along the entire upper edge of the gel. Up to 2.0 ml of sample containing several milligrams of proteins can be loaded using a discontinuous buffer system in the usual slab gel format. The sample capacity of the gel can also be increased by using thicker side-spacers to give a thicker gel. After electrophoresis the proteins are resolved into bands stretching transversely across the gel.

Localisation of Protein Bands

Once electrophoresis is complete the protein bands must be localised prior to excision. In general the yield of proteins from fixed, stained bands is much lower than from unfixed gels and so several methods have been devised for protein localisation in the latter.

(i)   Proteins may be localised by slicing the gel into segments (p. 56), eluting each into a suitable buffer by diffusion (p. 63) and then assaying for the protein of interest. For native proteins the assay is based upon the biological (e.g. enzymic) activity of the protein. Denatured proteins of interest separated by SDS-PAGE can be located by eluting into 0.01 M NH_4HCO_3, 0.05% SDS at 37°C overnight, followed by lyophilisation and then running a portion of each eluate in separate tracks of an analytical slab gel which is then stained and destained in the usual way.

(ii)   Another method is to cut two longitudinal strips from the sides of the slab gel and a narrow longitudinal strip from the centre, and to stain these with Coomassie blue whilst keeping the rest of the gel on a glass plate covered with cling film in a refrigerator. Following this, the stained side strips are lined up along the edges of the unstained gel (but not touching it) and used as guides to cut out bands of interest from the unstained gel. Using rod gels instead, one rod gel may be stained and destained and the protein R_f value used as a guide for cutting out bands from other identical but unstained rods. However, the method is usually applied to slab gels rather than rods because of the more uniform mobility achievable in this format. The method works well using slab gels provided the bands to be recovered are well separated from potential contaminants.

(iii)   Protein bands in gels lacking SDS can be detected by immersing the entire unstained gel in a 0.003% (w/v) solution of l-anilino-8-naphthalene sulphonate (ANS) in 0.1 M phosphate buffer, pH 6.8 (33). Within 5-10 min, fluorescent protein bands appear when viewed with a U.V. lamp. A band containing 20 μg protein is usually visible. The method is superior to the side-strip staining procedure, (ii) above, for bands which are not separated by more than a few millimetres. A variation on this methodology, to reduce diffusion in large-pore gels, is to use ANS in 75% saturated ammonium sulphate.

(iv) An increasingly popular and extremely sensitive method for localising polypeptides after fractionation by SDS-PAGE is to first mix the bulk sample with an aliquot of sample polypeptides which have been reacted with dansyl chloride. During electrophoresis the dansylated polypeptides comigrate exactly with their un-modified counterparts and serve as markers for these by monitoring the gel after electrophoresis with a U.V. lamp (62,63). Since dansylation alters the charge of the polypeptides, the method is usually only useful for electrophoresis in the presence of SDS which masks the small charge differences created by the treatment. The added advantage of the method is that one can also easily monitor the progress of polypeptide recovery from the gels during electrophoretic or diffusion elution (p. 63) simply by viewing the gel under U.V. light.

(v)   Protein bands can also be localised in unstained gels by virtue of the phosphorescence of tyrosine and tryptophan residues following U.V. irradiation (64) but the method has been little used.

(vi) Also little used at present are a number of other methods for localising polypeptides after SDS-PAGE without the use of dyes e.g. SDS-polypeptide complexes can be visualised by precipitation either by chilling the gel at 0-4°C, or by incubation with potassium ions or cationic surfactant. References to these and other 'direct detection' methods are given in Appendix I.

(vii) Some enzymes can be localised *in situ* by incubating the gel in an appropriate reaction mixture (pp. 254-263, Appendix I). The enzyme activity usually produces some change in colour or fluorescence which leads to its localisation. Since only the specific enzyme is visualised, the excised bands may well be contaminated with other proteins, especially since the coloured or fluorescent reaction products tend to diffuse to yield broad zones.

(viii) Finally, protein bands may be visualised by staining with Coomassie blue. Since the staining conditions will inactivate native proteins, this is only suitable for proteins which are not required in an active form for later analysis. The staining and destaining periods should be as brief as possible to allow visualisation of the bands but to minimise stain and fixative penetration into the gel. The stained gel is rinsed with water to remove excess fixative and then the relevant protein bands excised. This method of protein localisation and elution forms the basis of the Cleveland method of peptide mapping (Chapter 6).

Elution of Proteins

If the aim of gel fractionation has been to purify sufficient protein for antibody preparation there is usually no need to separate protein from the polyacrylamide since the latter does not interfere with antibody production. The gel segments are simply homogenised with Freund's adjuvant and injected directly (see Chapter 7). Antibodies raised against SDS-polypeptide complexes often precipitate the native protein as well as the denatured polypeptide.

For other purposes it is usually necessary to isolate the protein free of the gel matrix. Theoretically the simplest method would be to solubilise the gel (p. 58). However not only are these solubilising agents potentially harmful to proteins but also only the gel crosslinkages are labile, leaving the protein contaminated with long chain polyacrylamide molecules. Unfortunately no suitable methodology presently exists to remove this contaminant easily and quantitatively.

The two usual methods of protein recovery are elution by diffusion and by electrophoresis. The main advantages of electrophoretic elution are its speed and the fact that the protein is usually eluted in a much smaller volume than by diffusion. Diffusion elution is simpler but protein recoveries, especially in the absence of SDS, may be poor.

Elution by Diffusion

The gel slices are macerated by chopping finely with a scalpel or homogenisation.

This increases the gel surface area so that, when a suitable buffer is added, elution can occur more readily. Unfortunately there is a conflicting need to keep the volume as small as possible to minimise recovery problems after elution whilst using a large enough volume of buffer to obtain a good yield. Elution with about three gel volumes of buffer overnight with mixing, for example by putting the sample in a stoppered, siliconised tube fixed to a vertical rotating disc, and then a repeated extraction the next day, is a reasonable protocol. The process can be speeded up using higher temperatures if the protein is stable under these conditions. Thus SDS-polypeptide complexes can be eluted overnight at 37°C into buffer containing 0.05% SDS. The degree of protein recovery by diffusion elution is extremely variable with native proteins but can be as good as 70% if the elution buffer contains 0.05-0.1% SDS. In all cases gel fragments are easily removed by centrifugation. SDS can be removed by the use of urea or guanidine hydrochloride [see p. 82 , section (iv)].

Electrophoretic Elution

A number of devices have been designed for electrophoretic elution of proteins from polyacrylamide gel slices, most of which rely on electrophoresis of proteins out of the gel into a chamber which makes electrical contact with the reservoir buffer via a dialysis membrane. Proteins are eluted and then retained in the elution chamber. The apparatuses range from those requiring elaborate workshop facilities for their manufacture (65) to simple adaptations of the standard rod gel methodology (62,66), whilst some apparatus is available commercially (e.g. ISCO). Suitable apparatus and its use is also described in Chapter 3, p.150 .

MODIFICATIONS TO THE BASIC TECHNIQUES

Molecular Weight Analysis of Oligopeptides

Oligopeptides with a molecular weight below about 10,000 are not well resolved by SDS-PAGE even using 15% polyacrylamide gels (*Figure 6*). However, the separation of oligopeptides can be considerably improved by using 12.5% polyacrylamide gels prepared with a high ratio of bisacrylamide crosslinker to acrylamide (1:10), and the inclusion of 8M urea in the SDS-containing gel buffer (67). Under these conditions, the electrophoresis of cyanogen bromide cleavage products of certain proteins (e.g. cytochrome c and myoglobin) yields a good linear relationship between $\log_{10}$ oligopeptide molecular weight and mobility over the molecular weight range of 2,500 - 17,000 (*Figure 19*). Unfortunately, using native small peptides, considerable deviations from this linear relationship often occur which are believed to be due to the greater importance of the intrinsic charge and conformation of small peptides to their mobility under these conditions. Thus the molecular weights of diverse oligopeptides can still only be approximated by this method. Methodological details are given in *Table 10*.

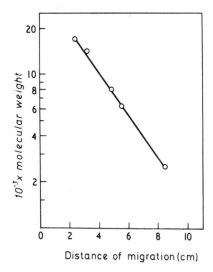

Figure 19. Mobility of oligopeptides in SDS-urea polyacrylamide gel electrophoresis according to the Swank and Munkres technique (67). Electrophoresis was carried out in a slab gel format as described in *Table 10*. The oligopeptides, in order of decreasing molecular weight, were horse heart myoglobin (mol. wt. 16,950), horse heart myoglobin cyanogen bromide fragments I + II (14,404), I (8,159), II (6,214), and III (2,512). The molecular weights of horse heart myoglobin and its cyanogen bromide cleavage fragments are calculated from the sequence data given in ref. 129.

Separation of Special Classes of Proteins

Although the electrophoretic methods described earlier in this chapter will separate the vast majority of cellular proteins, particular classes of proteins tend to be insoluble under the usual buffer conditions and so need special electrophoretic systems for their fractionation. These are described below.

Histones

Histones are small, highly basic proteins. Five major species can be resolved in most cases; H1, H2A, H2B, H3 and H4, and a number of variants of these can occur via charge modification through acetylation or phosphorylation, as well as ADP-ribosylation. The major problem associated with electrophoretic analysis of these proteins is their insolubility in the absence of denaturing agents. Using SDS as denaturant, these proteins can be analysed by SDS-PAGE under which conditions fractionation occurs mainly on the basis of size; therefore SDS-PAGE is useful for histone size fractionation, but is not able to analyse histone charge variants. A combined charge and size fractionation via polyacrylamide gel electrophoresis requires the use of denaturants able to prevent histone aggregation whilst preserving the charge differences between protein species. The usual method of achieving this has been the use of an acetic acid-urea buffer system (68) but even here is is often difficult to completely resolve histones H2A, H2B and H3 or to identify minor modified forms. The resolution of these histones can be improved markedly by including a non-ionic detergent (usually Triton X-100 or Triton DF-16) in the acid-urea buffer system

Table 10.   Urea-SDS-PAGE for Determination of Oligopeptide Molecular Weight.[a]

Acrylamide-bisacrylamide:	12.5g acrylamide and 1.25g bisacrylamide are dissolved in and made to 50 ml with water, then filtered through a Whatman No.1 filter. Stable at 4°C.
Gel buffer stock:	1% SDS, 1.0M H_3PO_4 adjusted to pH 5.0 with Tris base.
Reservoir buffer stock:	0.1% SDS, 0.1M H_3PO_4 adjusted to pH 6.8 with Tris base.
Sample buffer:	1% SDS, 8M urea, 1% 2-mercaptoethanol, 0.01M H_3PO_4 adjusted to pH 6.8 with Tris base.

1. The gel mixture is prepared by mixing:

acrylamide-bisacrylamide	15.0 ml
gel buffer stock	3.0 ml
urea	14.4 g
6% ammonium persulphate (freshly prepared)	0.3 ml
water to 30 ml final volume.	

2. The solution is deaerated for 1 min, 20 μl TEMED is added, and rod gels are poured without delay.[b]

3. The oligopeptides are dissolved in sample buffer and bromophenol blue is added to 0.002% (w/v) final concentration. Next, the samples are heated to denature the oligopeptides then cooled to room temperature prior to use. In the original study (67) the heating step was 60°C for 10 min but recent analyses have used 100°C for 3 min. Molecular weight marker peptides can be prepared by cyanogen bromide cleavage but are also available commercially.

4. Electrophoresis is at 6-8V/cm for rod gels until the bromophenol blue nears the end of the gel.

5. After electrophoresis, gels are stained in 0.25% Coomassie blue R250 in methanol:water:glacial acetic acid (5:5:1). Destaining is by diffusion using 12.5% isopropanol, 10% acetic acid; electrophoretic destaining is not recommended since some small peptides migrate under these conditions.

[a]Based on the method given in ref. 67.

[b]Slab gels are preferable to rod gels for comparison between samples but the highly-crosslinked resolving gel used in this method is sufficiently brittle that sample-well divisions often break during preparation. Therefore a low concentration gel (3.75%) is polymerised on top of the 12.5% gel, but with the same buffer composition, and the sample wells are formed in this 3.75% polyacrylamide gel. Slab gels with the dimensions given on p.21 require 100V overnight.

(69,70). These are the main gel systems used for histone fractionation and are described in more detail below. Finally, some histone modifications are acid-labile (for example, some phosphorylated species) and cannot be examined in the acid-urea system but can be fractionated by polyacrylamide gel electrophoresis using a neutral pH buffer system (71). Hardison and Chalkley (71) have recently reviewed all these methods for polyacrylamide gel electrophoresis of histones.

SDS-PAGE of histones. A number of protocols exist for SDS-PAGE of histones which differ in the proportion of bisacrylamide used in the polyacrylamide gel and the exact buffer conditions, for example the presence or absence of urea as well as SDS. Thomas and Kornberg (72) use the basic system of Laemmli (7) (*Table 2,*

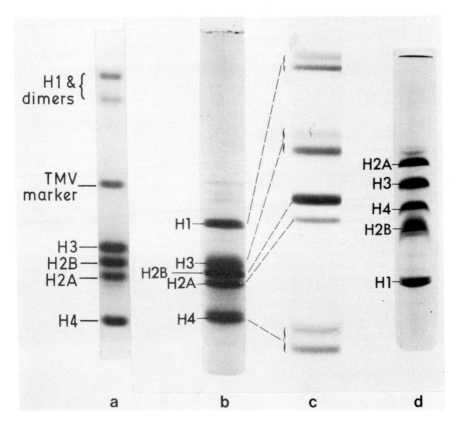

Figure 20. Separation of histones by polyacrylamide gel electrophoresis. (a) SDS-PAGE (slab gel track reproduced from ref. 72 with permission); (b) 2.5M urea-0.9M acetic acid (9 cm rod gel); (c) 2.5M urea-1.0M acetic acid (25 cm rod gel); (d) 2.5M urea-0.9M acetic acid-0.4% Triton X-100 (9 cm rod gel). The data in (b), (c), and (d) are reproduced from ref. 71 with permission.

p. 27) but with three modifications; the concentration of Tris buffer in the resolving gel (18% polyacrylamide) is increased to 0.75M, the ratio of acrylamide:bisacrylamide is increased to 30:0.15, and the electrode buffer comprises 0.05 M Tris, 0.38 M glycine, 0.1% SDS, pH 8.3. All other details of reagent and gel preparation are as described on p. 35. Slab gels are 1.5 mm thick and either 15 cm or 30 cm long. Samples used for electrophoresis can be either purified histones or intact chromatin. Electrophoresis is at 30mA for about 6h (15 cm long gels) or 4W for 24h (30 cm long gels) until the bromophenol blue tracking dye almost reaches the gel bottom. Staining with Coomassie blue and destaining is as described on p. 44. An example of the separation achieved is shown in *Figure 20a*.

Although histones separate mainly on the basis of size when analysed by SDS-PAGE, they migrate slower than would be expected on the basis of their known molecular weights (73). This is presumably due to a reduced overall-negative charge as a result of the high proportion of basic amino acids in histones. The practical consequence of this phenomenon is that the molecular weights of histones will be over-

Table 11. Acetic Acid-Urea Buffer System for Histone Separations.[a]

Solution A:	Acrylamide-bisacrylamide, 60.0g acrylamide and 0.4g bisacrylamide are dissolved in and adjusted to 100 ml with water. This is deionised by stirring for 30 min with 3g Amberlite MB-1 mixed-bed resin and filtered prior to use. Stable at 4°C.
Solution B:	4% (v/v) TEMED, 43.2% (v/v) glacial acetic acid in water. Stable at 4°C.
Solution C:	40 mg ammonium persulphate added to 20 ml deionised 4M urea. (Made fresh prior to use).

1. One volume of solution B is mixed with five volumes of solution C and deaerated. Two volumes of solution A are also deaerated but in a separate vacuum flask.

2. The deaerated solutions are combined, the gels poured, and overlayered with cold 0.9M acetic acid. The gels are left for an additional 30-60 min after polymerisation for optimal resolution of histones.

3. After filling the buffer reservoir with 0.9M acetic acid, the gels are pre-electrophoresed at 130V until constant current is obtained. This pre-electrophoresis is essential for good histone fractionation.

4. The histone sample, free of DNA and dissolved in 2.5M urea, 0.9M acetic acid at 1mg/ml, is then loaded. Electrophoresis is from anode (+) to cathode (−) at 130V for 9 cm long rod gels and takes about 3-4 h at room temperature. It is useful to include methyl green as tracking dye since the blue component of the dye migrates just ahead of the fastest migrating histone (H4) in this system and so gives an indication of the progress of the electrophoretic run. For very long (25 cm) gels, run at 200V constant voltage at 4°C for about 48 h.

5. After electrophoresis, the gels are stained with 0.1% Coomassie blue R250 in methanol:water: acetic acid (5:5:1) and destained electrophoretically or by diffusion in 10% acetic acid, 25% methanol.

[a]Based on the methods given in refs. 68, 71.

estimated if these are calculated solely by comparison of histone mobility with that of standard, non-basic proteins in SDS-PAGE.

Acid-urea gels. The banding pattern of histones separated by charge and size fractionation in an acetic acid-urea gel system depends on both the pH and the urea concentration but a useful system for routine analysis is 0.9M acetic acid, 2.5M urea (pH 2.7), in 15% polyacrylamide (68). No stacking gel is used. In this system the five major histones can be separated on a 9 cm resolving gel (*Figure 20b*) whereas some minor bands resulting from covalent modifications or sequence heterogeneity are visible only after electrophoresis on much longer (25 cm) gels (*Figure 20c*). Details of the acid-urea gel system are given in *Table 11*.

One of the disadvantages of the acetic acid-urea technique is that these solvents do not fully dissociate histones from DNA so that these proteins must be first purified free of DNA prior to electrophoresis (see Chapter 5). Alternatively, this problem can be overcome by including the cationic detergent cetyltrimethylammonium bromide (CTAB) in the sample buffer (74).

Triton-acid-urea gels. Addition of a non-ionic detergent, usually Triton X-100, to the acid-urea gel system results in the formation of micelles between the detergent and the hydrophobic regions of the histones. This results in complexes of histone and detergent and consequently a reduction in electrophoretic mobility of the individual histones proportional to their hydrophobicity (*Figure 20d*). Using this technique, it is possible to resolve histones which differ in the substitution of only a single, neutral, amino acid when this occurs in a hydrophobic region and results in a change in detergent binding (69). In most cases, it is usual to keep the detergent concentration constant at 6 mM Triton (0.4%) and to vary the concentration of urea (an inhibitor of detergent binding). The exact concentration of urea used depends on which histone components need to be maximally resolved. The main disadvantage of the Triton-acid-urea system is artifactual histone heterogeneity due to the oxidation of methionine residues; oxidised histones have altered Triton-binding characteristics. To avoid this potential artifact, the samples have to be isolated and stored in the presence of reducing agents and the polyacrylamide gels (which are polymerised using oxidising agents as catalysts) have to be scavenged with reducing agents after pre-electrophoresis. Full details of the method are given by Zweidler (70).

Nuclear Non-histone Proteins

These proteins have a pronounced tendency to aggregate and if one attempts electrophoretic separations using non-dissociating conditions many of the proteins remain as an aggregate at the top of the gel. Usually the inclusion of neither SDS nor urea alone is sufficient to ensure complete disaggregation of the proteins, rather both must be present in the gel and the sample solution. In practice it has been found that optimal resolution can be obtained using discontinuous SDS-PAGE as described in *Table 2* with the modification that urea is added to the gel to a final concentration of 4M by the addition of urea to the appropriate solution (75). For most complex mixtures a 12-15% polyacrylamide gel has proved to be suitable.

When preparing samples for electrophoresis, all procedures involving precipitation should be avoided. Samples should ideally be concentrated by lyophilisation after dialysis against 0.1% (w/v) SDS. Alternatively, concentration of the proteins in SDS or urea solutions can be carried out using polyethylene glycol or Sephadex G-200 (p. 36).

The protein sample is applied to the gel in 1% (w/v) SDS, 5% (v/v) 2-mercaptoethanol in 8M urea. Usually, because of the possible problem of cyanate ions in the urea, it is advisable to add 5 mM Tris HCl (pH 6.8). Similarly, heating of the sample should be avoided if at all possible since the heating of samples greatly enhances the rate of formation of cyanate ions. The presence of 8M urea is sufficient to increase the density of the sample to ensure that there is no difficulty in applying the sample to the gel without the addition of sucrose or glycerol.

The gels can be stained with Coomassie blue as described elsewhere (p. 44). After destaining, most non-histone fractions are revealed to be extremely complex mixtures. For this reason two-dimensional gel electrophoresis gives much better estimates of the complexity of such samples and has now been accepted as the preferred analytical method (see Chapter 5).

Ribosomal Proteins

Ribosomal proteins are basic proteins which are not readily solubilised in non-dissociating buffers. They can be successfully analysed on a size basis using the SDS-phosphate buffer system with a 10% acrylamide gel e.g. ref. 76. The SDS-discontinuous buffer system also may be used and gives good resolution especially with a 6-18% linear gradient gel. In each case the ribosomes or ribosomal subunits can be dissolved directly in the SDS-containing sample buffer without removing the RNA.

A combined size and charge separation of ribosomal proteins requires the presence of urea to maintain solubility. The separation can be carried out in a continuous buffer system (for example 0.9M acetic acid, 6.0M urea; ref. 77) or in the low pH discontinuous system of Reisfeld *et al.* (ref. 10; *Table 4*) modified to include 8M urea (78). However, although SDS-PAGE is still widely used for one-dimensional electrophoretic analysis of ribosomal proteins, most detailed studies of individual ribosomal proteins which include a charge separation now use two-dimensional polyacrylamide gel electrophoresis (see Chapter 5).

Membrane Proteins

SDS, when used in excess, solubilises all, or almost all, membrane protein components of prokaryotic and eukaryotic cells. Hence SDS-PAGE is now the most widely used method for investigating the complexity of membrane protein mixtures and polypeptide molecular weight determination. The protocols for SDS-PAGE of membranes are the same as described earlier for other proteins, although 8M urea is best included in the sample buffer to ensure membrane protein denaturation (e.g. ref. 79). In addition, plasma membrane proteins exposed on the cell surface can be identified by labelling their tyrosine residues with ^{125}I using lactoperoxidase (80) followed by SDS-PAGE of whole cell extracts then autoradiography. Several other reagents for cell-surface protein radiolabelling are also available (81).

The one drawback is that SDS extraction of membranes usually inactivates the solubilised protein. In contrast, extraction of membranes with Triton X-100 instead of SDS often preserves the protein's biological activity although the disadvantage here is that not all membrane proteins are extractable with this detergent. After extraction, the protein is fractionated using the SDS-discontinuous buffer protocol described earlier (p. 27) but replacing 10% (w/v) SDS in the recipes with 10% (v/v) Triton X-100. Further details of membrane protein preparation and electrophoresis using Triton X-100, plus assays for a number of membrane marker enzymes, are given by Dewald *et al.* (82). Other detergents such as sodium deoxycholate (82), Lubrol (83) and Sarkosyl (83) have also been used in combination with polyacrylamide gel electrophoresis for analysis of membrane proteins.

Other electrophoretic systems exist which aim to solubilise all membrane proteins but, unlike SDS-PAGE, allow fractionation on a charge and size basis. A common early method involved solubilisation with phenol:acetic acid:urea followed by electrophoresis in acid gels containing 5M urea and 35% acetic acid (84). Chloral hydrate will also solubilise membrane proteins very effectively and allow a charge and size fractionation by subsequent electrophoresis in polyacrylamide gels containing this reagent (85). These methods are little used at the present time.

Finally, one can carry out 'charge-shift electrophoresis' of membrane proteins in polyacrylamide gels. The basis of the method is that integral membrane proteins possess hydrophobic domains which anchor them to the membrane lipid whilst peripheral membrane proteins or soluble proteins do not. Thus hydrophilic proteins bind little or no Triton X-100 whereas membrane proteins possessing hydrophobic domains bind large amounts. Integral membrane proteins therefore show altered mobility in Triton X-100 ionic detergent mixtures (when they form charged Triton X-100-ionic detergent-protein complexes) compared to Triton X-100 alone, whereas the mobility of hydrophilic proteins is unaffected. The original work of Helenius and Simons (86) utilised agarose gel electrophoresis. However the technique can also be applied to polyacrylamide gel electrophoresis. Thus Bordier *et al.* (87) found that by replacing the SDS in the SDS-discontinuous buffer system with 0.5% Triton X-100 plus 0.25% sodium deoxycholate, the mobility of four membrane protein complexes was greatly increased compared to the situation when Triton X-100 was the only detergent present, showing that the membrane proteins contain hydrophobic domains and so are probably integral membrane proteins. The magnitude of the shift in mobility should be a measure of the size of the hydrophobic domain in the membrane protein.

Concentration Gradient Gels

Uses of Concentration Gradient Gels

Polyacrylamide gels which have a gradient of increasing acrylamide concentration (and hence decreasing pore size) with increasing migration distance are now being extensively used instead of single concentration gels, both for analysis of the protein composition of samples and molecular weight estimation using SDS as the dissociating agent. Although step gradients (in which gels of different concentration are layered one upon the other) were used in early work, they can give artifactual multicomponent bands at the interface between two layers. It is now usual to use continuous acrylamide gradients. The usual limits are 3-30% acrylamide in linear or concave gradients with the particular range chosen depending upon the size of the proteins to be fractionated. During electrophoresis in gradient gels, proteins migrate until the decreasing pore size impedes further progress (88). Once this 'pore limit' is approached the protein banding pattern does not change appreciably with time although migration does not cease completely.

One of the main advantages of gradient gel electrophoresis is that the migrating proteins are continually entering areas of gel with decreasing pore size such that the advancing edge of the migrating protein zone is retarded more than the trailing edge, resulting in a marked sharpening of the protein bands. In addition, the gradient in pore size increases the range of molecular weights which can be fractionated simultaneously on one gel. Therefore a gradient slab gel used for SDS-PAGE will not only fractionate a complex protein mixture into sharper bands than is usually possible with a single concentration gel, but also allows the molecular weight estimation of almost all the components irrespective of their size. Non-dissociating buffers have also been used for analysis of native proteins on gradient gels but less frequently (see below).

Molecular Weight Estimation using Gradient Gels

SDS-PAGE (SDS-denatured proteins). With uniform concentration polyacrylamide gels, there is a linear relationship between $\log_{10}$ molecular weight and R_f value or distance migrated by the SDS-polypeptide complex (p. 15). However, with linear concentration gradient gels, the linear relationship is between $\log_{10}$ molecular weight and $\log_{10}$ polyacrylamide concentration (%T) (89, 90, 91). This relationship would be expected to hold even in the case of non-linear gradients, but the calculation of %T is facilitated by using linear gradients. The procedure is to load the unknown sample proteins and a suitable set of calibration proteins onto a single linear gradient slab gel, and then carry out electrophoresis until the tracking dye reaches the last centimetre or so of gel. After staining, the position of the protein bands is measured and %T calculated for each. For example, a polypeptide which has migrated halfway through a 5-20% polyacrylamide gel will have reached a %T of 12.5%. A plot of $\log_{10}$ molecular weight versus log%T for the calibration proteins then yields a standard curve from which the molecular weight of the sample proteins can be determined. It is essential that molecular weight marker proteins should be included with each gel run and a standard curve drawn for that particular gel.

The range of polypeptide molecular weights over which there is a linear relationship between $\log_{10}$ molecular weight and log%T depends on the polyacrylamide gradient conditions chosen. Examples are 7-25 %T (C_{Bis} = 1%), mol. wt. ~ 14,300-330,000 (91); 5-20 %T (C_{Bis} = 2.6%), mol. wt. ~ 14,300-210,000 (*Figure 21*); 3-30 %T (C_{Bis} = 8.4%), mol. wt. ~ 13,000-950,000 (90). Each of these represents a considerable improvement over the very restricted range available for molecular weight determination using any one uniform concentration polyacrylamide gel (p. 15). Furthermore, there is some evidence (90) that glycoproteins do not behave anomalously during SDS-PAGE using gradient gels, unlike their behaviour in single concentration gels (p. 17), possibly because the gradient of pore size within gradient gels causes molecular sieving to predominate over the anomalous charge effect caused by reduced binding of SDS by glycoproteins. Gradient gels may therefore become the gel medium of choice for glycoprotein molecular weight determination.

Despite the above advantages of gradient gels it is important to note that these cannot match the resolution of two protein components obtainable with gels of a uniform, optimal polyacrylamide concentration (see Chapter 2). The approach should therefore be to use a gradient gel for SDS-PAGE initially to determine the complexity of a protein mixture and to obtain an estimate of the molecular weight of the components. If the polypeptides are well resolved this is all that may be required. However, if all the proteins of interest fall into a narrow molecular weight range, SDS-PAGE should then be carried out with an appropriate uniform concentration gel to obtain optimal resolution of the components.

Native proteins. As with SDS-PAGE, the fractionation of native proteins in polyacrylamide gradient gels enables a much larger range of protein sizes to be fractionated than gels of uniform concentration, and the protein bands are sharpened by the gel gradient. Several investigators (e.g. refs. 92, 93) have used polyacrylamide gradient gels to determine the molecular size of native proteins based on the distance migrated, but other workers find that this is only accurate when the proteins under

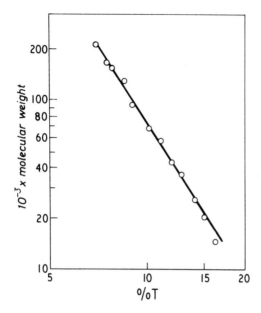

Figure 21. Calibration curve of $\log_{10}$ polypeptide molecular weight versus $\log_{10}$ %T for a 5-20% linear gradient slab gel. The marker polypeptides, in order of decreasing molecular weight, were myosin (mol. wt. 212,000), RNA polymerase β' (165,000) and β (155,000) subunits, β-galactosidase (130,000), phosphorylase a (92,500), bovine serum albumin (68,000), catalase (57,500), ovalbumin (43,000), glyceraldehyde-3-phosphate dehydrogenase (36,000), chymotrypsinogen A (25,700), soybean trypsin inhibitor (20,100), and lysozyme (14,300).

study are monomers and their homologous oligomers (89) and suggest that the molecular weight of a native protein electrophoresed in a linear polyacrylamide gradient gel is best estimated by measuring the rate of its migration through the gel (90). As with uniform concentration gels, molecular weight estimation for native proteins using gradient gels is little used compared to polypeptide molecular weight estimation using buffers containing SDS with these gels.

Preparation of Concentration Gradient Gels

Both linear and concave gradient gels can easily be produced in either slab or rod formats in the laboratory, although gradient slab gels are far more widely used than the corresponding rod gels. Multiple rod or slab gradient gel formation requires the use of a purpose-built gel forming tower (p. 76). However, gradient slab gels are often only required one or two at a time and so, although multiple gels may be made using a tower and stored, it is often convenient to produce them singly. This is easily done provided that a peristaltic pump and a suitable gradient maker are available (see below).

In addition to a gradient in acrylamide concentration, a density gradient of sucrose or glycerol is often included to minimise mixing by convective disturbances caused by the heat evolved during polymerisation. Some workers avoid the latter problem by including a gradient of polymerisation catalyst to ensure that polymerisation occurs

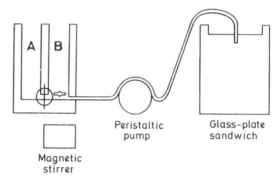

Figure 22. Apparatus for the formation of gradient polyacrylamide slab gels. A and B refer to the reservoir and mixing chambers of the gradient maker, respectively, which is supported about 1 cm above a magnetic stirrer using a clamp stand. Mixing chamber B contains a magnetic follower and is connected to reservoir A by a tunnel controlled by a two-way tap. The Tygon tubing used to connect the gradient maker via the peristaltic pump to the glass-plate sandwich is 0.075 cm or 0.15 cm I.D. Most simply, the Tygon tubing itself is cut at a 45° angle and then taped to the centre of the rear (unnotched) plate with the cut edge facing this plate, the tubing being pushed a centimetre or so between the plates to 'clamp' the cut end in position. If the tubing O.D. is too large to allow this, a suitably cut piece of fine-bore Teflon tubing or a syringe needle inserted into the end of the Tygon tubing will suffice instead.

first at the top of the gel (low acrylamide concentration) progressing to the bottom.

Linear gradient slab gels. Linear gradient makers are available commercially or can be constructed in the laboratory workshop from Perspex. The two chambers must be of exactly equal cross-section (or it will not be possible to produce linear gradients) and are joined by an inter-connecting tunnel controlled by a two-way tap. A typical set-up for producing linear gradient slab gels using this apparatus is shown in *Figure 22*. The outlet from the gradient maker is connected via fine-bore Tygon tubing to a peristaltic pump and thence to the glass-plate sandwich which has been previously assembled as described on p. 33. The Tygon tubing is cut at an angle of 45° and the cut edge taped in position facing the rear (unnotched) plate at the centre of the assembly (*Figure 22*). According to this procedure, the most concentrated acrylamide solution enters the glass-plate sandwich first and runs down the inside of the unnotched glass-plate to reach the gel bottom. As the level of acrylamide mixture rises in the sandwich, the acrylamide concentration steadily decreases. Surprisingly little mixing occurs if this is carried out carefully. Some commercial apparatuses and some homemade devices provide for a gradient of acrylamide to be introduced at the base of a single slab gel holder. In this case, the least concentrated acrylamide mixture enters first.

The exact composition of the acrylamide mixtures used for gradient gel preparation will depend on the concentration range of the gradient and the buffer system used. Most gradient gels are used for SDS-PAGE with the SDS-discontinuous buffer system (*Table 2*). Two acrylamide mixtures, corresponding to the lowest and highest concentrations in the gradient, are prepared according to the details in this table but with the amount of polymerisation catalyst reduced, to allow time to pour the gradient before polymerisation occurs, and with the highest acrylamide concentration

Table 12. Gel Mixtures for a 5-20% Gradient Gel.

5% acrylamide mixture:[a]
    5.0 ml acrylamide-bisacrylamide (30:0.8)
    3.75 ml resolving gel buffer stock; 3.0M Tris-Cl (pH 8.8)
    0.3 ml 10% SDS
    0.7 ml 1.5% ammonium persulphate
    20.25 ml water

20% acrylamide mixture:[a]
    20.0 ml acrylamide-bisacrylamide (30:0.8)
    3.75 ml resolving gel buffer stock; 3.0M Tris-Cl (pH 8.8)
    0.3 ml 10% SDS
    0.7 ml 1.5% ammonium persulphate
    4.5g sucrose (equivalent to 2.5 ml volume)
    2.75 ml water

[a]TEMED (10 μl) is added to initiate polymerisation just before pouring the gel (see the text).

mixture containing 15% (w/v) sucrose (4.5g per 30 ml gel mixture) to give a stabilising density gradient. Thus for a 5-20% linear gradient gel (2.6% bisacrylamide) using the SDS-discontinuous buffer system, the mixtures are as shown in *Table 12*.

The gel mixtures are degassed and TEMED (10 μl per 30 ml gel mixture) is then added to each. The low concentration mixture is added to reservoir A (*Figure 22*) and the connecting tube between the chambers of the gradient maker opened to fill it and then closed. Any gel solution which flowed into mixing chamber B (*Figure 22*) is returned to reservoir A. An equal volume of high concentration mixture is now added to chamber B. Each volume is calculated to correspond to half the volume of the final resolving gel. For the standard notched plate assembly (p. 21), this requires 15 ml acrylamide mixture per chamber. Chamber B can be mixed using a magnetic stirrer in which case care must be taken to prevent heat from the magnetic stirrer causing premature polymerisation. To avoid this, the gradient maker should be supported (using a clamp) about 1 cm above the stirrer. The chambers of the gradient maker are connected and the peristaltic pump and stirrer turned on. The flow rate of the gel mixture into the glass-plate sandwich should be about 3.0 ml/min. Following the quantitative delivery of the gel mixture into the sandwich, the outlet tubing from the gradient former is dipped into a flask of buffer with the same composition as in the resolving gel (3.75 ml 3.0 M Tris-HCl, pH 8.8, 0.30 ml 10% w/v SDS, 25.95 ml water) and the flow rate reduced to 0.5 ml/min. This serves to overlayer the resolving gel. Immediately the gradient has been poured, the gradient maker is flushed out with water to prevent acrylamide polymerising in the apparatus. After polymerisation of the slab gel the overlay is removed by tilting the gel and a stacking gel is polymerised in place, complete with sample wells, as described previously (p. 35).

Some workers prefer to use riboflavin as the polymerisation catalyst instead of ammonium persulphate when pouring gradient gels, since it allows more time for manipulations without the fear of premature gel polymerisation. In this case, riboflavin at a final concentration of 0.5 mg/ml of gel mixture replaces the ammonium persulphate and the gradient gel is polymerised, after pouring, by exposure to a fluorescent light for about 30 min. Then the stacking gel may be added. In practice, the sucrose gradient is sufficient to stabilise the acrylamide gradient during polymerisation and so a gradient of polymerisation catalysts to ensure that the slab gel polymerises at the top first is unnecessary. However, where convective mixing is a problem, the catalyst concentrations should be adjusted so as to cause the low concentration acrylamide mixture to gel in about 25-30 min and the high concentration mixture in about 40-45 min.

Concave gradient slab gels. Concave gradient polyacrylamide slab gels can be produced using the same gradient maker used for linear gradient gels. For a concave '5-20%' gradient gel of 30 ml total volume, for example, 7.5 ml of 20% acrylamide mixture is placed in the mixing chamber B and this is then stoppered with a rubber bung. The pressure inside this chamber is equalised with atmospheric pressure by momentary insertion of a hypodermic syringe needle. Reservoir A receives 22.5 ml of 5% acrylamide mixture. The slab gel is then poured as for linear gradient gels (see above). During pouring, the volume of chamber B remains constant and is continuously diluted by the incoming solution from reservoir A. This generates an exponentially decreasing gradient of acrylamide and yields a gel with the concave profile shown in *Figure 23a*, ranging from 20% at the gel bottom to about 6% at the gel top. A stacking gel may be polymerised in place if required.

Sample Preparation and Electrophoresis

Sample preparation is the same as for uniform concentration gels. Electrophoresis conditions will depend on the buffer system and gel concentrations comprising the gradient, but for the 5-20% linear gradient gel described above using the SDS-discontinuous buffer system, electrophoresis is at 25 mA constant current for about 5h, or 90V constant voltage overnight, or until the tracking dye nears the gel bottom. Analysis of gels after electrophoresis is the same as with uniform concentration polyacrylamide gels. The protein bands visible after staining are equally sharp with either linear or concave gradient gels (*Figure 23b* and *23c*, respectively) but the band distribution depends on the type of gradient chosen, concave gradient gels tending to separate polypeptides of high molecular weight better than linear gradient gels but at the expense of reducing band separation between lower molecular weight polypeptides. In the author's laboratory, linear gradient slab gels are preferred for one-dimensional SDS-PAGE of complex mixtures of polypeptides covering a wide range of sizes (e.g. *Figure 23d*) since this gradient profile is better suited than concave gradient gels to polypeptide molecular weight estimation (p. 72).

Large Numbers of Gels

Provided a multiple channel peristaltic pump and a gradient former of sufficient volume are available it is possible to produce several gradient gels simultaneously. Larger numbers of gradient slab gels can be produced using a purpose-built gel tower, several types of which are available commercially, although each of these is available for one slab gel size only. However, a suitable apparatus can be built cheaply in the laboratory from Perspex. Essentially it consists of a Perspex box to hold the slab gel assemblies, with the base of the box tapering down to form a funnel. The gel mixture enters the apparatus at the base of the funnel, which serves to decelerate the

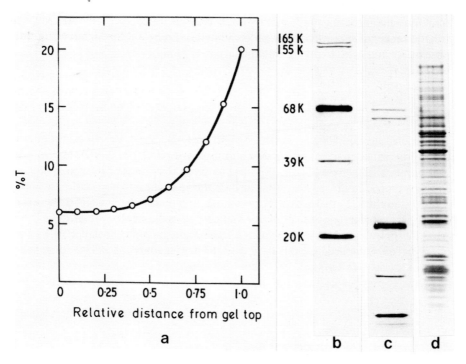

Figure 23. Use of gradient polyacrylamide gels. (a) profile of a '5-20%' concave ('exponential') gradient gel prepared as described in the text. (b) and (c), polypeptide band distribution after fractionation of RNA polymerase α, β, and β' subunits, bovine serum albumin, and soybean trypsin inhibitor on a linear 5-20% and concave '5-20%' polyacrylamide gradient slab gel, respectively, using the SDS-discontinuous buffer system. The molecular weights of the polypeptides are shown to the side of the gel tracks. (d) an example of a complex polypeptide mixture analysed using a linear gradient polyacrylamide slab gel; the sample was a subfraction of [^{35}S]-methionine-labelled proteins from *Dictyostelium discoideum* spores analysed on a 6-18% linear gradient polyacrylamide slab gel using the SDS-discontinuous buffer system. For other examples of complex polypeptide fractionation using gradient slab gels see *Figures 2* and *15*.

flow of liquid and to expand the horizontal cross-section of the liquid equal to that of the bottom of the gel holders. Details of gel tower construction and use are given elsewhere (88, 94, 95). The greatest practical problem in using a gel tower to produce gradient slab gels is to avoid convection currents during the exothermic polymerisation reaction of such a large volume of acrylamide. For this reason, it is essential that the polymerisation catalyst concentrations are arranged so as to cause polymerisation to occur first at the top and then to proceed down the gel.

In addition to producing gradient slab gels, a gel tower can be used to prepare large numbers of uniform concentration slab gels or rod gels.

Micro-rod and Micro-slab Gels

A variety of micro-gel systems has been developed for analysis of small amounts of protein. The extreme situation is the formation of polyacrylamide gels in capillary tubes which can analyse the amounts of protein found in single cells. Both single concentration and gradient polyacrylamide gels may be produced in the $1\mu l$ to $100 \mu l$ volume range using appropriately sized capillaries. Using 5 μl volume capillary gels Neuhoff (96) has detected 1 ng of albumin after staining. Micro-slab gels have also been used for protein analysis but here the highest sensitivity obtained is 20 ng for a single protein.

Micro-rod Gels

The detailed methodology for producing and analysing capillary gels, both uniform concentration and gradient gels, is described by Neuhoff (96). Basically, for uniform concentration gels, the capillaries are filled to about two-thirds of their total volume by being dipped in the gel mixture. They are then pressed into a plasticine cushion, about 2 mm thick, to seal the capillary bottom, and overlaid with water using a fine Pyrex glass capillary pipette of smaller diameter than the capillary used to hold the gel. After polymerisation, the overlay is removed and the protein solution (0.1-$1.0 \mu l$ of 1-3 mg/ml) is applied by capillary pipette, any free space in the capillary then being filled with electrode buffer. After filling, that length of capillary penetrated by the plasticine is snapped off and the gel subjected to electrophoresis. Grossbach (97) avoids the use of plasticine as a gel sealer and by using a micromanipulator with 50 μl capillary gels is able to polymerise both a resolving gel and a stacking gel in place prior to sample addition. Gradient gels may be produced in capillary tubing using either capillarity (98) or a gradient maker (99).

Despite these methods, capillary rod gels have been little used in recent years, possibly because some experience is needed to obtain good gels (particularly in water overlayering of individual gels) and the resolution of protein components is poor, at least in the lower volume capillary gels, compared to the standard rod gel. Furthermore, quantitation of stained protein bands can only be achieved using a micro-densitometer rather than the more generally available scanning spectrophotometer. However Condeelis (100) has described the use of intermediate-size gels made in 100 μl and 250 μl volume Drummond microcaps which can analyse proteins in the nanogram range (*Figure 24*). Any of the continuous and discontinuous gel buffer systems already described in this chapter could be used but Condeelis (100) found no increase in resolution when using a discontinuous buffer system with gels of this size. Details of the protein loading capacity and sensitivity of these micro-gels are given in *Table 13*. Perhaps the most useful of these micro-gels will be the 250 μl size which can be used to detect 2 ng of a single protein and yet protein bands can be quantitated using a scanning spectrophotometer as used with standard-size gels.

Micro-slab Gels

A number of micro-slab polyacrylamide gel apparatuses have been described, the smallest format being glass plates cut from microscope slides e.g. 7.5 cm x 2.5 cm (101), 7.6 cm x 3.8 cm (102). These are able to detect 20 ng of protein in a sharp band. However such small gel sizes cause serious problems of usage. With electrophoretic migration of samples along the long axis of these gels (101) only three samples can be

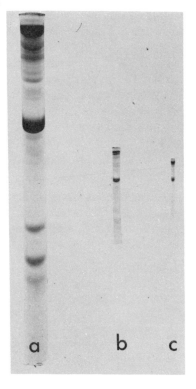
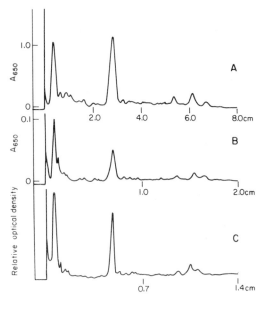

Figure 24. Use of micro-rod polyacrylamide gels. An actomyosin extract from rabbit skeletal muscle was electrophoresed on (a) a standard 0.5 cm diameter rod gel loaded with 20 μg total protein, (b) a 0.15 cm micro-rod gel loaded with 0.5 μg total protein, (c) a 0.07 cm micro-rod gel loaded with 0.1 μg total protein. The gels were electrophoresed using the SDS-discontinuous buffer system and then stained with Coomassie blue. A, B, and C show densitometer scans of the three gels (a), (b), and (c), respectively. Reproduced from ref.100 with permission.

co-electrophoresed, whilst electrophoretic migration along the shorter axis is too restricted for the separation of components comparable to that of the standard gel format. Not surprisingly therefore, micro-slab gels have been little used. Recently Matsudaira and Burgess (95) reported an intermediate-size slab gel format (8.2 cm x 9.2 cm) which enables the analysis of 21 samples with good resolution yet with a sensitivity such that 20 ng of a protein can be detected although 100 ng is required for a strongly-stained band.

Agarose-acrylamide Composite Gels

Polyacrylamide gels cannot be prepared at a total monomer concentration less than about 2.2%. Above this concentration, but below about 3%, gelation does occur but the gels are very difficult to handle, being almost viscous at the lower concentrations. For the vast majority of protein electrophoretic analyses it is unlikely that the investigator will need to use acrylamide gels at this low concentration. However, such gels have been used, with agarose added to 0.5% to give mechanical strength, for

Table 13. Properties of Micro-rod Gels.[a]

Gel format	Maximum loading of protein (μg)	Minimum detectable per band (μg)	Sample volume (μl)
Micro-rod gel in 250 μl Drummond microcap	2.0	2×10^{-3}	0.25-2.0 (0.5 optimal)
Micro-rod gel in 100 μl Drummond microcap	0.5	3×10^{-4}	0.1-0.5 (0.2 optimal)

[a]Data from ref. 100 with permission.

fractionation of ribosomes and polyribosomes (103, 104) and some viruses. Particles as large as polyribosomes comprising 8 ribosomes enter the gel with mobilities that approximate to the $\log_{10}$ of the particle weight. A detailed description of the preparation and electrophoresis of composite gels is available from Miss S.L. Bunting and Dr. A.C. Peacock (see ref. 104).

HOMOGENEITY AND IDENTITY

The detection of only one protein band by zone electrophoresis in polyacrylamide gels under a single set of experimental parameters does not prove homogeneity. Since the use of non-dissociating buffer conditions in polyacrylamide gel electrophoresis causes fractionation on the basis of both size and charge, it is possible to have two proteins which differ in each of these properties but which, under the electrophoresis conditions chosen, migrate at the same rate to yield a single band. Therefore, native proteins should always be analysed by electrophoresis at several pH values and gel concentrations before homogeneity or identity of protein samples is claimed. Identity testing by Ferguson plot analysis is covered in Chapter 2. Since SDS-PAGE fractionates only on the basis of molecular size, here again the number of protein bands observed does not necessarily equate with the number of distinct polypeptides in the sample mixture; any of the bands may be comprised of multiple components of identical molecular weight. Because of these problems, an increasingly common method to check on protein homogeneity is to use two-dimensional polyacrylamide gel electrophoresis (see Chapter 5) where proteins can be separated on the basis of charge in the first dimension followed by a size fractionation according to SDS-PAGE in the second. A single spot after two-dimensional polyacrylamide gel electrophoresis is a good indication of homogeneity, although not absolute proof (Chapter 5).

Assuming that the protein band of interest resolved by one-dimensional polyacrylamide gel electrophoresis represents a single pure protein, information about the protein structure and its identity or non-identity with other proteins can be obtained by a number of methods.

(i)   The amino acid composition of stained protein bands can be determined without elution from the gel and using only 5-10 nmoles protein (105, 106). Unfortunately the large amount of ammonia produced by hydrolysis of the polyacrylamide can prevent determination of basic amino acids. Hence it is better to elute the protein before analysis. One of the problems still encountered is that polyacrylamide gel often contains impurities which become increasingly important as the sensitivity of amino acid composition studies increases. As would be expected, the degree of interference depends on the volume of gel eluted. Some correction for this can be made by analysis of an equivalent volume of blank gel slices.

(ii) Polypeptides separated by polyacrylamide gel electrophoresis can also be characterised by peptide mapping. Given the small amount of polypeptide usually available, the sensitivity of peptide detection can be increased by elution of polypeptide from the gel followed by radioiodination prior to tryptic digestion (107, 108). An even simpler method has been described by Elder *et al.* (109) whereby polypeptides separated by SDS-PAGE are both radioiodinated and treated with trypsin whilst still in the gel. The tryptic peptides are then easily eluted from the gel and analysed by two-dimensional chromatography followed by autoradiography. Entire multicomponent systems can be analysed using only microgram amounts of total protein. The method works on fixed and stained protein bands and even on dried gels after years of storage, and requires only a few days work. If a protein mixture is too complex for adequate resolution of proteins it is best analysed by two-dimensional polyacrylamide gel electrophoresis (Chapter 5) and then the Elder method of mapping applied. The only disadvantage of this peptide mapping method methodology is that the protein analysed must contain amino acid residues susceptible to some form of radioiodination (see Appendix II). Another rapid method of peptide mapping involves partial proteolytic cleavage of proteins separated by polyacrylamide gel electrophoresis, without elution of protein, followed by a second SDS-PAGE separation during which the peptide products separate on a size basis (110). This method is described in detail in Chapter 6.

(iii) *N*-terminal amino acid sequences of proteins separated by polyacrylamide gel electrophoresis can be obtained by several protocols all of which rely on the sequential removal of *N*-terminal amino acids by reaction with phenylisothiocyanate (PITC), namely the Edman degradation. Originally Wiener *et al.* (111) described the elution of proteins into buffer containing SDS followed by a modified dansyl-Edman degradation. Although relatively simple to perform, this technique permits only small numbers of residues to be identified. A considerable improvement is the automatic solid-phase procedure described by Bridgen (112, 113). The protein is eluted from the gel, covalently attached to a di-isothiocyanate-activated glass support, and then subjected to the Edman degradation using [^{35}S]-PITC. The radioactive amino acid derivatives are identified by thin-layer chromatography followed by autoradiography. Using this protocol, 10 μg lysozyme (700 pmol) were sufficient to determine the sequence of 20 *N*-terminal amino acid residues (113). Recent advances in solid-phase sequencing are reviewed by Laursen and Machleidt (114). Proteins may also be labelled *in vivo* prior to polyacrylamide gel electrophoresis and sequencing (115). Although theoretically capable of very high sensitivity, this approach is only applicable to the restricted number of proteins where *in vivo* labelling to high specific activity is feasible. The limitations of these and other sequencing strategies for proteins separated using polyacrylamide gel electrophoresis are discussed by Bhown *et al* (116).

(iv) Although proteins are easily eluted from gels electrophoresed in non-dissociating buffers and the biological activity of distinct protein bands compared, this is also sometimes possible with denatured polypeptides separated by SDS-PAGE: several enzymes have been successfully renatured after elution from the gel and removal of SDS using urea with Dowex resin (117) or by electrophoresis through a urea-containing polyacrylamide gel (66). As an extension of this, Manrow and Dottin (118) have described a two-dimensional gel system whereby the protein mixture is fractionated by SDS-PAGE in the first dimension in a rod gel, and then by elec-trofocusing in the presence of urea and Nonidet P40 in a slab gel in the second dimen-sion. The urea and non-ionic detergent in the second-dimension gel remove SDS from the polypeptides so that, when the slab gel is then equilibrated with buffer to remove the urea, the polypeptides renature and can be located by assaying for biological activity. Thus, the method combines the high resolution associated with two-dimensional gel electrophoresis with the ability to identify the separated poly-peptides on a functional basis. An alternative protocol, which is better than the use of urea for the renaturation of at least some proteins, has been described recently by Hager and Burgess (119). The protein band is visualised in the gel after SDS-PAGE by KCl 'staining', then the band is cut out, crushed, and the polypeptide is eluted by diffusion in a buffer containing 0.1% SDS. The polypeptide is concentrated (and most of the SDS removed) by acetone precipitation. Renaturation of the polypeptide occurs after the precipitate is dissolved in guanidine hydrochloride and then diluted. The percentage of original activity recovered depends on the renaturation protocol chosen and on the particular protein under investigation.

(v) Many detailed characterisations of individual proteins have only been possible by a combination of immunoprecipitation, using antibodies raised against the pro-tein, and SDS-PAGE. Proteins eluted from polyacrylamide gels can be used as an-tigens for immunisation or entire gel slices may be macerated and injected (p.63). If SDS-PAGE has been used to obtain the polypeptide, it is usually not necessary to remove the SDS prior to injection and the resulting antibodies frequently precipitate both native and denatured forms of the protein. The monospecific antibody can then be used to precipitate the protein from a complex mixture either by direct immuno-precipitation using carrier protein (120) or indirect immunoprecipitation using say, monospecific rabbit IgG followed by goat anti-rabbit IgG (121). Contaminating pro-teins are then removed by repeated washing with detergents such as Triton X-100 (122) or centrifugation through sucrose containing detergent (120). A more recent method involves the use of *Staphylococcus aureus* protein A, a bacterial coat protein with high affinity for the Fc portion of certain immunoglobulins including IgG (123, 124). Addition of killed *S. aureus* binds the immune complexes which are then purified by pelleting the bacteria. A further refinement is to use protein A-Sepharose instead of bacteria to reduce non-specific protein binding. After purification by any of these methods, immune complexes are solubilised by heating in SDS-containing sample buffer and subjected to SDS-PAGE. Polypeptides specifically precipitated by the antibody are identified by electrophoresis of a control sample exposed to preimmune serum in a parallel track. Further identification of polypeptide antigens after electrophoresis is possible either by molecular weight estimation relative to standard polypeptides or by comigration with purified antigens. If the polypeptide under study has a molecular weight which corresponds to that of the antibody light chain (mol. wt. ~ 23,500) or heavy chain (mol. wt. ~ 50,000), comigration in SDS-PAGE can be avoided by omitting the thiol reagent during sample dissociation or by use of papain-treated antibody respectively.

The technique works best when the protein antigen to be precipitated exceeds about 0.1% of the total protein although lower concentrations of specific proteins have been successfully analysed. Among the more common uses for this methodology are:-

(a) the relatedness of two proteins from different sources can be analysed by testing for immunological cross-reaction using antibody raised against one of the proteins followed by SDS-PAGE. Alternatively, SDS-PAGE may be carried out first and then the immunological cross-reactivity of resolved protein bands analysed (see p. 60).

(b) immune precipitation followed by SDS-PAGE is now the most common method of following *in vivo* or *in vitro* synthesis of specific proteins (e.g. ref. 125). The protein is labelled using a radioactive amino acid (usually [^{3}H]-leucine or [^{35}S]-methionine), immunoprecipitated, and fractionated by SDS-PAGE. After electrophoresis, the specific protein band is excised and its radioactivity determined. One big advantage of the combined immunoprecipitation/SDS-PAGE approach over immunoprecipitation alone is that SDS-PAGE separates specifically-precipitated protein from non-specific protein contaminants which may be carried through the immunoprecipitation procedure, thus reducing background counts considerably.

(c) the technique can also be used to isolate microgram amounts of immunologically related proteins for further analysis. For example, Platt *et al.* (126) used antibody raised against *E. coli lac* repressor to isolate a mutant protein which was compared to the wild-type protein by *N*-terminal amino acid sequence determination.

ARTIFACTS AND TROUBLESHOOTING

(i) Lack of polymerisation is usually due to incorrect concentrations of the prepared reagents or the omission of a reagent from the gel mixture, or impurities in the reagents. The simplest remedy is usually to discard solutions and prepare a fresh batch using pure reagents. High concentrations of thiol reagents will also inhibit polymerisation.

(ii) Polymerisation should occur according to the time periods specified in this chapter, usually 10-30 min for single concentration gels. Too fast or too slow polymerisation can most easily be corrected by varying the concentrations of the polymerisation catalysts.

(iii)   Cracking of the gel during polymerisation (usually only high concentration gels) is often due to excessive heat production by the polymerisation reaction itself and is remedied by using cooled solutions. For rod gels, siliconising the tubes may also help. This problem is usually not encountered at most gel concentrations when rod or slab gels are made individually. Gel cracking during electrophoresis occurs due to excessive current input overheating the gel, and is remedied by using less current over a longer period of time.

(iv)   Provided that the ionic strength of non-dissociating buffers is high enough to prevent aggregation of native proteins, insoluble material in the sample usually represents denatured protein and must be removed by centrifugation prior to electrophoresis. Insolubility in SDS-containing buffers indicates too little SDS, too little reducing agent, or too low a pH, especially after TCA precipitation of proteins.

(v)   Failure of the sample to form a layer at the well bottom when applied to slab gels indicates either the accidental omission of sucrose or glycerol from the sample buffer or the use of a sample comb where the teeth do not form a snug fit with the glass plates, allowing gel to polymerise between the teeth and glass plates which then interferes with sample loading. The remedy for the latter problem is to use a better fitting comb, but in the short term, excess gel can be removed from the wells using a syringe needle connected to a water aspirator.

(vi)   Detachment of slab gels from the glass plates during gel electrophoresis usually indicates inadequately cleaned plates. Low concentration gels sometimes detach from rod gel tubes even though these are clean; the problem is overcome by attaching a piece of nylon mesh to the bottom of the tube (p. 28).

(vii)   Sometimes the top of a rod gel collapses inward upon itself and away from the tube wall during electrophoresis, causing deformation of the gel and allowing sample to migrate between the gel and the tube wall. A similar phenomenon is observed with slab gel tracks where the base of the sample well appears to be dragged downwards in the direction of electrophoresis. Both these effects can be caused by trapping of high molecular weight, highly charged species at the gel surface and is particularly common with high concentrations of nucleic acid in the sample (low concentrations are usually tolerable). If problematical, nucleic acid concentrations can be reduced by one of several methods (see Chapter 5, p.194).

(viii)   Poor staining after SDS-PAGE is often cured by increasing the volume of staining solution to dilute out the SDS present.

(ix)   A metallic sheen on gels after staining with Coomassie blue usually indicates that solvent has been allowed to evaporate causing dye to dry on the gel at that point. Slight films of Coomassie blue sometimes observed on the gel surface after destaining can be removed by a quick rinse in 50% methanol or by gently swabbing the gel surface with methanol-soaked tissue paper.

(x)   Blue blotches near gel borders are usually fingerprints caused by gel handling without gloves!

(xi)   Protein bands observed in all tracks of a slab gel or all rod gels may indicate contamination of the sample buffer. Contaminated reservoir buffer produces a continuous stained region from the gel origin to near the buffer front, even in tracks which have not been loaded with sample.

(xii)   A high background of protein staining along individual rod gels or slab gel tracks with indistinct protein bands usually indicates extensive sample proteolysis [see (xviii) for remedy]. However, this phenomenon has also been observed with the SDS-discontinuous system when using certain impure grades of SDS; analysis of the same samples with purified SDS gave sharp bands.

(xiii) Variations in staining density along the width of a stained band often indicates an uneven gel surface, resulting in sample accumulating at the low points prior to electrophoresis. Uneven gel surfaces can be caused by insufficient care or experience in overlayering the gel or as a result of vibration during gel polymerisation.

(xiv)   The inability of a substantial portion of the protein to enter the resolving gel may cause a heavily stained band at the gel origin. Provided that the gel concentration is sufficiently low that one would have expected the proteins to be able to enter the gel matrix, the problem may be caused by aggregated protein in the sample prior to electrophoresis [see (iv) above] or, in the case of non-dissociating discontinuous buffer systems, precipitation of the proteins due to the formation of highly concentrated zones during electrophoresis in the stacking gel. If the latter is correct, one would be advised to use less concentrated samples with a continuous buffer system.

(xv)   Protein streaking along individual rod gels or slab gel tracks is often accompanied by protein at the gel origin and indicates protein precipitation followed by dissolution of the precipitates during electrophoresis. For remedies see (iv) and (xiv) above.

(xvi)   Cracking of single concentration gels during drying under vacuum will occur if the vacuum is accidentally released before the gel is properly dry, or if thick gel slices or slabs (>1.5 mm) are being used, or if the gel has been allowed to swell appreciably before drying.

(xvii) Artifactual blackening of X-ray film during fluorography may be caused by inadequate removal of DMSO. With future gels, ensure sufficient soaking in water before drying the gel.

(xviii) Irreproducibility of the protein band pattern is usually caused by problems of sample preparation rather than by polyacrylamide gel electrophoresis. Reduction in the staining intensity or complete loss of individual components may be indicative of proteolysis, as is the appearance of previously unobserved fast migrating bands. Working at low temperature and the use of protease inhibitors during sample preparation may eliminate the problem. If the problem occurs with SDS-PAGE, check that the sample is being heated to at least 90°C for 2 min during dissociation.

ACKNOWLEDGEMENTS

Thanks are due to Drs D.Rickwood, E.J.Wood, and S.J.Higgins for critical reading of this manuscript, to Sandra Gray and Susan Whaley for dedicated typing of near-illegible drafts, to David Wilkinson for some of the data presented, and finally to my wife, Irene, for continual good advice and encouragement, and for detailed proofreading.

REFERENCES

References of General Interest

Gordon,A.H. (1975) Electrophoresis of Proteins in Polyacrylamide and Starch Gels (Laboratory Techniques in Biochemistry and Molecular Biology; Work,T.S. and Work.E., eds) North-Holland, Amsterdam, Oxford, Vol. 1, pt. 1.

Smith,I. (ed.) (1975) Chromatographic and Electrophoretic Techniques, Vol. 2. Zone Electrophoresis, William Heinemann Medical Books Ltd., London.

Allen,R.C. and Maurer,H.R. (eds) (1974) Electrophoresis and Isoelectric Focusing in Polyacrylamide Gel, Walter de Gruyter, New York.

Maurer,H.R. (1971) Disc Electrophoresis and Related Techniques of Polyacrylamide Gel Electrophoresis, Walter de Gruyter, Berlin and New York.

References in the Text

1. Gordon,A.H. (1975) see general references.
2. Smith,I. (1975) *in* Chromatographic and Electrophoretic Techniques (Smith,I., ed.), Vol. 2. Zone Electrophoresis, William Heinemann Medical Books Ltd., London, p. 153.
3. Shapiro,A.L., Vinuela,E. and Maizel,J.V. (1967) Biochem. Biophys. Res. Commun. **28**, 815.
4. Weber,K. and Osborn,M. (1969) J. Biol. Chem. **244**, 4406.
5. Ornstein,L. (1964) Ann. N.Y. Acad. Sci. **121**, 321.
6. Davis,B.J. (1964) Ann. N.Y. Acad. Sci. **121**, 404.
7. Laemmli,U.K. (1970) Nature (London) **277**, 680.
8. Neville,D.M. (1971) J. Biol. Chem., **246**, 6328.
9. Williams,D.E. and Reisfeld,R.A. (1964) Ann. N.Y. Acad Sci. **121**, 373.
10. Reisfeld,R.A., Lewis,V.J. and Williams,D.E. (1962) Nature (London) **195**, 281.
11. Ferguson,K.A. (1964) Metabolism **13**, 21.
12. Rodbard,D., Chrambach,A. and Weiss,G.H. (1974) *in* Electrophoresis and

Isoelectric Focusing in Polyacrylamide Gel (Allen,R.C. and Maurer,H.R., eds), Walter de Gruyter, Berlin and New York, p. 62.

13. Hedrick,J.L. and Smith,A.J. (1968) Arch. Biochem. Biophys. **126**, 155.

14. Rodbard,D. and Chrambach,A. (1974) *in* Electrophoresis and Isoelectric Focusing in Polyacrylamide Gel (Allen,R.C. and Maurer,H.R., eds), Walter de Gruyter, Berlin and New York, p. 28.

15. Segrest,J.P. and Jackson,R.L. (1972) *in* Methods in Enzymology (Ginsburg,V., ed.), Academic Press, New York and London, Vol. 28B, p. 54.

16. Studier,F.W. (1973) J. Mol. Biol. **79**, 237.

17. Swaney,J.B., Vande Wonde,G.F. and Bachrach,H.L. (1974) Anal. Biochem. **58**, 337.

18. Payne,J.W. (1973) Biochem. J. **135**, 867.

19. Carpenter,F.H. and Harrington,K.T. (1972) J. Biol. Chem. **247**, 5580.

20. Rice,R.H. and Means,G.E. (1971) J. Biol. Chem. **246**, 831.

21. Diezel,W., Kopperschläger,G. and Hofmann,E. (1972) Anal. Biochem. **48**, 617.

22. Chrambach,A., Reisfeld,R.A., Wyckoff,M. and Zaccari,J. (1967) Anal. Biochem. **20**, 150.

23. Reisner,A.H., Nemes,P. and Bucholtz,C. (1975) Anal. Biochem. **64**, 509.

24. Fenner,C., Traut,R.R., Mason,D.T. and Wilkman-Coffelt,J. (1975) Anal. Biochem. **63**, 595.

25. Kohler,P.O., Bridson,W.E. and Chrambach,A. (1971) J. Clin. Endocrinol. **32**, 70.

26. Maizel,J.V. (1971) *in* Methods in Virology (Maramorosch,K. and Koprowski,H., eds), Academic Press, New York, Vol. 5, p. 179.

27. Switzer,R.C., Merril,C.R. and Shifrin,S. (1979) Anal. Biochem. **98**, 231.

28. Oakley,B.R., Kirsch,D.R. and Morris,N.R. (1980) Anal. Biochem. **105**, 361.

29. Ragland,W.L., Pace,J.L. and Kemper,D.L. (1974) Anal. Biochem. **59**, 24.

30. Eng,P.R. and Parker,C.O. (1974) Anal. Biochem. **59**, 323.

31. Barger,B.O., White,F.C., Pace,J.C.,Kemper,D.L. and Ragland,W.L. (1976) Anal. Biochem. **70**, 327.

32. Ragland,W.C., Benton,T.L., Pace,J.L., Beach,F.G. and Wade,A.E. (1978) *in* Electrophoresis '78 (Catsimpoolas,N., ed.), Elsevier/North-Holland Publishing Co., Amsterdam, Vol. 2, p. 217.

33. Hartman,B.K. and Udenfriend,S. (1969) Anal. Biochem. **30**, 391.

34. Jackowski,G. and Liew,C.C. (1980) Anal. Biochem. **102**, 34.

35. Laskey,R.A. and Mills,A.D. (1975) Eur. J. Biochem. **56**, 335.

36. Laskey,R.A. and Mills,A.D. (1977) FEBS Lett. **82**, 314.

37. Bonner,W.M., and Laskey,R.A. (1974) Eur. J. Biochem. **46**, 83.

38. Laskey,R.A. (1980) *in* Methods in Enzymology (Grossman,L. and Moldave,K., eds), Academic Press, New York, Vol. 65, p. 363.

39. Chamberlain,J.P. (1979) Anal. Biochem. **98**, 132.

40. McConkey,E.H. (1979) Anal. Biochem. **96**, 39.

41. Walton,K.E., Styer,D. and Gruenstein,E. (1979) J. Biol. Chem. **254**, 795.

42. Kronenberg,L.H. (1979) Anal. Biochem. **93**, 189.

43. Peterson,J., Tipton,H.W. and Chrambach,A. (1972) Anal. Biochem. **62**, 274.

44. Goodman,D. and Matzura,H. (1971) Anal. Biochem. **42**, 481.

45. Paus,P.N. (1971) Anal. Biochem. **42**, 372.

46. Hansen,J.N., Pheiffer,B.H. and Boehnert,J.A. (1980) Anal. Biochem. **105**, 192.

47. Spath,P.J. and Koblet,H. (1979) Anal. Biochem. **93**, 275.

48. Anderson,L.E. and McClure,W.O. (1973) Anal. Biochem. **51**, 173.

49. Basch,R.S. (1968) Anal. Biochem. **26**, 184.

50. Ames,G.F. (1974) J. Biol. Chem. **249**, 634.

51. Aloyo,V.J. (1979) Anal. Biochem. **99**, 161.

52. Martin,A.F., Prior,G. and Zak,R. (1976) Anal. Biochem. **72**, 577.

53. Airhart,J., Kelley,J., Brayden,J.E. and Low,R.B. (1979) Anal. Biochem. **96**, 45.

54. Auerbach,S. and Pederson,T. (1975) Biochem. Biophys. Res. Commun. **63**, 149.

55. Olden,K. and Yamada,K.M. (1977) Anal. Biochem. **78**, 483.

56. Burridge,K. (1978) *in* Methods in Enzymology (Ginsburg,V., ed.), Academic Press, New York, Vol. 50, p. 54.

57. Renart,J., Reiser,J. and Stark,G.R. (1979) Proc. Nat. Acad. Sci U.S.A. **76**, 3116.

58. Chua,N.H. and Blomberg,F. (1979) J. Biol. Chem. **253**, 3924.

59. Showe,M.K., Isobe,E. and Onorato,L. (1976) J. Mol. Biol. **107**, 55.

60. Saltzgaber-Müller,J. and Schatz,G. (1978) J. Biol. Chem. **253**, 305.

61. Sri Venugopal,K.S. and Adiga,P.R. (1980) Anal. Biochem. **101**, 215.

62. Stephens,R.E. (1975) Anal. Biochem. **65**, 369.

63. Schetters,H. and McLeod,B. (1979) Anal. Biochem. **98**, 329.

64. Mardian,J.K.W. and Isenberg,I. (1978) Anal. Biochem. **91**, 1.

65. Tuszynski,G.P., Damsky,C.H., Fuhrer,J.P. and Warren,L. (1977) Anal. Biochem. **83**, 119.

66. Hanaoka,F., Shaw,J.C. and Muelter,G.C. (1979) Anal. Biochem. **99**, 170.

67. Swank,R.W. and Munkres,K.D. (1971) Anal. Biochem. **39**, 462.

68. Panyim,S. and Chalkley,R. (1969) Arch. Biochem. Biophys. **130**, 337.

69. Franklin,S.G. and Zweidler,A. (1977) Nature (London) **266**, 273.

70. Zweidler,A. (1978) *in* Methods in Cell Biology (Stein,G., Stein,J. and Kleinsmith,L.J., eds), Academic Press, New York, Vol. 27, p. 223.

71. Hardison,R. and Chalkley,R. (1978) *in* Methods in Cell Biology, (Stein,G., Stein,J. and Kleinsmith,L.J., eds), Academic Press, New York, Vol. 17, p. 235.

72. Thomas,J.O. and Kornberg,R.D. (1975) Proc. Nat. Acad. Sci. U.S.A. **72**, 2626.

73. Panyin,S. and Chalkley,R. (1971) J. Biol. Chem. **246**, 7557.

74. Shmatchenko,V.V. and Varshavskey,A.J. (1978) Anal. Biochem. **85**, 42.

75. MacGillivray,A.J., Cameron,A., Krauze,R.T., Rickwood,D. and Paul,J. (1972) Biochim. Biophys. Acta **277**, 384.

76. Bickle,T.A. and Traut,R.R. (1974) *in* Methods in Enzymology (Moldave, K. and Grossman,L., eds), Academic Press, New York, Vol. 30, p. 545.

77. Kanda,F., Ochiai,H. and Iwabuchi,M. (1974) Eur. J. Biochem. **44**, 469.

78. Traub,P., Mizushima,S., Lowry,C.V. and Nomura,M. (1971) *in* Methods in Enzymology (Moldave,K. and Grossman,L., eds), Academic Press, New York and London, Vol. 20, p. 391.

79. Siu,C.H., Lerner,R.A. and Loomis,W.F. (1977) J. Mol. Biol. **116**, 469.

80. Hubbard,A.L. and Cohn,Z.A. (1972) J. Cell Biol. **55**, 390.

81. Gahmberg,C.G. (1977) *in* Dynamic Aspects of Cell Surface Organisation, Cell Surface Reviews (Poste,G. and Nicolson,G.L., eds), Elsevier/North-Holland Publishing Co., Amsterdam, Vol. 3, p.371.

82. Dewald,B., Dulaney,J.T. and Touster,O. (1974) *in* Methods in Enzymology (Fleischer,S. and Packer,L., eds), Academic Press, New York, Vol. 32, p. 82.

83. Newby,A.C. and Chrambach,A. (1979) Biochem. J. **177**, 623.

84. Zahler,W.L. (1974) *in* Methods in Enzymology (Fleischer,S. and Packer,L., eds), Academic Press, New York, Vol. 32, p. 70.

85. Ballou,B. and Smithies,O. (1977) Anal. Biochem. **80**, 616.

86. Helenius,A. and Simons,K. (1977) Proc. Nat. Acad. Sci. U.S.A. **74**, 529.

87. Bordier,C., Loomis,W.F., Elder,J. and Lerner,R. (1978) J. Biol. Chem. **253**, 5133.

88. Margolis,J. and Kenrick,K.G. (1967) Nature (London) **214**, 1334.

89. Lambin,P.C. (1978) Anal. Biochem. **85**, 114.

90. Lambin,F. and Fine,J.M. (1979) Anal. Biochem. **98**, 160.

91. Poduslo,J.F. and Rodbard,D. (1980) Anal. Biochem. **101**, 394.

92. Anderson,L.O., Borg,H. and Mikaelson,M. (1972) FEBS Lett. **20**, 199.

93. Kopperschläger,G., Diezel,W., Bierwagen,B. and Hofman,E. (1969) FEBS Lett. **5**, 221.

94. Leaback,D.H. (1975) *in* Chromatographic and Electrophoretic Techniques, Vol. 2. Zone Electrophoresis (Smith,I., ed.), W.Heinemann Medical Books Ltd., London,p. 250.

95. Matsudaira,P.T. and Burgess,D.R. (1978) Anal. Biochem. **87**, 386.

96. Neuhoff,V. (1973) *in* Micromethods in Molecular Biology (Neuhoff,V., ed.), Springer Verlag, New York, p. 1.

97. Grossbach,U. (1974)*in* Electrophoresis and Isoelectric Focusing in Polyacrylamide Gel (Allen,R.C. and Maurer,H.R., eds), Walter de Gruyter, Berlin and New York, p. 207.

98. Ruchel,R. (1974) *ibid.* p. 215.

99. Dames,W. and Maurer,H.R. (1974) *ibid.* p. 221.

100. Condeelis,J.S. (1977) Anal. Biochem. **77**, 195.

101. Maurer,H.R. and Dati,F.A. (1972) Anal. Biochem. **46**, 19.

102. Amos,W.B. (1976) Anal. Biochem. **70**, 612.

103. Dahlberg,A.E., Dingman,C.W. and Peacock,A.C. (1969) J. Mol. Biol. **41**, 139.

104. Dahlberg,A.E. (1979) *in* Methods in Enzymology (Moldave,K. and Grossman,L., eds), Academic Press, New York, Vol. 59, p. 397.

105. Kyte,J. (1971) J. Biol. Chem. **246**, 4157.

106. Huston,L.L. (1971) Anal. Biochem. **44**, 81.

107. Bray,D. and Brownlee,S.M. (1973) Anal. Biochem. **55**, 213.

108. Raison,R.L. and Marchalonis,J.J. (1977) Biochemistry **16**, 2036.

109. Elder,J.H., Pickett,R.A., Hampton,J. and Lerner,R.A. (1977) J. Biol. Chem. **252**, 6510.

110. Cleveland,D.W., Fischer,S.G., Kirschner,M.C. and Laemmli,U.K. (1977) J. Biol. Chem. **252**, 1102.

111. Wiener,A.M., Platt,T. and Weber,K. (1972) J. Biol. Chem. **247**, 3242.

112. Bridgen,J. (1976) Biochemistry **15**, 3600.

113. Bridgen,J., Snary,D., Crumpton,M.J., Barnstaple,C., Goodfellow,P. and Bodmer,W. (1976) Nature (London) **261**, 200.

114. Laursen,R.A. and Machleidt,W. (1980) *in* Methods of Biochemical Analysis (Glick,D., ed.), J.Wiley and Sons, New York, Vol. 26, p. 201.

115. Ballou,B.T., McKean,D.J., Freedlender,E.F. and Smithies,O. (1973) Proc. Nat. Acad. Sci. U.S.A. **73**, 4487.

116. Bhown,A.S., Mole,J.E., Hunter,F. and Bennett,J.C. (1980) Anal. Biochem. **103**, 184.

117. Weber,K. and Kuter,D.J. (1971) J. Biol. Chem. **246**, 4505.

118. Manrow,R. and Dottin,R.P. (1980) Proc. Nat. Acad. Sci. U.S.A. **77**, 730.

119. Hager,D.A. and Burgess,R.R. (1980) Anal. Biochem. **109**, 76.

120. Maurer,R.A., Stone,R. and Gorski,J. (1976) J. Biol. Chem. **251**, 2801.

121. Gorecki,M. and Zeelon,E.P. (1979) J. Biol. Chem. **254**, 525.

122. Schutz,G., Beato,M. and Feigelson,P. (1974) *in* Methods in Enzymology (Moldave,K. and Grossman,L., eds), Academic Press, New York, Vol. 30, 701.

123. Kessler,S.W. (1975) J. Immunol. **115**, 1617.

124. Ivarie,R.D. and Jones,P.P. (1979) Anal. Biochem. **97**, 24.

125. Ma,G.C.L. and Firtel,R.A. (1978) J. Biol. Chem. **253**, 3924.

126. Platt,T., Weber,K., Ganem,D. and Miller,J.H. (1972) Proc. Nat. Acad. Sci. U.S.A. **69**, 897.

127. Banker,G.A. and Cotman,C.W. (1972) J. Biol. Chem.**247**, 5856.

128. Frank,R.N. and Rodbard,D. (1975) Arch. Biochem. Biophys. **171**, 1.

129. Dautrevaux,M., Boulanger,Y., Han,K. and Biserte,G. (1969) Eur. J. Biochem. **11**, 267.

CHAPTER 2

"Quantitative" and Preparative Polyacrylamide Gel Electrophoresis

ANDREAS CHRAMBACH and DAVID RODBARD

INTRODUCTION

Polyacrylamide gel electrophoresis (PAGE) is the only method at present capable of separating molecules on the basis of their physical differences both in molecular size and net charge (1). When the nature of the differences between molecular species are not known, and consequently it is not known whether a size or a charge fractionation are desirable, this is the separation method of choice at this time. For many applications involving comparative analysis of protein mixtures the techniques described in Chapter 1 will suffice. However, when there are only a few components, an exact determination of the optimum pore size (by Ferguson plot analysis) for separating the species of interest from its most closely migrating contaminants is possible. This analysis by polyacrylamide gel electrophoresis then provides the information whether both size and charge fractionation, or either one of these, promises to be most effective.

When one attempts to use the relative mobility (R_f values) of proteins as a physico-chemical tool for the determination of size and charge of the protein, or for the determination of the optimum pore size for resolution of the protein from another, it is essential that reproducible gels are used. We have designated this particular application of polyacrylamide gel electrophoresis "quantitative PAGE" (1,2,3). Polyacrylamide gels become reproducible if one rigidly standardises the polymerisation rate by controlling the purity of reagents, the concentration of monomers, catalysts, and inhibitors (oxygen), as well as temperature, as has been discussed in detail previously (2). In this chapter, only the polymerisation procedure which resulted from these considerations will be given.

The points of distinction between PAGE and SDS-PAGE are procedurally minor. Therefore it appears sufficient for the purposes of this review to provide a single procedure for PAGE, which is also applicable to SDS-PAGE, unless specified otherwise at the appropriate points of the text.

A SURVEY OF APPARATUS FOR QUANTITATIVE PAGE

The central operations in the strategy of quantitative PAGE, that is the physical characterisation of proteins and the determination of the optimally resolving gel concentration (T_{opt}) for the resolution of any pair of components require simultaneous electrophoresis of these in gels varying in gel concentration. This requirement is best met by a gel tube apparatus. Contemporary forms of PAGE tube apparatus differ from that described by Ornstein and Davis (4), at the time PAGE originated, in one essential aspect only, that is, temperature control through a water-jacketed lower buffer reservoir. Basically, this apparatus consists of a plastic or glass beaker (upper buffer reservoir) with rubber grommets lining holes in the floor of the beaker. Glass tubes containing the gels are held in these grommets. One end of the gel tube protrudes into a second beaker, the lower buffer reservoir, and the other end into the upper buffer reservoir. Present designs offer a great deal of versatility in providing the upper reservoirs with grommets for tubes of 0.3, 0.6 and 1.8 cm I.D. The efficiency of heat dissipation during polymerisation and electrophoresis can be improved by maintaining a high rate of coolant flow (1.5-3.5 1/min) and by continuous magnetic stirring of the lower reservoir buffer. Relatively minor conveniences have been added, such as vertical levelling devices, a safety interlock and condensate drainage (2). To ensure that this water-jacketed tube apparatus of polyacrylamide gel electrophoresis (*Figure 1,* available in Pyrex from RND Optical Systems Inc) is also suitable for mechanically-labile gels, the following design and procedural features have been introduced.

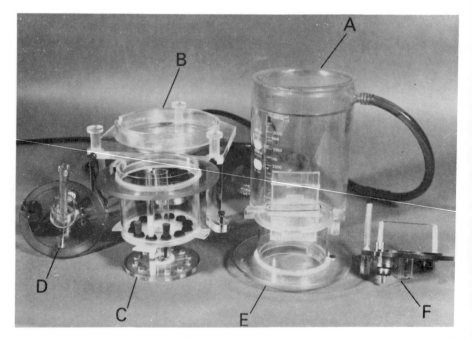

Figure 1. Gel tube apparatus. A) lower buffer reservoir; B) support stand with condensate drainage tube and screw adjustment for vertical alignment of the apparatus; C) upper buffer reservoir for gel tubes of 0.6 cm I.D.; analogous reservoirs for 0.3 and 1.8 cm I.D. tubes are also part of the apparatus; D) electrode cover with upper and lower electrodes and safety interlock; E) and F) analogous upper buffer reservoir and electrode cover for gel slabs (in inverted position). Technical drawings A) to D) are provided upon request.

(i) *Hydrostatic equilibration of the gel tube.* The upper buffer reservoir rests in the lower buffer reservoir, and the levels of the cathodic and anodic buffers are equalised in both reservoirs.

(ii) *Equalisation of the polymerisation temperatures with those of electrophoresis.* This is intended to suppress thermal expansion and/or contraction of the gel and consequent weakening of wall adherence. Such equalisation is simply achieved by polymerising gels within the electrophoresis apparatus maintained at its operating temperature. Usually polyacrylamide gel electrophoresis of proteins is conducted at 0-4°C to preserve native properties. SDS-PAGE is also carried out at low temperature (5°C). The rationale in this case is not to prevent denaturation (which is by necessity complete in the presence of alkylsulphonate detergents), but to improve the reproducibility of polymerisation, as reflected by the variation of R_f (5).

(iii) *Mechanical support of the gel.* After polymerisation and removal of the Parafilm tube support, a square sheet of Nylon mesh (100 micron mesh size has been found suitable) is slipped over the bottom of the gel tube and held in place with a 2 to 3 mm section of 8 mm I.D. Tygon tubing; dialysis tubing is not suitable for this purpose since there is a localised substantial voltage drop across it which would give rise to electrical heating at the bottom of the gel.

(iv) *Maximising the polarity of the glass surface in order to promote adherence of hydrophilic polyacrylamide.* This is achieved by soaking gel tubes in methanolic KOH or dichromate cleaning solution, and washing thoroughly with water until neutralised. It should be noted that washing with detergent would have the opposite effect.

(v) *Coating the gel tube.* For PAGE using mechanically-labile gels the tubes are coated with linear 1% polyacrylamide (Gelamide 250, Union Carbide or Polysciences).

Amongst the three most up-to-date commercial models of tube gel apparatus (RND Optical, Miles, and Hoefer Scientific), the apparatus manufactured by Miles Research Products is the most economical model, although it still lacks some of the features of the more expensive all-Pyrex apparatus, such as interchangeable upper buffer reservoirs for tubes of various diameters or slabs, a levelling indicator attached to the cover, and apparatus support plate with vertical screws, both for vertical alignment of the tubes and to collect condensate from the walls of the apparatus. Some of these features have been incorporated in the apparatus made by Hoefer Instruments, but this has other design problems. To operate at 0-2°C it requires a cooling bath set a −7°C; the plate which holds the rubber grommets is so thin that it is easy to push grommets through the plate while inserting or withdrawing gel tubes; coolant tubing needs to be disconnected every time a gel tube is removed in the course of an experiment (unless one dispenses with the apparatus cover); and finally, a levelling device and an apparatus support plate (see above) are lacking.

A special gel tube apparatus with a set of six or eight separated upper and lower buffer chambers, for the purpose of simultaneous gel electrophoretic analysis in several buffer systems, at different pH's, ionic strengths (see below), concentrations of reducing agents, cofactors, etc., is available in glass from RND Optical Systems. Alternatively, it can be easily constructed (6) by cementing Perspex cylinders into both upper and lower buffer Perspex reservoirs. In this apparatus, hydrostatic equilibration of the gels has not yet been incorporated as a design feature. However, mechanical gel stability is enhanced by an increase in gel length, approximately 20 cm in the Pyrex unit depicted in *Figure 6* of ref. 7.

A vertical gel slab apparatus (e.g. Hoefer Scientific) is convenient in three special cases, assuming that sample volumes are small.

(i)   When the gel concentration has already been optimised and is being used for a comparison of migration distances or R_f values of proteins from different samples. Such direct comparison on a single gel has advantages as an identity test since it avoids "between-experiment" error sources (8). However, comparison of zones based on migration distances also requires that sample ionic strength, buffer composition, sample volumes, exact gel thickness and wall adherence are constant in all channels, to avoid voltage differences between channels.

(ii) A vertical slab apparatus is also convenient for constructing Ferguson plots on a large number of standard molecular weight proteins simultaneously, if their range of molecular weights is sufficiently limited to allow one to analyse all proteins in a relatively small number of pore sizes.

(iii) Finally, vertical slab apparatus is suited for distinction of protein subunits on the basis of molecular size by SDS-PAGE at a single gel concentration.

Horizontal slab apparatus (for example from LKB or Pharmacia) is restricted to continuous buffer systems, since it does not lend itself easily to polymerisation of parallel, contiguous stacking and resolving gels. In general, therefore, it is not useful for PAGE with multiphasic buffer systems. However, horizontal slab apparatus does offer the highest degree of mechanical stability of any apparatus type and therefore has advantages in the relatively rare application of PAGE to very large molecules or particles, where the mechanically-labile polyacrylamide gels crosslinked with 10 to 50% bisacrylamide (9), or agarose, are used.

CHOICE OF GEL

Polyacrylamide gels can be prepared at total monomer concentrations (%T) of between 3.5 and 40% and, within that range, provide effective pore sizes capable of retarding the electrophoretic migration of most proteins (1). When bisacrylamide is used as a crosslinking agent, the size of the molecular sieve decreases from 2 to 5% crosslinking (%C), while at larger %C values it increases progressively until, at 50 %C, molecular sieving effects appear to cease altogether (9). However, because mechanical stability declines as crosslinking with bisacrylamide is increased, it is advantageous in practice (for separations at 0-4°C) to replace bisacrylamide by other crosslinking agents, particularly N,N'-diallyltartardiamide (DATD) (10), or possibly to replace polyacrylamide by agarose when extremely large pore sizes are required. The effective pore size of DATD crosslinked polyacrylamide gels does not increase progressively with increasing %C above 10% and so gels at equivalent molar concentrations are more restrictive above 10 %T than those crosslinked with bisacrylamide.

Furthermore, DATD crosslinking leads to short average chain lengths and pools of unpolymerised crosslinker which increase as the polymer chain lengthens. These properties are the ones that most probably provide DATD gels with the practical advantages of relatively enhanced elasticity, mechanical stability and adherence to glass walls (10).

Acrylamide, bisacrylamide, persulphate, riboflavin and TEMED still are the practical reagents in nearly all applications of PAGE. These reagents allow one to polymerise, within 10 min, polyacrylamide gels at any practical pH, ionic strength and temperature within the wide range of gel concentrations listed above. In view of this fact, we will consider these reagents exclusively in our discussion, although there are alternative monomers, crosslinkers and polymerisation catalysts that remain untried, or have only seldom been used, which promise to provide separation science with a presently unknown variety of pore architecture. A few of these alternative reagents have been discussed previously (1,3,5).

Empirically, it has been found that analytical-scale polyacrylamide gels (0.6 cm diameter) polymerised within 10 ± 2 min, exhibit a maximum degree of conversion of monomers to polymer, and a maximum average chain length. These two characteristics of gels are antagonistic: an increase in the concentration of polymerisation catalysts increases the percentage conversion but decreases the average chain length. Thus, in practice, one should use the lowest catalyst concentrations capable of causing polymerisation within 10 min. In selecting these concentrations it appears, on the basis of rather sparse data (5), that the three free radical donors/acceptors, persulphate, riboflavin and TEMED, are freely interchangeable, although persulphate appears more highly effective under basic conditions while riboflavin is more effective at neutrality and at low pH. The polymerisation reaction appears better controlled if all three reagents are used, rather than just two. Thus, as a rule, photopolymerisation with riboflavin in the presence of persulphate is used (2,5).

In polymerising polyacrylamide, attention should be paid to hidden free radical donors or acceptors of the system, and to the need to compensate for their presence by a corresponding change in the concentrations of the three standard polymerisation catalysts. Oxygen is the most ubiquitous free radical trap but also low pH acts in similar fashion. Amines accelerate polymerisation similarly to TEMED. Apparatus materials such as soft-glass and plastics appear to be inhibitory, while rubber appears to act as a polymerisation catalyst (2). All of these factors can be neglected as long as they can be kept constant and are allowed to contribute in a reproducible fashion to an overall polymerisation rate of 10 min for a gel of 0.6 cm diameter.

The "10 minute rule" becomes particularly important when one is trying to make reproducible gels. This is because the lower the percentage conversion of monomers to polymer, the less reproducible the pore size, measured in terms of R_f, becomes (5). Variance of pore size also appears directly related to average chain length, since R_f values in "10 min gels" formed at 0-4°C are more reproducible than those formed at 25°C (11).

There are three applications of PAGE in which gels may be polymerised without concern for the reproducibility of pore size.

(i) The "non-restrictive" gels required for stacking gels, for isotachophoresis (Chapter 3), or for gel electrofocusing (Chapter 4): as long as these gels do not significantly retard the protein of interest, it does not matter what their real gel concentration is.

(ii) SDS-PAGE. As long as all proteins of the system under investigation, unknowns as well as standards, share an identical surface charge density (Y_0 value of the Ferguson plot, see below) and as long as standard molecular weight proteins and unknowns are being subjected to SDS-PAGE at the identical gel concentration, the mean pore size is irrelevant over a wide range.

(iii) When the migration distances of a large number of proteins are compared directly on a single gel: in this case an "optimal pore size" (T_{opt}) cannot be meaningfully defined and has to be replaced by an "average pore size" (see below).

CHOICE OF BUFFER SYSTEM

Multiphasic Versus Continuous Zone Electrophoresis

Polyacrylamide gel electrophoresis of proteins can be carried out using either a continuous or discontinuous (multiphasic) buffer system. Polyacrylamide gel electrophoresis in multiphasic buffer systems (multiphasic zone electrophoresis, MZE), has the one major advantage over PAGE in a single, continuous buffer (continuous zone electrophoresis, CZE) of being able to concentrate the sample into a narrow starting zone prior to resolution. This is due to the formation of a moving boundary.

A moving boundary is generated when an electric field is applied to a mixture between a relatively rapidly migrating charged species (the leading constituent) and a relatively slowly migrating species (the trailing constituent). In practice the moving boundary is more rapidly formed by first separating the two species by a stationary phase boundary, stabilised by a density difference, for example an upper liquid phase containing the trailing constituent, and a lower gel phase holding the leading constituent *(Figure 2)*. Once an electric field is imposed on the system, with a polarity which

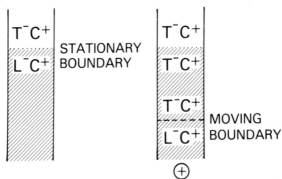

Figure 2. Moving boundary across leading (L) and trailing (T) charged constituents in the electric field; migration toward the anode. C denotes the common constituent. Shading represents a dense medium (the gel) relative to the medium, depicted as clear. The solid lines represent a gel tube, dashed line the moving boundary and dotted line the stationary phase boundary. Left: prior to electrophoresis. Right: in the electric field.

allows the leading constituent to migrate ahead of the trailing one, a moving boundary arises. At the steady-state this is characterised by equal migration velocities of leading, trailing and intermediate charged species, by an ordered alignment of these species within the moving boundary, and by "regulation" not only of all migration velocities, but of voltage, pH, temperature and protein concentration within the space traversed by the moving boundary.

This phenomenon whereby the protein of interest is either concentrated in the moving boundary ("selectively stacked") while other ("unstacked") proteins remain *outside* the stack, or by which the protein of interest is selectively excluded from the stack while the contaminants are confined to the moving boundary will be referred to as steady-state stacking (SSS). Concentration of the sample proteins by steady-state stacking in MZE is, as a rule, of vital importance in analysis of biological materials, so that we are justified in discussing here PAGE linked to MZE exclusively. However, under conditions where a concentrated sample of 1 mg/ml protein or more is available, and where it can be applied at such a concentration and at a very low ionic strength (less than 0.01 M), one is justified in applying CZE, since under those conditions the initial zone width and resolution of CZE does not differ significantly from those of MZE (12).

Other differences between PAGE in multiphasic or single buffers are relatively minor. Optimisation of pH in MZE requires the analysis of single zones per gel, or of single gel slices for activity; in CZE, gels at various pH have to be analysed in their entirety. A clear disadvantage of MZE is that access to its buffer systems requires a computer output, mastering of some rationales and terminology, and availability of a number of unusual buffers (7). A minor difference between CZE and MZE also exists with regard to the zone width of the reference zone for calculation of R_f values; in MZE, this zone is a moving boundary capable of concentrating protein and/or tracking dye into a very sharp zone which remains sharp idependently of migration distance. In CZE , the width of a dye zone is much larger, and it increases in proportion to migration distance. This wider reference zone usually introduces a relatively larger measurement error into R_f values, defined in CZE as migration distance relative to a dye.

Multiphasic Buffer Systems Available: The Jovin Output

Moving boundaries have been treated theoretically in two, presently non-interconverted, scientific languages by Jovin (13) and by Everaerts (14). Both theories have been incorporated into computer programs. The Jovin program has been used to generate an explicit buffer systems output (the "Jovin output"), which exhaustively describes the physical properties of moving boundaries at any pH. This "Jovin output" is available at nominal cost from the National Technical Information Service (NTIS) on microfiche (Public Board Numbers 258309 to 259312), together with a catalogue (Number 196090) listing the principal properties of more than 4,000 buffer systems capable of setting up moving boundaries.

In contrast, the program of the Everaerts theory by Routs has not been designed to generate a simple recipe book of buffer systems with steady-state moving boundaries; it does not exhaustively display the physical properties of the various buffer phases;

and, to date, it has been applied mostly to small molecular weight compounds, in liquid phases and using a capillary apparatus pioneered by Everaerts (14). However, compared to that of Jovin, this treatment has the advantage that it can be used to compute sets of *sequential* moving boundaries, each capable of providing desired mobility ranges, that is selective "mobility compartments" for single proteins or groups of proteins with similar constituent mobilities. Recently, a graphic way to use these sequential moving boundaries for protein fractionation was presented by Svendsen (15), but cannot be treated here prior to publication of its programs and some experimental application.

Therefore, this chapter will be restricted to the language of the Jovin theory, and to the format of the "Jovin output". *Table 1* summarises the relevant terminology. *Figures 3-7* depict the "Jovin output" for one representative "multiphasic" buffer system, No. 2052. Because of their multiple columns of numbers, many symbols and Greek letters, the figures and the table appear hopelessly confusing. Nonetheless, if at this point the reader courageously overcomes the urge to close the book, he will be guided, step-by-step, through the maze of symbols and will recognise that only very few parameters and symbols on these figures matter in practice, and that by merely finishing to read the chapter, he can become familiar with them.

Table 1. Terminology of the Jovin Output.

Constituents	Phases			Stacking Limits
	As Set	Operative	Function	
(1) = Trailing	ALPHA, α = (1)		Upper Buffer	Lower, RM(1,ZETA)
(2) = Leading	BETA, β = (2)	ZETA, ζ = (4)	Upper Gel Buffer (Stacking gel)	Upper, RM(2,BETA)
(3) = Leading	GAMMA, γ = (3)	PI, π = (9)	Lower Gel Buffer (Resolving gel)	Lower, RM(1,PI)
(6) = Common	EPSILON, ϵ = (11)		Lower Buffer	Upper, RM(2,LAMBDA)

SYSTEM NO.	CONSTITUENTS 1 2 3 6	PHI(4) RANGE	PHI(2) RANGE	PHI(9) RANGE	PHI(3) RANGE	RM(1,4) RANGE	RM(2,2) RANGE	RM(1,9) RANGE
JOV 2049	35 18 18 83	6.00	4.79	7.50	6.04	-0.123	-0.42	-0.41
		4.41/ 9.41	4.10/ 9.22	4.41/ 9.41	4.10/ 9.22	-0.004/-0.44	-0.28/-0.79	-0.004/-0.44
JOV 2050	35 19 19 83	6.00	4.77	7.50	6.09	-0.123	-0.28	-0.41
		4.41/ 9.41	4.40/ 9.29	4.41/ 9.41	4.40/ 9.29	-0.004/-0.44	-0.15/-0.63	-0.004/-0.44
JOV 2051	36 84 99 5	6.78	4.28	7.50	7.01	-0.102	-1.28	-0.28
		6.78/ 8.15	3.91/ 7.87	6.78/ 8.15	3.87/ 7.84	-0.101/-0.41	-1.28/-1.28	-0.101/-0.41
JOV 2052	36 15 99 5	6.63	4.13	7.50	7.01	-0.077	-0.44	-0.28
		5.34/ 8.13	3.19/ 7.98	6.89/ 8.13	5.66/ 7.82	-0.005/-0.40	-0.12/-0.66	-0.122/-0.40
JOV 2053	36 18 18 97	6.00	4.45	7.50	5.14	-0.021	-0.27	-0.28

Figure 3. Representative page of the catalogue (NTIS PB No. 196090) summarising the Jovin output. Phase designations are defined in *Table 1*.

```
DATE = 03/26/70     COMPUTER SYSTEM NUMBER = JOV-CHR    2052
POLARITY = - (MIGRATION TOWARD ANODE)      TEMPERATURE =   0  DEG. C.

CONSTITUENT 1 = NO.   36 , ACES
CONSTITUENT 2 = NO.   15 , LACTIC ACID
CONSTITUENT 3 = NO.   99 , CHLORIDE -
CONSTITUENT 6 = NO.    5 , 4-PICOLINE
```

	PHASES					
	ALPHA(1)	ZETA(4)	BETA(2)	PI(9)	LAMBDA(8)	GAMMA(3)
C1	0.0400	0.0400		0.0254		
C2			0.0484		0.0307	
C3						0.0443
C6	0.0240	0.0240	0.0323	0.3075	0.3128	0.3264
THETA	0.599	0.599	0.669	12.116	10.191	7.361
PHI(1)	0.164	0.164		0.591		
PHI(2)			0.663		1.000	
PHI(3)						1.000
PHI(6)	0.274	0.274	0.992	0.049	0.098	0.136
RM(1)	-0.077	-0.077		-0.278		
RM(2)			-0.438		-0.660	
RM(3)						-1.626
RM(6)	0.195	0.195	0.704	0.035	0.070	0.096
PH	6.63	6.63	4.13	7.50	7.17	7.01
ION.STR.	0.0066	0.0066	0.0321	0.0150	0.0307	0.0443
SIGMA	0.748	0.748	4.241	1.708	4.056	9.997
KAPPA	189.	189.	1004.	418.	962.	2333.
NU	-0.103	-0.103	-0.103	-0.163	-0.163	-0.163
EV	0.024	0.024	0.026	0.047	0.064	0.088

Figure 4. Physical properties of a representative buffer system, No. 2052, in the form provided on page 2 of the description of each buffer system by the Jovin output.

```
                RECIPES FOR BUFFERS OF PHASES ZETA(4),BETA(2),GAMMA(3),PI(9)
                                      1X       4X       4X       4X
   CONSTITUENT                     PHASE 4  PHASE 2  PHASE 3  PHASE 9

      ACES              GM           7.29                              1.85
      LACTIC ACID       GM                    1.74
      1N HCL            ML                             17.74
      4-PICOLINE        GM           2.23     1.20     12.16          11.45
      H20 TO                       1 LITER   100 ML   100 ML         100 ML

   AT FINAL CONCENTRATION =
      PH(25 DEG.C.)                  6.27     4.09     6.86           7.25
      KAPPA(25 DEG.C.)               523.    1964.    4387.          1043.
```

Figure 5. Buffer system recipe of a representative buffer system, No. 2052, in the form provided on page 2 of the description of each buffer system by the Jovin output.

101

STACKING AND UNSTACKING RANGES

PHASE ZETA(4) OR PI(9)				PHASE BETA(2) OR LAMBDA(8)					PHASE GAMMA(3)			
RM(1)	PHI(1)	C(1)	C(6)	PH	RM(2)	PHI(2)	C(2)	C(6)	PH	C(3)	C(6)	PH
-0.005	0.010	0.0400	0.0005	5.34	-0.12	0.182	0.0484	0.0088	3.19	0.0	0.0	0.0
-0.028	0.060	0.0400	0.0045	6.15	-0.17	0.265	0.0484	0.0128	3.40	0.0	0.0	0.0
·0.052	0.110	0.0400	0.0117	6.43	-0.27	0.415	0.0484	0.0201	3.69	0.0	0.0	0.0
-0.075	0.160	0.0400	0.0228	6.62	-0.42	0.641	0.0484	0.0312	4.09	0.0	0.0	0.0
-0.099	0.210	0.0400	0.0385	6.76	-0.61	0.923	0.0484	0.0469	4.92	0.0	0.0	0.0
-0.122	0.260	0.0400	0.0597	6.89	-0.65	0.990	0.0484	0.0681	5.83	0.1008	0.1292	5.66
-0.146	0.310	0.0400	0.0876	6.99	-0.66	0.996	0.0484	0.0959	6.21	0.0846	0.1421	6.04
-0.169	0.360	0.0400	0.1237	7.09	-0.66	0.998	0.0484	0.1320	6.45	0.0728	0.1600	6.29
-0.193	0.410	0.0400	0.1701	7.18	-0.66	0.998	0.0484	0.1785	6.64	0.0639	0.1830	6.48
-0.216	0.460	0.0400	0.2298	7.27	-0.66	0.999	0.0484	0.2382	6.80	0.0570	0.2117	6.64
-0.240	0.510	0.0400	0.3068	7.36	-0.66	0.999	0.0484	0.3152	6.95	0.0514	0.2476	6.79
-0.263	0.560	0.0400	0.4070	7.44	-0.66	0.999	0.0484	0.4154	7.09	0.0468	0.2925	6.93
-0.287	0.610	0.0400	0.5392	7.53	-0.66	1.000	0.0484	0.5476	7.22	0.0430	0.3499	7.06
-0.310	0.660	0.0400	0.7177	7.63	-0.66	1.000	0.0484	0.7261	7.36	0.0397	0.4248	7.20
-0.334	0.710	0.0400	0.9663	7.73	-0.66	1.000	0.0484	0.9747	7.49	0.0369	0.5262	7.33
-0.357	0.760	0.0400	1.3290	7.84	-0.66	1.000	0.0484	1.3374	7.64	0.0345	0.6705	7.48
-0.381	0.810	0.0400	1.8957	7.97	-0.66	1.000	0.0484	1.9040	7.79	0.0324	0.8914	7.63
-0.404	0.860	0.0400	2.8849	8.13	-0.66	1.000	0.0484	2.8933	7.98	0.0305	1.2710	7.82

Figure 6. Buffer subsystems table of a representative buffer system, No. 2052, in the form provided on page 3 of the description of each buffer system by the Jovin output.

RESTACKING PARAMETERS

		PHASE PSI(5)					PHASE TAU(6)				
CT7	IS	RM(7)	PHI(7)	C(7)	C(6)	PH	C(7)	C(6)	PH	PHI(7)	KAPPA
23	0.009	-0.154	0.376	0.0233	0.3054	7.74	0.0368	0.0207	6.94	0.088	90.
24	0.008	-0.126	0.451	0.0180	0.3001	7.77	0.0284	0.0124	6.82	0.086	60.
25	0.004	-0.070	0.219	0.0198	0.3019	8.05	0.0312	0.0152	7.24	0.042	34.

Figure 7. Restacking buffer table of a representative buffer system, No. 2052, in the form provided on page 3 of the description of each buffer system by the Jovin output.

QUANTITATIVE PAGE PROCEDURE

Strategy

Since PAGE separates molecules on the basis of their differences in molecular size and in net charge, it is rational to optimise both the pH of electrophoresis and the pore size of polyacrylamide for the purposes of separation. Furthermore, since the pH of PAGE in multiphasic buffer systems, together with ionic strength and other parameters important for native protein stability, can be optimised using a single gel for each condition and evaluating a single zone or gel slice (the stack), while pore size optimisation requires the analysis of many gels and many slices per gel, it is labour-saving to make pH optimisation the initial step in the fractionation strategy of PAGE (2,7), and to tackle the optimisation of pore size subsequently. Once the pH and pore size are optimised for a specific separation, subsequent analytical and preparative PAGE can be carried out efficiently.

To date, the approach has been to assess the optimum pH for separation based on molecular net charge ("charge fractionation") according to the rationale that the lower the net charge on the macromolecules to be separated, the larger the relative net charge difference, and the more efficient the "charge fractionation". To this end, a systematic search is undertaken to determine the minimum pH for protein anions, and maximum pH for protein cations, at which the species of interest can be stacked. This systematic search can be carried out in a single experiment, or in very few experiments depending on how exacting a choice of pH is required to bring about separation. Optimisation of other parameters important for protein stability at the optimum or near-optimum pH, such as ionic strength, the temperature of electrophoresis, the redox conditions in the gel, the incorporation into the gel of metal ions, or chelating agents, or detergents is also best carried out at the level of stacking gels since this has the enormous advantage over the same optimisation performed on resolving gels that only a single gel slice needs to be analysed for each particular condition investigated. Using the gel tube apparatus with internal partitions (6), as many different conditions can be investigated in a single experiment as the number of partitions allows, making it possible to scan through all the relevant conditions within a few workdays.

Optimisation of pH

STEP 1: Choice of Buffer Systems which give rise to Slow Moving Boundaries in the Desired pH Range

The systems catalogue (NTIS PB Number 196090) is used to select the desired stacking pH's [*pH(ZETA) of Table 1* or *PH(4) of Figure 3*]. When viewing the *PH(4)* values in the catalogue (*Figure 3,* column 7), the *"RANGE"* of possible *PH(4)* values, represented by two numbers separated by a slash (/), which appear directly under each *PH(4)* value, should be disregarded at this stage.

The catalogue lists buffer systems in numerical order. Positively charged proteins at $0°C$ require systems numbered 1-691; 692-1578 refers to these at $25°C$; negatively charged proteins at $0°C$ require systems numbered 1579-2969 and 2970-4269 refers to these at $25°C$. Thus, one should select the desired "quadrant" among those four cases, and start to read the *PH(4)* column (*Figure 3,* column 7) at an extreme of pH, with the purpose of locating a *SYSTEM NO.* (columns 1-2 of *Figure 3*) corresponding to the desired *PH(4)* which displays both a low value for the lower stacking limit [*RM(1,ZETA) or RM(1,4)* in column 11 of *Figure 3*] and a desirable set of buffers (*"CONSTITUENTS"* in columns 3-6 of *Figure 3*).

The value of *RM(1,ZETA),* also designated the "lower stacking limit", represents the "constituent mobility" relative to Na^+ *(RM)* that the trailing buffer *(CONSTITUENT 1 of Table 1)* exhibits in the operative stacking phase (*PHASE 4 or ZETA*). Constituent mobility is the mobility of the charged species multiplied by its mole fraction in the total concentration of both the charged and uncharged species. In practical terms, the value of *RM(1, ZETA)* represents the lowest protein mobility which can be stacked in the particular moving boundary. At this point, systems with a low value of *RM(1,ZETA)* should be used, because this will allow the protein of interest to be stacked as its migration velocity declines, that is as it approaches its pI. "Low values" of *RM(1,ZETA)* are those around 0.050; the further the pH from the isoelectric point, the higher this value may be for the stacking of a protein.

Thus, to select suitable pH systems from the catalogue, columns 7 and 11 of *Figure 3* should be screened simultaneously. It will be noticed that systems with low *RM(1,4)* are relatively rarely encountered in such a search. A number of such buffer systems have been tabulated in the literature *(Figure 8* of ref. 7, the upper gel buffers of Appendix C of ref. 2 and Appendix 1 of ref. 16). However, most of these tabulated buffer systems describe moving boundaries with a large mobility range [*RM(2,BETA)—RM(1,ZETA)*], and therefore a steep pH gradient across the stack and little buffering capacity within it. Systems with better buffering capacity within the stack can be generated with the Jovin program (Newby,A.C. *et al.,* in preparation) using the limited number of buffer constituents available at the time when the representative "Jovin output" was produced (the latter buffer constituents are listed in *Figure 6* of ref. 2, in *Figure 4* of ref. 17, and on the first page of the Systems Catalogue). More such systems should become available once the number of buffer constituents with known pK's and ionic mobilities is increased.

Let us assume that no buffer systems with a sufficiently low value of *RM(1,ZETA)* can be found among any of the buffer systems at the desired pH. In that case, a "subsystem" of any of the numbered buffer systems may be required. The *"RM(1,4) RANGE"* listed under each *RM(1,4)* value in column 11 of *Figure 3* indicates the lowest possible value of *RM(1,ZETA),* which can be obtained with the sets of four buffer constituents characteristic for each system (columns 3-6, *Figure 3*). However, the *PH(4) RANGE* (column 7, *Figure 3*) corresponding to the *RM(1,4) RANGE* (column 11, *Figure 3*) frequently indicates that the price to pay for a minimum *RM(1,ZETA)* is a *pH(ZETA)* at an undesirable value. Thus, in practice, a reasonable compromise must be struck between the choice of *RM(1,ZETA)* and *pH(ZETA).* The catalogue is of no help in finding this compromise, since it lists only the minimum and maximum values of *RM(1,4)* and *PH(4),* and no intermediate values of either one of these parameters. Thus recourse has to be taken to page 3 of each system of the "Jovin output" (columns 1 and 5 of *Figure 6*), found on the microfiche output available from NTIS. In choosing *RM(1,ZETA)* attention should be paid to the fact that with decreasing *RM(1,ZETA),* the upper stacking limit *RM(2,BETA)* (column 6 of *Figure 6*) simultaneously decreases. It may decrease so much that one may no longer be able to stack all the proteins in a mixture, that is, some of their constituent mobilities may exceed *RM(2,BETA).* Thus, to achieve an acceptable range of stacking limits, one may have to settle for an *RM(1,ZETA)* value larger than the desired minimum.

To designate unequivocally the selected "subsystem" of any of the systems of the "Jovin ouput", certain terminological conventions have been adopted (2). Subsystems describing a moving boundary across phases *BETA* and *ZETA* are given an arabic number, in the order of their listing on page 3 of the ouput for each system (*Figure 6,* top to bottom). Thus, in the example shown in *Figure 6,* the subsystem of MZE buffer system 2052 with a value of *RM(1,4)* = −0.122 would be designated 2052.6.[1]

Stacking of SDS-polypeptide complexes allows for relatively high values of lower stacking limits [*RM(1,ZETA)* or *RM(1,PI)*] since the net charge of SDS-polypeptide

---

[1]Similarly, subsystems describing a moving boundary across phases *LAMBDA* and *PI* are given Roman numerals, in order of their appearance in vertical sequence in the *PHASE GAMMA* column of *Figure 6* (columns 11-13). Thus, the system with *RM(1,9)* = 0.122 would be designated as 2052.0.I, unless *RM(1,4)* was changed simultaneously from the value shown on page 2 of the output (*Figure 4*). It is possible to stack proteins in *PHASE PI* as well as in *PHASE ZETA* although most of the computed *PI/LAMBDA* boundaries exhibit high values of *RM(1,PI)*, i.e. are likely to unstack, but not to stack proteins.

complexes is vastly augmented by the bound negatively charged detergent. A lower stacking limit [*RM(1,ZETA*] for SDS-polypeptide complexes of 0.2 at pH's above neutrality is representative (11).

In many cases, there are several buffer systems operative at the desired pH and capable of providing a low value of *RM(1,ZETA)*. To choose among these, the following properties of buffer *CONSTITUENTS 1, 2* and *6* in columns 3, 4 and 6 of the catalogue *(Figure 3)*, that is, the trailing, leading and common constituents of each system, should be taken into consideration: buffering capacity [closeness of *pH(ZETA)* to p*K*], solubility, optical properties, compatibility with protein activities, commercial availability in pure form, and price.

STEP 2: Preparation of Buffers

To determine experimentally the minimum pH at which anionic proteins can be stacked (or the maximum pH at which cationic proteins may be stacked), the stacking gel buffers ("upper gel" buffers), and cathodic and anodic reservoir buffers ("upper" and "lower" buffers) for each of the selected buffer systems must be prepared. Page 2 of the "Jovin output" for each system provides a *"RECIPE"* section *(Figure 5)* which lists in full detail the composition of the stacking gel buffer (fourfold concentrated *PHASE 2*) and the upper buffer (by arbitrary convention prepared as a *PHASE 4*).[2] The upper buffer is composed of *CONSTITUENTS 1* and *6* (see *Table 1*) at any concentration at a pH favourable to the protein of interest. This is important to prevent denaturation of the sample, which is in contact with the upper buffer across a density-stabilised sample phase boundary, during the relatively short time before the power is turned on. Equally, the constitution of the lower buffer is irrelevant as long as it contains *CONSTITUENT 6*. Since the protein is at no time in contact with the lower buffer, its pH is also irrelevant. The composition of *CONSTITUENT 5* has been arbitrarily set as K^+ or Cl^-, depending on the polarity of the system. Any other counterion can be used as long as it does not give off electrolytic products of such charge as to pass through the gel. Conventionally, the lower buffer is prepared as 0.0625 M common constituent of the system *(CONSTITUENT 6)* and 0.05 M HCl (for anionic proteins) or 0.05 M KOH (for cationic proteins).

To prepare the upper gel buffer for any particular subsystem, the final molar concentrations of each of the two buffer constituents of *PHASE BETA*, that is *CONSTITUENT 2 (C2)* and *CONSTITUENT 6 (C6)*, are obtained from the subsystems table on page 3 of the "Jovin output" for each system (*Figure 6*, columns 8 and 9). The upper gel buffer is prepared at fourfold its final concentration, since the buffer is by convention one-fourth of the polymerisation mixture of polyacrylamide (see *Figure 8*). When using a subsystem, the upper buffer given in the *RECIPE* section *(Figure 5, PHASE 4)* for the particular buffer system may be used. The lower buffer remains unaltered.

---

[2]The upper buffer composition suggested in the Buffer Recipe section of the "Jovin output" is made on the basis of the arbitrary convention to equalise phases *ALPHA* and *ZETA* in composition, thus eliminating one stationary phase boundary from the system.

Polyacrylamide Gel System No. 2052

	Vol. Ratio Stock Soln.	LOWER GEL (GAMMA)			UPPER GEL (BETA)			UPPER BUFFER (ALPHA)			LOWER BUFFER (EPSILON)		
		Components/100 ml Stock Solution	pH (25°C 1/4 dil.)	κ	Components/100 ml Stock Solution	pH (25°C 1/4 dil.)	κ	Components/Liter	pH (25°C)	κ	Components/Liter	pH (25°C)	κ
BUFFER	1	1N HCl 17.74 ml 4-picoline 12.16 gm	6.86	4387.	Lactic 1.74 gm 4-picoline 1.20 gm	4.09	1964.	Aces 7.29 gm 4-picoline 2.23 gm	6.27	523.	1N HCl 50.0 ml 4-picoline 5.8 gm	5.46	4912.
CATALYST	1	KP ___ mg RN ___ mg			KP ___ mg RN ___ mg								
MONOMER SOLUTION	2	%T a) 10.0 b) 20.0 c) 40.0			%T 6.25								
		%C ___			%C 20								
		TD/100 ml gel (μl) a) b) c)			TD/100 ml gel (μl)								

pH (PI) = 7.50
pH (ZETA) = 6.63
RM(1,ZETA) = 0.077
RM(2,BETA) = 0.438
RM(1,PI) = 0.278
°C (electrophoresis) = 0
mm Hg (polymerization) = 10
tracking dye =

Figure 8. Representative procedural instruction sheet for PAGE, using MZE buffer system No. 2052. No catalyst concentrations are given since concentrations providing polymerisation at any one %T, %C, pH and ionic strength within 10 min will vary amongst laboratories. Abbreviations used in this figure are TD = TEMED, KP = potassium persulphate, RN = riboflavin, and k = specific conductance (μmhos/cm).

The systems generated by the "Jovin output" exhibit various ionic strengths in the operative stacking phase. However, for the purpose of comparing various stacking pH's it is imperative to equalise the ionic strength in the operative stacking phase *(PHASE ZETA)*. A convenient value for the ionic strength of the various buffer systems under comparison is 0.01 M. Equalisation is simply achieved by noting the value of *ION.STR.(ZETA)* on page 2 of the systems output *(Figure 4)*, by calculating the factor needed to bring the operative ionic strength to 0.01, and by applying this factor to both *CONSTITUENTS 2* and *6* given in the buffer *RECIPE* section of 4 × *PHASE 2* (stacking gel buffer) *(Figure 5)*. Accordingly, the published buffer systems (e.g. *Figure 8* of ref. 7) have been multiplied by the necessary factor to bring the ionic strength to 0.01 M. It should be noted, however, that page 2 of the systems ouput *(Figure 4)* providing ionic strength values for the various subsystems *(Figure 6)* can only be generated by computation, using the Jovin program (NTIS, PB Number 196092).

STEP 3: Preparation of Stacking Gels

The stacking gel serves solely as an anticonvective medium for the stack. Because proteins migrate slowly compared with the buffer constituents setting up the moving boundary, they should not be retarded in their migration in the stacking gel any more than necessitated by a gel concentration capable of effectively reducing zone diffusion. In other words, the gel should be both anticonvective and "non-restrictive". Furthermore, it needs to adhere well to the walls of the apparatus, since without such "wall adherence" current does not flow uniformly across the cross-sectional area of the gel but preferentially along the interface between gel and vessel, giving rise to distorted band geometry and loss of resolution. In severe cases, wall separation causes seepage and migration of the protein between the gel and walls.

Among "open-pore" gel media, the 3.125 %T, 20 %C_{Bis} gel originally introduced by Ornstein for use in MZE on polyacrylamide (4) is comparatively "non-restrictive" but lacks mechanical stability and wall adherence unless supported by a sturdy resolving gel, as in PAGE. Therefore, the somewhat more restrictive 5 %T, 15 %C_{DATD} gel, with its relatively good wall adherence and elastic properties, appears preferable for use in initial studies to optimise electrophoretic pH using stacking gels. The degree of restrictiveness as a function of the molecular weight of the protein can be gauged by a criterion of stacking; any protein capable of migrating within the moving boundary (with high pH and low displacement rate) at a particular gel concentration may be considered "unrestricted" (7,10). *Figure 1* of Chapter 3 depicts the gel concentrations of "non-restrictive" 15% DATD-crosslinked gels for proteins of various molecular weights (10).

For each buffer system to be tested, one requires either a separate PAGE apparatus *(Figure 1)* or separated chambers within a single apparatus (6,7). The apparatus is equilibrated to 0-4°C by circulating coolant at 1.5 to 3.5 1/min from a refrigerated, thermostated bath. Pyrex tubes 12.0 cm long, 0.8 cm O.D., 0.6 cm I.D. are cut with a triangular file (avoiding glass dust inside the tube), fire polished with care so as not to reduce the internal diameter, and sealed with stretched Parafilm by winding the Parafilm repeatedly across the bottom of the tube. The requisite number of tubes is inserted into the upper buffer reservoir where they are held vertically by the rubber grommets *(Figure 1)*. The remaining open grommets in the upper buffer reservoir are sealed with rubber stoppers. The upper buffer reservoir is then lowered into the two-thirds filled lower buffer reservoir *(ibid., Part A)*. For photo-polymerisation the apparatus is placed concentrically between two lamp units with six 20-watt daylight fluorescent tubes contained in two units of semi-circular cross-section (Buchler Instruments).

The apparatus should contain sufficient lower and, in the upper reservoir, upper buffer, to surround the tubes with fluid at a controlled temperature. To improve the dissipation of both heat of polymerisation and electrical heating, the lower buffer should be stirred magnetically.

Upper gel buffer, a stock solution of acrylamide and crosslinking agent (10 %T, 15 %C_{DATD}) and a stock solution of initiators (frequently 0.06% potassium per-sulphate, 2×10^{-3}% riboflavin (for rationales in the choice of catalyst concentrations, see p.116) are assembled and cooled in ice (for gel polymerisation and electrophoresis at 0°C). The polymerisation mixture is prepared in a 25 ml amber bottle, containing a small stirring bar, by mixing 5 ml acrylamide-crosslinking agent stock solution, 2.5 ml of upper gel buffer and 2.5 ml initiator stock solution (see *Figure 8*). The bottle is immediately placed into an evacuator (vacuum pump with electronic vacuum controller) over an ice-water bath and it is deaerated for 5 min to 10 mm Hg, 0°C (*Figure 9;* also see refs. 2,18). Alternative partial deoxygenation methods can be applied if they are capable of providing a reproducible partial pressure of oxygen, for example, flushing with argon at a constant gas flow rate and time (2); an uncontrolled oilpump or aspirator does not provide reproducible deaeration.

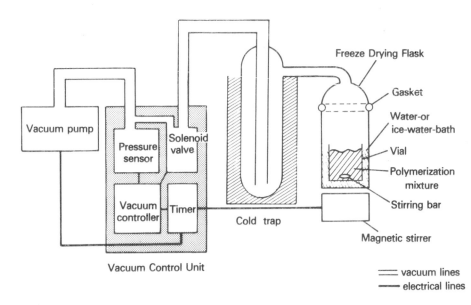

Figure 9. Schematic diagram of the polymerisation mixture deaerator (18).

The final polymerisation catalyst, TEMED, is added (usually 5 μl per 10 ml polymerisation mixture) and the mixture is pipetted without delay into the gel tube. Polymerisation lamps should not be turned on at this time. The polymerisation mixture is gently overlayered with water, using a lubricated (Lubriseal) 1 ml glass syringe fitted with a 23 gauge, 2.5 inch needle, or a 100 μl Hamilton syringe, taking care not to disturb the gel surface. Overlayering is complete when the liquid layer above the gel is 1-2 mm deep. The polymerisation lamps are turned on for 30 min. Then the upper buffer reservoir is inverted, the water overlay is gently absorbed onto paper tissue, and the Parafilm seals are cut off by sliding a single edged razor blade along the tube. The upper buffer reservoir with gel is then placed into the lower reservoir.

STEP 4: Preparation of the Sample

The protein load per analytical gel (0.6 cm I.D.) will be of the order of five to several hundred micrograms for the purposes of stacking if the protein is being detected by staining, as is usual for PAGE when one attempts to stack both the protein of interest and its contaminants for the purpose of later separation by electrophoresis in a resolving gel. If the protein is detected by activity or isotope analysis, the protein load may be as low as the limit of detectability allows. However, whenever less than microgram amounts of protein are loaded, it is necessary to guard against non-specific protein adsorption to the matrix by incorporating microgram amounts of another protein into the sample (19). Apparently, the added protein can be of any size, as long as it carries the same net charge as the protein of interest in the chosen buffer systems. It is usually selected so that it migrates ahead of the protein of interest and can be readily separated from the isolated protein by gel filtration.

Usually, the sample is made in upper buffer, containing enough sucrose or glycerol or other uncharged, dense additive to allow one to form a distinct phase boundary between upper buffer and sample phase, and a suitable tracking dye, that is, a charged dye with a mobility in the range defined by the upper and lower stacking limits. When the leading constitutent *(CONSTITUENT 2)* in the selected buffer system is a strong acid or base, theoretical considerations would call for making the sample not in upper buffer, but in upper gel buffer. However, the pH of upper gel buffer *(PHASE BETA)* is frequently too extreme for the maintenance of protein activities, or too close to its isoelectric point to safeguard the protein from isoelectric aggregation or precipitation. It can be assummed that a tracking dye is stacked within a particular set of stacking limits, when its zone boundaries are sharp and do not broaden in proportion to migration distance. To check that assumption, it is necessary to verify that the dye migrates either at the front edge of the trailing constituent *(C1)* or at the trailing edge of the leading constituent *(C2)* in the gel. Chemical detection of these constituents (see below) is suitable for this purpose.

Sample volume may equal the volume of the gel. It is, therefore, practically unlimited. The larger the sample volume, however, the longer it takes to stack the protein. The only real limitation on sample volume is therefore the length of gel and time needed to obtain protein concentration at the steady-state. If the sample is small, that is 200 μl or less per analytical-scale gel, it is easiest to layer the sample solution under the upper buffer, by directing the tip of an unconstricted micropipette, controlled by a rigid syringe-type micropipette control (Microchemical Specialties), well lubricated with Lubriseal, or a Hamilton syringe (100 μl) to just above the top of the stacking gel. If the volume of the sample is larger, it is pipetted onto the gel surface and subsequently overlayered with upper buffer. This overlayering can be easily done without disturbing the interface between the sample and upper buffer by delivering buffer dropwise to the top inner wall of a clean gel tube with a Pasteur pipette, allowing the buffer to run down the wall of the tube in a continuous stream. In a dirty tube, the buffer will coalesce to drops and "bomb" the interface. Using either procedure, the interface must be watched during overlayering to make sure it remains undisturbed during the operation.

STEP 5: *Electrophoresis*

To reach the steady-state as fast as possible one should employ the maximum voltage compatible with the electrical heat dissipation capacity of the apparatus. When the gel is completely surrounded by stirred liquid coolant (the lower buffer) at 0-4°C, as is suggested here, a regulated current of 1-2 mA for a standard 0.6 cm diameter gel can be tolerated if the protein load does not exceed 200 μg.

The duration of stacking has to be sufficient to attain the steady-state. This is recognised for analytical load levels (200 μg or less) by maximum zone sharpening of the stained protein stack. In practice, the zone sharpening observed for the tracking dye is a reasonable guide to the time for the stacking of protein if the sample volumes are small (not larger than 100 μl/gel). Electrophoresis times of 1-2 h are nearly always sufficient, unless extremely slow moving boundaries [*RM(1,ZETA)* values of 0.005-0.020 or *RM(2,BETA)* values of 0.1-0.2] are used.

To gauge the effect of pH on stacking, other parameters such as ionic strength, temperature and current density have to be equalised. It is therefore helpful to conduct electrophoresis in a single multi-chamber apparatus (6). The ionic strength of all operative stacking gel buffers *(PHASES ZETA)* is equalised for that purpose, usually to a value of 0.01 M (see above).

STEP 6: *Stacking Gel Analysis*

After turning off the power, the stacking gels at various pH's are withdrawn from the apparatus individually by stoppering the top of the tube with a rubber stopper, lifting the upper buffer reservoir and pulling the tube downward until the rubber stopper is seated in the grommet. The stopper is tightened and electrophoresis is resumed. Gels are removed from the tube by rimming with water, using a 23 gauge, 2.5 inch long, blunt hypodermic needle connected to a water tap, syphon or syringe. Since stacking gels are mechanically labile, this has to be done by holding the gel tube vertically 1-2 cm above the bench, allowing the water to emerge dropwise from the syringe. A continuous stream of water is likely to destroy the gel. After inserting the needle along the gel circumference to its full length from one end, the gel is inverted and rimmed from the opposite end until it slides out onto a clean surface such as Parafilm.

To determine whether the protein of interest is stacked at the various pH values, the stack needs to be recognised. When the pH of stacking is high or low, tracking dyes can usually be found which mark the stack visually as a characteristically sharp, coloured zone. If the lower stacking limit is very low (0.005 to 0.050), and/or if the pH is neutral, coloured proteins like haemoglobin, ferritin or cytochrome *c* can sometimes be used to mark the moving boundary visually. If that is not the case, the boundary can sometimes be located by specifically precipitating either the trailing constituent *(C1)* or the leading constituent *(C2)* of the stack, for example, phosphate by lanthanum acetate, sulphate by barium chloride, chloride by silver nitrate. In these cases, the gel is dipped into a solution of the precipitating agent, and the sharp boundary of precipitate on the gel is marked using a syringe needle dipped into India ink. The same gel can then be stained for protein. Finally, if all else fails, one can determine the pH across the gel, using the procedures for pH gradient determination in electrofocusing (Chapter 4, p. 180) and locate the boundary at the point of inflection between *pH (BETA)* and *pH (ZETA)*.

The stack is excised from the gel, suspended in assay buffer and assayed for activity. Alternatively, the position of the stack relative to gel length is marked, and the gel is fixed and stained. If the protein is stacked, staining reveals the position of the stack as an extremely sharp, slightly almond-shaped zone. Its position relative to gel length should be identical to that of the stacked dye in order to confirm stacking. When in doubt, the constancy of stack width and appearance with varying time of electrophoresis should be verified experimentally. The detection of a stained stack does not reveal, of course, whether the protein of interest is stacked, unless we are dealing with a homogeneous sample. One can only infer from the absence of unstacked stained protein zones, which would usually migrate as relatively diffuse bands behind the stack, that the protein of interest is stacked, assuming that it is present in the sample in stainable amounts.

In those cases in which one relies on staining methods to reveal the stack position, it is important to fix the protein zones efficiently and rapidly in order to maintain the typical extremely sharp zone boundaries characteristic for the stack. Thus, it is advantageous to employ TCA solutions, rather than acetic acid solutions in aqueous methanol, for fixation. We prefer the staining procedure of Diezel *et al.* (20), since it provides a background-free gel pattern within 30 min (see *STEP 11*).

Choice of Stacking Limits

Once the pH of stacking is optimised, both selective stacking of the protein of interest (SSS) [and, even more so, resolution within the stack (ITP); see Chapter 3] depend on tailoring the stacking limits as narrowly as possible; unwanted slowly migrating species can be retarded so as to migrate behind the stack. The technique is exactly the same as used for the optimisation of pH, except that the various stacking gels are now made in the subsystems of the single buffer system of optimum pH. Since initially we had assumed a minimum constituent mobility for the protein of interest, the value of *RM(1,ZETA)* is now *increased* to a maximum value compatible with the stacking of the protein. Thus, increasing *RM(1,ZETA)* values are selected from column 1 of the subsystems table for the chosen system *(Figure 6)*. The upper gel buffers corresponding to those subsystems are prepared as the four-fold concentrates of the final concentrations of *CONSTITUENTS 2* and *6* in columns 8 and 9 of the same table *(PHASE BETA).* If so desired, subsystems intermediate to those listed in the table may be prepared by graphic interpolation of a plot of *RM(1,ZETA)* versus the final concentration of *CONSTITUENT 6* (the parameter *C6* in column 9) and, where *CONSTITUENT 2* is variable, a plot of *C2* (in column 8) versus *RM(1,ZETA)*. Necessarily, the maximisation of the lower stacking limit, *RM(1,ZETA),* causes a minor change in pH, in the direction opposite to that used for optimising the pH in the previous section. However, the relatively small pH variation concomitant with maximisation of the lower stacking limit will usually have an only negligible effect on protein migration rates. The gel is made, electrophoresis is carried out, and the gel is analysed as described for the selection of pH above.

Choice of Optimal Conditions for Protein Stability and Resolution

If the protein of interest is being detected on the basis of its activity and if the recovery of protein activity within the stack has been less than quantitative, it may be improved at the optimum, or near-optimum, pH and at the maximum *RM(1,ZETA)* by changing the ionic strength of the stacking gel buffer, the temperature of electrophoresis, the redox conditions in the gel, sweeping it electrophoretically with charged reducing agents, or by incorporation into the stacking gel of metal ions such as Ca^{++} or Mg^{++}, or of chelating agents, sucrose, glycerol, or detergents compatible with the protein activity. In each case, the introduction of one of these factors into the stacking gel is monitored by the identical technique described above for the optimisation of pH. Thus, the stack is excised and assayed for activity. Only a few specific procedural points need to be noted explicitly in this regard:

Ionic strength. To vary the operative ionic strength of stacking, only the concentration of the stacking gel buffer needs to be changed. Upper and lower buffers remain constant. Gels varying in ionic strength should not be subjected to electrophoresis in the same apparatus, since the potential gradient varies between them in inverse proportion to ionic strength. Even when the total power across all gels of variable ionic strength is regulated, that is, the sum of the various voltages multiplied by the sum of the various currents, the power varies to an unknown degree between gels of varying ionic strength. By conducting the comparison of gels at various levels of ionic strength in separate apparatuses, the power could be kept constant and would be known in each case. However, power supplies capable of regulating at less than 1 watt are not yet available, and 1 watt per 0.6 cm gel appears excessive in practice. Thus, comparative stacking experiments at various ionic strengths are best carried out using either wattage or voltage regulated at a level that, at the beginning of electrophoresis, gives rise to 0.25 watt in the gel with the highest ionic strength, under the conditions of electrophoresis described above. At constant voltage the wattage will be less in all cases of lower ionic strength, and it will decrease with time of electrophoresis in all cases, since at constant voltage, current decreases with time. Thus, the 0.25 watt per gel compatible with the heat exchange capacity of the apparatus will not be exceeded at any time. Of course, a maximum tolerable wattage value of 0.25 watt per gel is arbitrary because it not only refers to a particular heat dissipation capacity of a particular apparatus, but also to the heat sensitivity of the protein under investigation. The value should therefore be adjusted in each individual case.

The width of the stack is inversely proportional to ionic strength. Since the protein concentration within the stack is regulated, this means that ionic strength is directly proportional to protein concentration within the stack. Thus, ionic strength is the prime regulator by which one can, at will, modify the stack length and protein concentration in the stack. A practical lower limit of ionic strength appears to be 0.0020 M (21) in representative buffer systems of the "Jovin output". The upper limit of ionic strength is set by time, since in view of the limited heat dissipation capacity, power cannot be increased in proportion to ionic strength.

Temperature. Since the "Jovin output" specifies stacking systems at 0°C and 25°C, it is easy to compare protein activities in the stack at these two temperatures. At least one case is known in which the biological activity of a protein could only be maintained by conducting the electrophoresis at a lower buffer temperature of between 0° and 2°C, requiring insulation of the apparatus with polyurethane foam (22). In other cases, protein solubilities were found to be enhanced at 0°C as compared to 25°C. However, protein gelation may occur at the high concentration in the stack at 0°C, but not at 25°C. Thus, it appears reasonable, within the limitations of the known temperature-activity profile for the protein of interest, to spend a few hours testing this relation by means of stacking gels.

Redox conditions. Polyacrylamide gels are highly oxidative due to the free-radical catalysis of the polymerisation reaction which presumably results in a relatively high concentration of peroxides, which remain in the gel after polymerisation. These, being uncharged, cannot be removed by pre-electrophoresis. Therefore, they can only be removed by sweeping the gel with electrophoretically-mobile, charged reducing agents ($HS-R-COO^-$ or $HS-R-NH_4^+$). In a representative case, an analytical-scale 10 %T gel at pH 10 required 5 micromole equivalents of thiol groups for reduction (23). Experimentally, a gel may be reduced by layering electrophoretically-mobile, charged, reducing agent over the surface of the stacking gel and under the sample. To maintain phase boundaries, the layer of reducing agent should contain 30-40% sucrose. The charged reducing agent sweeps the gel ahead of the protein, thus providing net reducing conditions. Recommended loads of reducing agent are 5 to 20 μmoles per stacking gel of 0.6 cm diameter. In buffer systems with negative polarity (anionic protein), thioglycolic acid, brought to the pH of the operative stacking gel (*PHASE ZETA*) with the common constituent (*C6*), is the reducing agent commonly used. Since it is a very strong acid, it should be neutralised with solid or undiluted *CONSTITUENT 6,* except for the fine adjustment of pH.

Metal ions and chelating agents. When required for protein activity, ionic additives such as metal ions or chelating agents may be included in the upper or lower buffer reservoir, depending on polarity of migration of the additive, at sufficient steady-state concentrations. Concentrations of the order of 0.01 M can usually be tolerated without perturbing the stacking. Several levels should be tested.

Sucrose, glycerol, ethylene glycol. The precipitation and denaturation of proteins under conditions of gel electrophoresis in general and of stacking in particular (in view of the high protein concentration reached during stacking) can sometimes be counteracted by modifying the solvent, such as by adding sucrose, glycerol or ethylene glycol to the polymerisation mixture. A range of 20-60% solutions is usually tested for a beneficial effect on recovery of protein activity from the stack. As a first approximation, upper and lower buffers without these additives may be used for this purpose, although then the additive concentrations in the gel progressively decrease during electrophoresis. Because the viscosity of the gel is increased by non-ionic adducts, the potential across the gel is increased, thus requiring a decrease in current in order to prevent undue electrical heating.

Detergents. Gel electrophoresis of water-insoluble hydrophobic proteins is carried out in the presence of detergents. Frequently protein activities can be maintained in non-ionic detergents, cholic acid-type charged detergents or "uncharged" amphoteric detergents (e.g. "Zwittergent 3-14", Calbiochem). The most suitable detergent species appear to be those in which the amphoteric group is linked to a cholic acid-type rigid hydrocarbon backbone (24). Since even these "benign" detergents at very high concentrations may denature the protein, usually the maximum tolerable detergent concentration must be found experimentally. This is done most easily by stack analysis on gels at various detergent concentrations. In the case of charged detergents, these must be added to the cathodic or anodic reservoir, depending on the polarity of the detergent, as well as to the gel. In addition, proteins can sometimes be solubilised at lowered detergent concentrations, that is with higher recovery of activity, when detergents are mixed. Thus, it has been shown that a mixture of deoxycholate and Lubrol is advantageous in the case of adenylate cyclase (25).

Optimisation of Pore Size: The Ferguson Plot

Quantitative PAGE has two objectives: determination of the appropriate fractionation method (the "optimal pore size", T_{opt}) to provide optimal resolution and physical characterisation. Both objectives are experimentally met by the Ferguson plot ($\log_{10}R_f$ versus %T), that is by conducting PAGE at several gel concentrations and measuring the R_f of the protein in each of the gels. The R_f range desired is 0.25 to 0.85. Measurements of R_f outside this range are so imprecise that the weighting function, used to evaluate the Ferguson plot statistically, largely suppresses them (26,27). The workload of this experiment is one day for a single investigator.

STEP 7: Choice of Resolving Gel Buffer System, Resolving Gel and Polymerisation Catalyst Concentrations

Unstacking of the protein in a resolving gel to provide R_f values in the range of 0.25-0.85 is brought about by two measures taken individually or concurrently. Firstly, it is possible to increase the lower stacking limit of the resolving phase *(PHASE PI)*, designated *RM(1,PI)* or *RM(1,9)* in the "Jovin output", by an increase (anionic systems) or a decrease (cationic systems) in pH. Secondly, the gel concentration can be increased to a level which provides mobilities in the desired range. An increase in the gel concentration and change from a "non-restrictive" highly crosslinked gel to a 2-5 % C_{Bis} crosslinked gel at elevated %T leads to additional retardation of the protein to a constituent mobility less than the value of *RM(1,PI),* that is, it leads to unstacking. Traditionally, "disc" electrophoresis has adopted both measures at once.

The resolving gel buffer (*PHASE GAMMA*). The resolving gel buffer suggested by the physical properties table (*Figure 4*) and the recipe table (*Figure 5*) of the "Jovin output", although likely to unstack the protein of interest in a "non-restrictive" gel, neglects the presence of a gel and cannot be expected to provide R_f values between 0.25 and 0.85 at convenient gel concentrations in a particular application. Thus, in practice it is reasonable to select a convenient gel concentration arbitrarily, for example, one between 5 and 12 %T, 2 %C_{Bis}, for a resolving gel made with the buffer suggested by the "Jovin output" (fourfold concentrated *PHASE 3* of *Figure 5*). If, after electrophoresis, R_f values turn out to be excessively high, a lower gel buffer with increased $RM(1,9)$ should be selected from the subsystems table (column 1 of *Figure 6*). Columns 11 and 12 of the same table provide the recipe, expressed as the final molar concentrations in *PHASE GAMMA* of the two buffer constituents of the lower gel buffer [designated $C(3)$ and $C(6)$ in the table; the top of *Figure 4* defines the chemical identity of *CONSTITUENT 3* and 6 in the particular system]. Correspondingly, if R_f values on gels of between 5 and 12 %T turn out to be excessively low, a subsystem of decreased $RM(1,PI)$ can be selected by the same procedure.

The resolving gel. Once a tentative choice of resolving gel buffer has been made by the rationales discussed above, the parameters defining the resolving gel, that is, gel concentration (%T, %C) and polymerisation catalyst concentrations need to be selected.

The %T range is selected to provide R_f values in the range of 0.25 to 0.85. PAGE should be carried out on resolving gels at three or four gel concentrations, such as 5, 7.5, and 10 %T. Values of R_f are measured as described in *STEP 11* (see below), and rudimentary Ferguson plots are constructed on semi-log graph paper as described in *STEP 12*. The %T values yielding R_f values of 0.25 to 0.85 are read off the plot. Within the range of these values, the remaining gel concentrations are selected at even intervals. In total, for most accurate determinations of the Ferguson plot, seven, but at least three, gel concentrations should be selected.

The choice of %C, and of the crosslinking agent, depends on the aim of the investigation and on the molecular size of the proteins under study. If one aims at analytical information only, a 5% bisacrylamide crosslinked gel is the most restrictive at all gel concentrations. It has the drawback, however, that even a 10 %T gel is turbid (9) and that wall adherence is reduced because of the relative rigidity of this gel type. Thus, when preparative PAGE is anticipated, with its requirement for maximum wall adherence of large blocks of gel, 2 %C_{Bis} even on an analytical scale is preferable to maintain the identity of analytical and preparative conditions. Another consideration in the choice of %C relates to the interdependence of %C and the slopes of Ferguson plots. The flatter the Ferguson plots are, the wider the range of gel concentrations that can be used to construct molecular weight standard curves (11). Thus, when one aims mainly at a molecular weight analysis, 2 %C provides an advantage. Finally, if one is dealing with proteins of high molecular weight, or with a large molecular weight complex between proteins and non-ionic detergents (22,28), 15 %C_{DATD} is advantageous since it allows one to make relatively open-pore gels, while maintaining mechanical gel stability and good wall adherence. No matter which %C is chosen by any of these criteria, it is important to keep the value of %C constant at all gel concentrations in order to obtain linear Ferguson plots.

The choice of catalysts. Selection of catalysts to provide polymerisation within 10 min has to be made by systematic experimentation for each gel concentration. It is easiest, as a first approximation, to fix the concentrations of persulphate and of riboflavin at such a level, that polymerisation will not result within 10 min unless and until TEMED is added to the polymerisation mixture (2). Reasonable final concentrations for polymerisation at 0°C would be 0.015% potassium persulphate, 5 x 10^{-4}% riboflavin (see *Figure 8*). Keeping both catalysts constant at these concentrations, TEMED is then added to the polymerisation mixtures at each gel concentration in varying amounts ranging from 0.1 to 2 μl/ml of gel. Gels are photopolymerised for 10 min (using the procedure described above in *STEP 3*), the gel surface is inspected critically for straightness, (a convex surface indicates excessive catalyst while a concave surface indicates the opposite; the reverse is true if the problem is the average chain length), gels are inverted and, if polymerised, removed from the glass tubes and tested for mechanical stability. If necessary, the concentrations of persulphate (in basic systems) or riboflavin (in acid systems) are increased to achieve polymerisation within 10 min. The catalyst concentrations required usually decrease with increasing gel concentration (*Figure 8*).

STEP 8: Polymerisation of Resolving Gels

Gel tubes are prepared and inserted into the apparatus, and the gels are polymerised as described in *STEP 3*, except that bisacrylamide is used as a crosslinking agent and TEMED is added to the polymerisation mixture in inverse proportion to %T. This is the reason for providing space for various TEMED levels in *Figure 8*. In general, gel concentrations in PAGE need to be accurately controlled by careful pipetting, by checking pH and conductance of stock solutions (*Figure 8*), and by constant deaeration conditions of the polymerisation mixture (e.g. 5 min, 10 mm Hg at 0°C, or 20 mm at 25°C) (2). An appropriate deaeration apparatus (18) is depicted schematically in *Figure 9*. The volume of the resolving gel is usually 1.2 ml for a 0.6 cm diameter gel.

STEP 9: Polymerisation of Stacking Gels

The upper buffer reservoir (Part C, *Figure 1*) is lifted out of the lower one (Part A, *ibid.*) and inverted. Overlayering water is drained from the resolving gel surface onto tissue paper by shaking the reservoir. Alternatively, overlayering water may be removed by touching with a surgical cotton swab. The Parafilm seals at the bottom of the gels are removed gently. The stacking gel is polymerised on top of the resolving gel, using the procedure detailed in *STEP 3*. The volume of stacking gel should be not less than twice the sample volume. Usually, 0.5 ml are applied for a 0.6 cm diameter gel. If sample volumes are very large, for example 1-3 ml for a gel of this size, the tube length has to be increased to accommodate the sample plus twice its volume of stacking gel. In view of the mechanical lability of bisacrylamide crosslinked stacking gels, the layering water should be removed gently by adsorbing it onto tissue paper, or with a cotton swab.

STEP 10: Sample Application, Electrophoresis and Gel Removal from the Tube

The sample is prepared as described earlier (*STEP 4*). PAGE is conducted at a regulated current of 1-2 mA for a 0.6 cm diameter gel at 0-4°C, 2-4 mA per gel at 25°C. PAGE is discontinued and each gel tube is withdrawn from the upper buffer reservoir individually (see p. 110) when the tracking dye marking the moving boundary in the resolving gel has migrated to within 2-3 mm from the bottom of the gel. The gel is removed from the tube in one of three ways, depending on gel concentration.

(i)   A robust 2-5% bisacrylamide crosslinked 5 to 12 %T gel is rimmed with a continuous stream of water using a No. 23 gauge, 2.5 inch long, blunt hypodermic needle attached by tubing to a water tap. If after rimming from both sides the gel does not slide from the tube easily, it can be forced out by blowing into the tube.

(ii) At high %T, that is, 12-40 %T, the same gel type is peeled out of the tube under water, after shattering the glass with a hammer. Gentle progressive shattering of the gel tube with a heavy hammer on a stone surface starting at one end does not damage the gel. The same technique is applied for very long gels, irrespective of %T.

(iii) Finally, 3.5 to 5 %T gels of the same type, or 10-30 %C_{Bis} gels, are rimmed using a syringe needle, delivering the water dropwise as described earlier (p. 110). The stacking gel can be discarded, unless one wants to test for protein aggregates too large to enter the stacking gel or for unstacked migrating components.

STEP 11: Gel Staining, Slicing and Calculation of R_f

The bottom of the gel is marked by inserting a fine syringe needle, optionally dipped into India ink. The position of the stack may be marked in the same way. Alternatively, it may be measured in relation to gel length, either directly on the gel tube, or more easily from a Polaroid photograph of the gel, or by means of an electronic R_f measuring device (29), prior to the removal of the gel from the tube.

The gel pattern is obtained either by (i) staining for protein, (ii) transverse slicing of the gel and assay, or (iii) autoradiography, with or without prior longitudinal slicing of the gel. R_f's are measured as described in section (iv) below.

(i) *Staining.* The staining method for protein should provide a pattern rapidly, without a stained background, and employ an effective fixative, such as TCA, to precipitate every protein species and prevent its gradual resolubilisation. The procedures preferred by these criteria are those of Diezel *et al.* (20), and, in case of particularly acid-soluble proteins, one described previously from this laboratory (30). In either procedure, the proteins are first denatured in 12.5% TCA for 5 min, using a sufficient volume excess of fixative (40 ml for a 1.2 ml gel) to be able to neglect the dilution of the fixative by the gel. Then, in the Diezel procedure, 2 ml of 0.25%

aqueous Coomassie brilliant blue G-250 are added to the fixative. After 30 min, the gel is transferred to a screw-cap tube of 9 ml volume. The tube is filled completely with 5% acetic acid. The pattern is immediately apparent without background and is stable for at least a few months. It should be noted that, even after dilution with acetic acid solution, the final TCA concentration remains approximately 2%. In cases where this residual TCA concentration appears inadequate for continued fixation, the alternative procedure is to stain the protein, after the initial short fixation, in a saturated solution of Coomassie brilliant blue R-250 in 12.5% TCA for 30 min, followed by transfer to a tube and storage in 10% TCA (30). This procedure, besides avoiding poor fixatives such as acetic acid, has the added advantage that the dye remaining in 10% TCA, or added in small amounts to it, is progressively adsorbed onto the stained protein zones, giving rise to progressive darkening of zones upon storage in the dark. The procedure has two disadvantages. Firstly, the staining solution needs to be freshly prepared with care to avoid dye precipitation. This is done by mixing 1 ml 0.1% aqueous dye with 50 ml 25% TCA followed by 49 ml water and mixing again immediately. The stain is not stable for more than a day. The second problem is that gels shrink in 10% TCA after two weeks of storage.

Special staining problems arise in SDS-PAGE or PAGE in other detergents. These detergents compete with the acid fixative as well as the dye. Thus charged detergents such as SDS compete in the reaction of the protein with anionic dyes and preclude the use of cationic dyes. It is also necessary to disperse the micellular structure of any detergent to be able to diffuse monomeric detergent out of the gel while fixing the protein effectively. Present procedures attempt to achieve this by dissociating micellular SDS with acid alcoholic solutions (11). However, these cannot in general be trusted as effective protein fixatives. Furthermore, staining in acid alcoholic solutions produces a deep background stain which takes a long time to wash out (11). Both the rate of diffusion of monomeric SDS from the gel and the destaining of the gel seem significantly accelerated by adsorbing SDS and dye to activated charcoal (e.g. using the diffusion destainer available from Hoefer Scientific Instruments) using 12.5% TCA, 50% methanol (B. An der Lan, in preparation).

(ii) *Transverse gel slicing and assay.* Transverse gel slicing to give slices of 1 mm thickness should employ instrumentation and techniques for which the variance of slice thickness has been determined for the gel type, and within the gel concentration range of a particular experiment, and has been found adequate. Also, the adaptability of a gel slicer to gels of various diameters is a valid criterion for selection of a slicer. Finally, sectioning techniques which obliterate the position of the moving boundary front either through gel maceration, or through retention of an unsliced residual gel stump are less useful since they make it difficult to characterise zones by R_f. No commercial gel slicer exists as yet whose performance is known in terms of the variance of slice thickness at different %T, %C, and which would fulfil the needs for size adaptability and R_f measurement. A diaphragm-type slicer which meets these criteria and which can be assembled from commercially available parts has been reported by this laboratory (31) (*Figure 10*). For relatively elastic gels, crosslinked with 15%DATD, which are suitable for large molecular weight proteins, a commercial wire slicer with an electrovibrator (Hoefer Scientific) appears most suitable, at least at or above

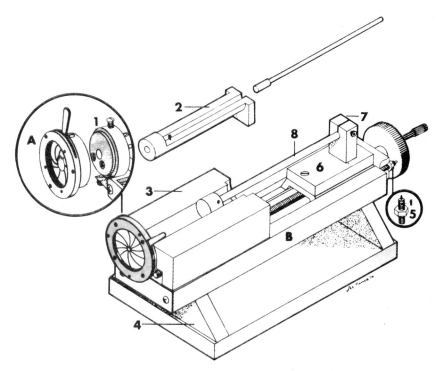

Figure 10. Schematic diagram of a gel slicer, adaptable for cylindrical gels of 0.3, 0.6 and 1.8 cm diameter (31). All parts except 3, 6, 7, 8 are commercially available. For part designations see ref.31.

5 %T. However, the frequency of the vibrator is excessive in the present commercial model for gels of this type and needs to be reduced, by means of a rheostat, in relation to %T, and in proportion to gel diameter.

The protein pattern is obtained either by isotope analysis on the gel slices, or by analysis of protein activity in extracts of the gel slices. In the former case, analysis for β-emitting radioisotopes requires slice solubilisation (3) if quantitative data are needed, while analysis for γ-emitters can be done directly on the gel slices. Gel solubilisation conditions depend on %T and on the choice of crosslinking agent (ref. 3; also see Chapter 1).

Activity assay on gel slices is usually carried out qualitatively by incubating slices in assay cocktail. Since the diffusion rate of proteins from the gel varies with conformation and degree of adsorption to polyacrylamide of individual proteins and with gel concentration, a time curve of diffusion time versus activity should be constructed to determine the minimal time required for elution of a maximum amount of activity.

(iii) *Autoradiography*. Detection of radioactive proteins by autoradiography is usually carried out on longitudinal slices of cylindrical gels or on gel slabs of 1-2mm thickness, after drying the slices onto filter paper (Chapter 1). Since at higher gel concentrations polyacrylamide gels crack upon drying, the gels are allowed to equilibrate for 0.5h with a solution of 5% polyethylene glycol (20,000 MW) and 5% glycerol

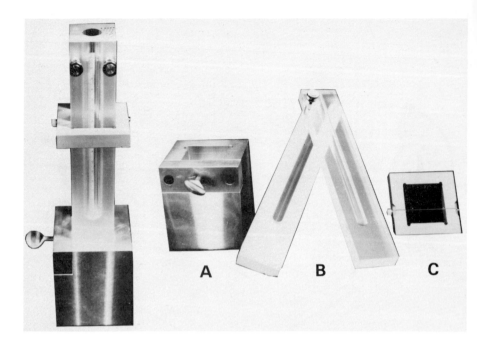

Figure 11. A longitudinal slicer for gels of 1.8 cm diameter capable of providing a guidestrip, consisting of A) an aluminium base; B) a Perspex (Plexiglas) gel holder; C) a wire frame, with asymmetric mounting of the gel cutting wire.

prior to drying. A longitudinal slicer developed in this laboratory (2,32) for 0.6 cm gels is constructed as shown in *Figure 11* (construction details are available on request). A commercial model exists (Miles, Elkhard, Ind.), but it is available for gels of 0.5 cm diameter only and lacks tightening screws and the heavy base needed for slicing on the bench surface. A commercial slice dryer (Hoefer Instruments) appears satisfactory. For proteins labelled with γ-emitters, autoradiographs can be prepared by direct exposure of the gel tubes to film (Kodak X-Omat R, XR1) using, if necessary, an intensifying screen (Dupont Lightning Plus). Gels contained in tubes of rectangular cross-section (2 x 4 mm I.D.) used with a suitable upper buffer reservoir appear to be ideal for this (B. An der Lan *et al.,* in preparation). Also, gel halves can be scanned directly on a commercial paper strip scanner if 5 x 10^5 cpm per protein band are available (33).

(iv) *Measurement of* R_f. Protein zones in the gel pattern are characterised by measuring their migration distances in the resolving gel relative to the moving boundary in front of the zones (relative mobility, R_f). In many cases the front moving boundary can be marked with a dye ("tracking dye") migrating between the particular stacking limits [*RM(1,PI)* and *RM(2,LAMBDA)*]. In other cases, a chemical detection method for either the trailing constituent (*C1*) or the leading constituent (*C2*) is available, which allows one to detect the moving boundary, for example sulphate by incubation of the gel in 0.1 M barium chloride, chloride in silver nitrate (see *STEP 6*).

Alternatively, labelled *CONSTITUENTS 1* or *2* may be detected by isotope analysis. For stained gel patterns, an electronic R_f measuring device has been described (29) and is commercially available (Hoefer Scientific). If the gel is made in a reproducible fashion, and electrophoresis is carried out under controlled conditions, the R_f is also highly reproducible (26). The measurement of the R_f of each zone, at each gel concentration, for the protein of interest and molecular weight standards, is the sole experimental work required for "quantitative PAGE". It provides the "R_f-%T sets" for computation of Ferguson plots and derived parameters in fully automated fashion using a series of computer programs, designated the *"PAGE-PACK"* (2,7,26,27,34).

STEP 12: Computation of Ferguson Plots

The Ferguson plot ($\log_{10} R_f$ versus %T) for the measured R_f-%T sets should be plotted on semi-log paper. If a minimum of gel concentrations over a wide range was used initially, as suggested above, such a plot will indicate the required range of gel concentrations for the desired R_f range of 0.25 to to 0.85. For optimal statistical definition of the line, seven R_f-%T sets in that range are desirable; three is the minimum required. Thus, in practice, *STEPS 8-11* are repeated at this stage to accumulate the desired number of points on the Ferguson plot. Also, the manual plot allows one to spot where experimental confirmation of certain points might be needed.

The computer programs for constructing the Ferguson plot and evaluating it, the *PAGE-PACK* (2,7,26,27,34), serve to provide adequate statistics, error-free mathematical manipulations and a convenient speed of interpreting the R_f-%T sets. The *PAGE-PACK* is available in the form of a magnetic tape, a listing and instructions for use of the programs from Biomedical Computing Technology Information Center, Room 1302, Vanderbilt Medical Center, Nashville, Tennessee 37232, USA; program identification number *MED-34 PAGE-PACK*. The programs are written in Fortran IV G. *Figure 12* lists the components of the *PAGE-PACK* and their arrangement in the form in which they are used at the NIH. Program names correspond to those given in the instructions. *Figures 13-23* illustrate representative input and output formats of the key programs of the *PAGE-PACK*.

Values of R_f at the various %T are entered into an input file (*Figure 13*), preferably using a remote terminal. Program *RFT1* employs this input to compute the slope of the Ferguson plot, K_R, the retardation coefficient (a measure of molecular size), and the y-intercept of the Ferguson plot, Y_0, (a measure of molecular net charge), their standard deviations, and the correlation coefficient for $(\log R_f)$/%T. These parameters are indicated by an arrow or bracketing in *Figure 14*. Furthermore, the joint 95% confidence envelope of K_R and Y_0 is computed in numerical terms that allow one to plot the envelope manually on semi-log paper, with Y_0 on the logarithmic scale (*Figure 15*). Finally, program *RFT1* provides the graph of the Ferguson plot together with the 95% confidence envelopes for the line (crosshatched) and for each observation (hatched) (*Figure 16*).

PAGE PACK

	DATAFILE	PROGRAM	IN	OUT
1.	PAGEPACK			INSTRUCTIONS
2.	PAGE.PROG			PROGRAMS in FORTRAN
3.	DATA01 card		R_f, %T	
4.		RFT1		K_R, Y_o
5.	DATA02 card		Prot.#, K_R	
6.		PLOTRUN		$\bar{R}$, MW
7.	DATA03 card		Prot.#, K_R R_f K_{AV}	
8.		GIANTRUN		$\bar{R}$, MW $\frac{MW}{\bar{R}}$
9.	DATA04 card		K_R, Y_o, $\bar{R}$	
10.		CHARGE		M_o, V
11.	DATA06 card		$[K_R, Y_o, \bar{R}]A$ $[K_R, Y_o, \bar{R}]B$	
12.		TOPT		T_{OPT}, T_{MAX}
13.	"MUNI"	IDENT	7 parameters	Significant distinction based on F-test criteria

Figure 12. Schematic overview of the *PAGE-PACK* computer files and programs.

Identity Testing

The appearance of a single component at the various gel concentrations used to construct a Ferguson plot in a buffer system optimised for pH, ionic strength, temperature, and other conditions providing maximal stability and activity, indicates molecular homogeneity. To the degree that the same molecular properties as those used to provide separation by PAGE are also used by other physical techniques, their application to test homogeneity appears unnecessary, particularly if the

```
18.      ── System
2.       ──%C
800.     ── Prot#
9.       ── #of cases
3.5        .959
3.75       .935
4.5        .730
5.         .640
6.         .533
7.         .470
8.         .390
9.         .320
10.        .240
18.
2.
801.
9.
3.5        .898
3.75       .840
4.5        .625
5.         .575
6.         .465
7.         .390
8.         .300
9.         .230
10.        .140
9999.
```

PROGRAM
RFT1

INPUT
File DATA01

Figure 13. Input format for file *DATA01*.

"independent method" exhibits lower resolving power measured as the number of theoretical plates, N, (37) than PAGE.

Questions of molecular identity between experiments carried out at different times or in different laboratories arise frequently. They can only be settled if the analyses have been performed under an identical set of PAGE conditions. The resolution of any identity question can then be based on either the shape of the concentration profile (peak) of each component in question, or on R_f, or on the Ferguson plots (26,35). In the latter, optimal, way the question is resolved by plotting the joint 95% confidence envelopes of K_R and Y_0 (*Figure 15*) of the two Ferguson plots in question on the same graph (semi-log paper, with Y_0 on the log axis). Envelope overlap of two components shows that they are indistinguishable under the particular conditions. Non-overlap demonstrates that they are distinguishable. Partial overlap raises problems. If envelopes of vastly different sizes are involved, the larger envelope should be reduced to the size of the smaller one by increasing the number of data points (R_f-%T sets). If the ellipse overlap still persists, and if components have been analysed under identical conditions, preferably simultaneously, the statistical method used to resolve the problem may be refined, using program *IDENT* to reduce the size of the K_R-Y_0 envelopes (26, *Figure 12*). Instructions for use of this program are not provided here to avoid complication of a presentation of the major aspects of the *PAGE-PACK* by secondary considerations; but both the program in *BASIC* and instructions are available upon request.

OUTPUT 1

TUESDAY   APRIL 22, 1980

SYSTEM   18.2.0.   CROSSLINKING   2.00   PROTEIN   800.00

T	RF	PREDICTED RF	LOG(RF)	WORKING LOG(RF)	WEIGHT
3.50	0.9590	0.93358	-0.01818	-0.01802	0.24112
3.75	0.9350	0.88710	-0.02919	-0.02858	0.23415
4.50	0.7300	0.76110	-0.13668	-0.13630	0.21158
5.00	0.6400	0.68720	-0.19382	-0.19275	0.19541
6.00	0.5330	0.56024	-0.27327	-0.27274	0.16165
7.00	0.4700	0.45673	-0.32790	-0.32772	0.12831
8.00	0.3900	0.37235	-0.40894	-0.40846	0.09792
9.00	0.3200	0.30356	-0.49485	-0.49424	0.07219
10.00	0.2400	0.24748	-0.61979	-0.61959	0.05174

```
UNWEIGHTED REGRESSION
MEAN X=   6.31   MEAN Y=   -0.2781      ANTILOG(MEAN Y)=  0.527147
Y INTERCEPT A =      0.27499E+00       STD.DEV. A =    0.022982
ANTILOG A =    1.88359                 ANTILOG (A+1 S.D. OF A)=    1.98595
SLOPE B =    -0.87709E-01              STD.DEV. B =     0.34406E-02
CORRELATION R =   0.994658
VARIANCE Y =       0.42497E-01
RESIDUAL VARIANCE OF Y = 0.51756E-03
RESIDUAL SUM OF SQUARES= 0.36229E-02
SUM XOBS. =     56.75                  SUM XOBS. SQARED =    401.562
SUM YLOG. =     -2.5026                SUM YLOG. SQARED =      1.0359
SUM XOBS.*YLOG. =  -19.61520
```

```
WEIGHTED REGRESSION
MAXIMUM LIKELIHOOD METHOD      10 ITERATIONS
```

$$\boxed{YO = 1.908236} \qquad \boxed{KR = 0.08871}$$

```
MEAN X=   5.35818  MEAN Y= -0.194691  MEAN (RF)=    0.63872
Y INTERCEPT C =      0.28063E+00      STD.DEV. C =    0.23708E-01
ANTILOG C =   1.908236
YO+1STD.DEV. =      2.01530           STD.DEV. YO =   0.10706    ◄———————
SLOPE D =   -0.08871                  STD.DEV. D =    0.41816E-02 ◄———
CORRELATION R =    0.998905
VARIANCE OF Y =      0.47821E-02
RESIDUAL VARIANCE OF Y = 0.83704E-04
STANDARD DEVIATION OF MEASUREMENT =      0.91490E-02
RESIDUAL SUM OF SQUARES= 0.58593E-03
SUM W*XOBS.=    7.46965               SUM W*XOBS. SQARED=   44.81082
SUM W*YLOG.=   -0.27141E+00           SUM W*YLOG. SQARED=    0.91098E-01
SUM W*XOBS.*YLOG.=  -1.87893          SUM OF WEIGHTS=    1.39406
SUM(WXX)=   4.78703   SUM(WYY)=  0.38257E-01   SUM(WXY)=   -0.42466E+00
RESIDUAL SUM OF SQUARES/SUM OF WEIGHTS =      0.42030E-03
```

Figure 14. Output format of program *RFT1*. The five essential output parameters are boxed in or noted by arrows.

Size and Charge Isomerism

The inspection of bands in PAGE within a gel, or between gels, does not provide any information concerning physical relationships between the proteins under each of the zones. In contrast, Ferguson plot analysis, and in particular the plot (*Figure 15*) of K_R versus Y_0 derived from it does provide such information. Thus, parallel Ferguson plots reveal identity in molecular size and a difference in net charge ("charge isomerism"), while Ferguson plots intersecting at or near the Y_0 axis suggest an oligomeric series (i.e. identity of surface net charge density) and a difference based on molecular size ("size isomerism") (see Chapter 1, p. 14 and ref.36). Crossed Ferguson plots reveal a difference between the species based both on charge and size differences (Chapter 1, p. 14; *Figure 19* of ref.7). The same information is conveyed

124

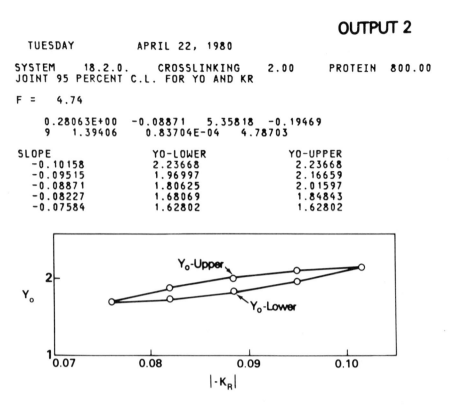

OUTPUT 2

TUESDAY   APRIL 22, 1980

SYSTEM   18.2.0.   CROSSLINKING   2.00   PROTEIN   800.00
JOINT 95 PERCENT C.L. FOR YO AND KR

F =   4.74

     0.28063E+00 -0.08871   5.35818 -0.19469
     9   1.39406   0.83704E-04   4.78703

SLOPE   YO-LOWER   YO-UPPER
  -0.10158   2.23668   2.23668
  -0.09515   1.96997   2.16659
  -0.08871   1.80625   2.01597
  -0.08227   1.68069   1.84843
  -0.07584   1.62802   1.62802

Figure 15. Output format of program *RFT1* continued. The joint 95% confidence envelopes of K_R and Y_0 are given in numerical form (top panel) and need to be plotted manually on semi-log graph paper as shown in the bottom panel.

by the plot of K_R versus Y_0, but with the statistical significance of these parameters added. Thus, joint 95% confidence envelopes of K_R and Y_0 displaced from one another along the K_R axis demonstrate size isomerism, while envelopes displaced along the Y_0 axis reveal charge isomerism. Since a single protein exists, as a rule, in several oligomeric and charge isomeric forms, the envelope pattern consisting of horizontally and vertically displaced species in this plot reveals molecular homogeneity. This homogeneity would remain unrecognised if one merely counted the number of zones on a gel. Even a series of Ferguson plots, graphically superimposed, but without superimposition of the statistical limits for each, could not clearly provide the evidence of identity or non-identity, and of the charge- or size-isomeric relatedness, which is directly available upon inspection of the "ellipses". Their computation is therefore one of the major justifications for applying the *PAGE-PACK* in preference to a non-computerised analysis of Ferguson plots.

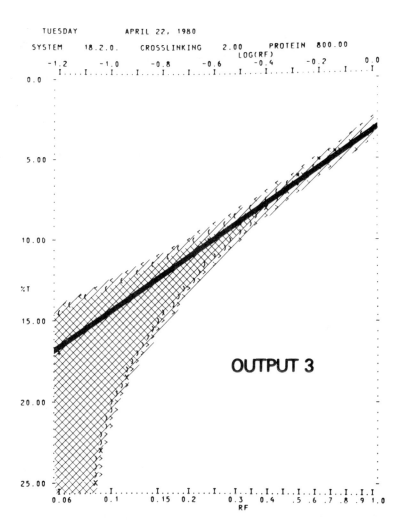

Figure 16. Format of the output of program *RFT1* continued. Plot of $\log R_f$ versus %T (Ferguson plot). The 95% confidence envelope across the line is given by cross-hatching, the corresponding envelope for a single observation is given by hatching.

Physical Characterisation

The values of K_R and Y_0, within their 95% confidence limits, are experimentally-derived measures of molecular size and net charge, that is, they are derived without any assumptions. Therefore, a protein species is best described in terms of its joint 95% confidence envelope for K_R and Y_0 which directly assesses its molecular size and charge in a particular buffer milieu. Since one can select various buffer milieus of different degrees of hydrophobicity, all of the major properties of macromolecules (namely charge, size, and hydrophobicity) find expression in these envelopes of PAGE. This fact makes it redundant to test molecular properties, or to exploit them

126

in fractionation, by methods other than PAGE which are sensitive to the same properties. Alternative fractionation methods sensitive to size or to charge, such as gel filtration, DEAE- or CMC-chromatography, are less efficient than PAGE in practice, yielding a lower number of theoretical plates (1,37). However, it should be noted that this redundancy of other methods of fractionation based on exploitation of the same molecular differences as PAGE holds only for analytical characterisation of systems containing a few components and not for multicomponent systems; determination of the T_{opt} (see below), and hence isolation of proteins by fractionation at the T_{opt}, are feasible only with simple mixtures of a small number of components. This is because Ferguson plots cannot be constructed on multicomponent systems since one cannot be sure which protein band (R_f) at one %T value corresponds to the same protein (and R_f) at another. Therefore multicomponent systems need to be prefractionated before it is possible to determine the optimally resolving gel concentration for any set of two components.

Very little is known about the physical meaning of K_R as a measure of molecular size, and about Y_0 as a measure of molecular net charge. From considerations of pore theory it appears that the aspect of molecular size expressed as K_R depends on the gel type. Thus, in the ideal case, K_R values found experimentally in gels made of fibres so long that the occurrence of fibre endings can be neglected, reflect molecular surface area. Correspondingly, K_R values in "point gels" in which the fibre length is so short that it can be neglected would be predicted to reflect molecular volume (7,37). These predictions, although plausible, still await experimental testing. At this time, they cannot be more than guidelines in the interpretation of K_R as a measure of molecular size, assuming that the usual polyacrylamide gels in PAGE, crosslinked with 2 to 5% bisacrylamide, are of the "infinite length or one-dimensional" type, while gels crosslinked with 10 to 50% bisacrylamide, or highly crosslinked DATD gels, approximate to the "zero length or zero-dimensional" type (1,7,37). In real gels, one would expect intermediates between the "one-dimensional" and "zero-dimensional" ideal gel types, depending on %C and polymerisation rate (catalyst concentrations, temperature), the latter being inversely proportional to average chain length.

The interpretation of Y_0 as a measure of molecular net charge is even more ambiguous than that of K_R (see previous relevant discussions: p.124 of ref.27, p.58 of ref.26, p.401 of ref.38). One problem is that the value of Y_0 varies with %C in a manner analogous to K_R except that at high (10-50%) values of %C it remains constant (9). Since molecular valence cannot vary with %C, it is uncertain which value of Y_0 to apply for the computation of valence. Secondly, it is possible that Ferguson plots in the range of 0-2 %T are non-linear, although this can neither be proved nor disproved, since gels of such concentration cannot be made. Finally, the computation of valence itself makes numerous assumptions which will be enumerated further below. In spite of all of these shortcomings of the translation of Y_0 to valence, V, it is frequently useful for interpreting and visualising Y_0 differences between species qualitatively in terms of net charge.

In contrast to the direct interpretation of the experimental parameters K_R and Y_0 as measures of molecular size and net charge respectively, the translation of K_R to conventional size parameters (molecular radius and weight) and of Y_0 to values of free electrophoresis mobility and molecular valence hinges on numerous assumptions. Therefore, it is important to distinguish molecular characterisation in terms of K_R and Y_0 which is more accurate, but is difficult to conceptualise, and molecular characterisation in terms of the translated parameters, which is less accurate but can be visualised in familiar terms. For instance, description of a molecule in terms of a molecular weight of 60,000 derived from PAGE data evokes an image, while a corresponding K_R of 0.04 (which is more accurate) would not do so until such time when the use of the K_R will become more common. It is for that reason that "translation" still remains useful at this time. It is carried out by means of the *PAGE-PACK* as follows:

STEP 13: Ferguson Plots of Molecular Weight Standards

Protein standards sufficiently homogeneous to exhibit a major band must be found. Although commercial molecular weight standards are available for SDS-PAGE, none are presently marketed for PAGE of native proteins. Standard molecular weight proteins are selected over a range of molecular weight encompassing the unknown, and resembling the unknown as much as possible in chemical composition. Glyco-, lipo- and nucleoproteins, for example, should not be mixed in this choice if possible and globular-type proteins should not be mixed with excessively asymmetric ones. Assuming such a choice of standards can be made, seven standard molecular weight proteins should suffice. However, in practice one can rarely afford to be so selective, since available proteins vary in all of these respects. This usually makes it necessary to select ten to twenty standards in an attempt to randomise their compositional and conformational differences. Furthermore, the selection must be narrowed to those proteins which stack in the buffer system optimised for the unknown protein species, although one can adjust the stacking limits of the stacking gel selectively within subsystems to stack the standards (*Figure 6*) and still compare unknown and standards in the same resolving gel. Ferguson plots on ten to twenty standard proteins of a wide range of molecular weight requires PAGE at a wide variety of pore sizes, and is therefore best done in gel tubes. This requires a great deal of work. With twelve to fourteen gels or two Ferguson plots per apparatus, a single operator can at best handle four proteins per day using two sets of apparatus. A standard curve therefore requires at least two to four days work. The procedure can be shortened by about a factor of two if the molecular weight range is narrow enough to allow for analysis of all proteins in four pore sizes. In this case, vertical gel slabs, each at a particular %T, are labour-saving. The procedure for arriving at the Ferguson plots of the standards is otherwise the same as described above.

For SDS-PAGE, Ferguson plots of standards and unknowns are needed to verify experimentally the equality or near-equality of relative free mobility Y_0 for all species. If any standard or unknown differs in Y_0 from the common Y_0 of proteins at or near %T = 0, due to inadequate saturation of the protein surface with SDS (inadequate reaction time in 1% SDS at 100°C), or due to covalent linkage with non-protein moieties, molecular weight in SDS-PAGE can only be derived in the same way as described below for PAGE, assuming a random-coiled conformation for SDS-polypeptide complexes, that is, by a standard curve of molecular weight (MW)

HEMAGGLUTININ I AND CONT PH 2.9

KR ◄——— PAGE, Gel filtration, SDS-PAGE   **PROGRAM GIANTRUN**

51. ◄— Prot #	**.0567** —— K_R	
6.	**.04674**	
45.	**.08044**	
114.	**.09134**	Standards INPUT **File DATA03**
11.	**.1096**	
60.	**.1331**	
800.	**.08871**	Unknown
801.	**.10887**	
999.		

OUTPUT 1

PROTEIN	NAME	WEIGHT	RADIUS	KR	
51.0	PAPAIN	20700.	1.82430	0.05670	1
6.0	CHYMOTRYPSIN	21600.	1.85040	0.04674	2
45.0	LACTOPEROXASE	93000.	3.01030	0.08044	3
114.0	GLYCERALD-3PDH	140000.	3.45000	0.09134	4
11.0	GAMMA-GLOBULIN	160000.	3.60700	0.10960	5
60.0	CATALASE	232000.	4.08260	0.13310	6

Figure 17. Format of the input for program *GIANTRUN*. Top panel: input file *DATA03*. Bottom panel: output listing of the input parameters.

versus K_R (*STEP 14*). If, in contrast, standards and unknowns do share a common Y_o at or near %T = 0, molecular weight can be derived from R_f at a single %T-value by means of the sigmoidal standard curve of $\log_{10}(MW)$ versus R_f, using either the central linear segment of that curve within a limited molecular weight range (program *GIANTRUN*, as in *STEP 14*, with an input of "*RF*" and R_f values for standards and unknowns in *Figure 17*), or non-linear curve fitting using a smooth, symmetrical, logistical curve (34).

STEP 14: Translation of K_R to Molecular Radius ($\bar{R}$) and Molecular Weight (MW)

The K_R values of unknown and standards are entered into a data file (*Figure 17*, top panel). The entry of "*KR*" in line 3 of the data file determines the type of data analysis, namely the computation of molecular geometric mean radius, $\bar{R}$, and molecular weight, MW, on the basis of a standard curve for globular proteins [$(K_R)^{1/2}$ versus $\bar{R}$] or for random-coiled proteins (K_R versus MW). The program then summarises the input data (*Figure 17*, bottom panel), computes $\bar{R}$ and MW for the unknowns on the basis of the standard curve for globular proteins, with the correlation coefficient for the standard curve (*Figure 18*) and provides a plot of the standard curve with 95% confidence envelopes for the line (cross-hatched) and for a single observation around the line (hatched) (*Figure 19*). An analogous computation and

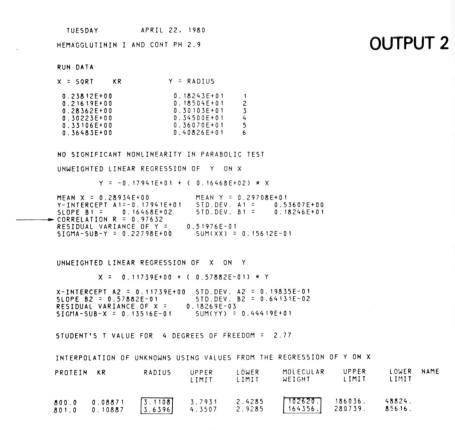

Figure 18. Output format for program *GIANTRUN* continued. The essential output data are boxed in or indicated by an arrow.

plot (not shown) follows for the case of a random-coiled protein. The following three assumptions are made in the translation of K_R to $\bar{R}$ or MW:

(i) a common globular or random-coil conformation for both unknown and standard proteins

(ii) a partial specific volume of 0.74

(iii) zero hydration for both unknown and standards.

To reduce errors due to these assumptions, an attempt should be made to select standards of the same conformational type as the unknown, and if it is a conjugated protein, to select standards with the same conjugated moiety. Systematic errors arising due to violations of these three assumptions decrease as one increases the number of standards.

The identical program (*Figures 17-19*) translates R_f to molecular weight for SDS-PAGE, within the limitations pointed out above. In this case the letters *RF* are entered into the input file (*DATA03, Figure 17*) in lieu of *KR*. R_f values are then entered instead of K_R values.

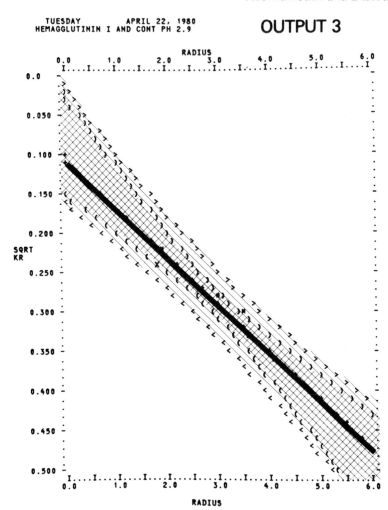

Figure 19. Output format for program *GIANTRUN* continued. Standard curve of $(K_R)^{1/2}$ versus $\bar{R}$ (nm) for globular proteins. The negative intercept of the line on the *RADIUS* axis is predicted to be a measure of the thickness of the polyacrylamide fiber (37). Symbols designating the line and its confidence limits are the same as in *Figure 16*.

STEP 15: Translation of Y_0 to Valence, V

The values of Y_0 and of $\bar{R}$ form the input parameters for computation of free electrophoretic mobility, $M_0(\frac{cm/sec}{volt/cm})$ and valence, V (net protons/molecule) (*Figure 20*). The assumptions involved in this computation are numerous: electric potential gradient surrounding the charged macromolecule according to the Debye-Hueckel theory; sphericity of the macro-ion; univalency of the small (buffer) ions; net charge less than 10; uniform surface charge; rigid macro-ion; nonconducting macro-ion; equality of dielectric constants between buffer and water; Newtonian viscosity. Other

PROGRAM CHARGE
Data File DATA04

```
        TUESDAY          APRIL   1, 1980

        THIS PROGRAM DEVELOPED AT

  NATIONAL INSTITUTE OF CHILD HEALTH AND HUMAN DEVELOPMENT
          -- REPRODUCTION RESEARCH BRANCH

  NATIONAL INSTITUTES OF HEALTH, BETHESDA, MD

        BY    DAVID RODBARD

  FORTRAN VERSION BY LOEL GRABER
```

```
  SYSTEM    18.00 CROSSLINKING 2.0 PROTEIN  801.0 YO =2.094193

  RM(1,9)   IONIC STRENGTH  TEMP.   MU(SODIUM+)      MU(FRONT)
   0.140        0.0150       0.      0.000245        0.34235E-04     INPUT

  FREE MOBILITY =                                  0.7169462E-04  ◄──────
  RADIUS =                                         0.3640001E-06
  COUNTERION RADIUS =                              0.2500000E-07      OUTPUT
  DEBYE-HUCKEL RECIPROCAL THICKNESS =              0.3978788E+07
  HENRY'S FUNCTION OF  1.44828  = X1 =             0.1043548E+01
  CHARGE (COULOMBS/MOLECULE) = Q =                 0.1957434E-10
  VALENCE (NET PROTONS/MOLECULE) = V1 =            0.1222637E+02  ◄──────

  MOBILITY AT I=.1  =                              0.4179173E-04
  HENRY'S FUNCTION OF  3.73944 = X1 =              0.1116080E+01
```

Figure 20. Input and output format of program *CHARGE*. The essential output parameters are designated by the arrows.

unsolved problems with regard to the computation of valence from Y_0 and the interpretation of Y_0 have been discussed above.

Optimally Resolving Pore Size

The optimisation of the pore size for the separation of any pair of proteins with known K_R, Y_0 and $\bar{R}$ is the most important practical justification for using "quantitative" PAGE. The theoretical basis of pore size optimisation and graphic solutions have been described previously (39).

STEP 16: Computation of Optimal Pore Size

The characteristic parameters K_R, Y_0, and $\bar{R}$ for each of the two proteins to be separated are entered into program *T-OPT* (*Figure 21*), together with the estimated time of electrophoresis and its temperature. Program *T-OPT* computes the gel concentration for optimal resolution between the two species (*Figure 22*) and plots resolution against gel concentration (*Figure 23*). The output also gives explicit advice to use electrofocusing, or isotachophoresis, when a resolution optimum occurs at zero gel concentration. In the example shown in *Figures 22* and *23,* such an optimum does exist but is predicted to provide a little less resolution, in this case, than the optimum at 10.8 %T. In many other cases, however, a true optimum exists at %T = 0,

PROGRAM TOPT

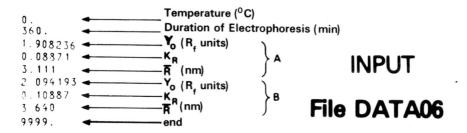

```
 0.                    ←————————  Temperature (°C)
 360.                  ←————————  Duration of Electrophoresis (min)
 1.908236              ←————————  Y₀ (R_f units)          ⎫
 0.08871               ←————————  K_R                      ⎬ A
 3.111                 ←————————  R̄ (nm)                   ⎭
 2 094193              ←————————  Y₀ (R_f units)          ⎫
 0.10887               ←————————  K_R                      ⎬ B
 3 640                 ←————————  R̄ (nm)                   ⎭
 9999.                 ←————————  end
```
INPUT

File DATA06

Figure 21. Input format for program *T-OPT*. The species to be separated are designated A and B.

```
TUESDAY          APRIL 22,1980                      OUTPUT 1

THIS PROGRAM DEVELOPED AT

NATIONAL INSTITUTE OF CHILD HEALTH AND HUMAN DEVELOPMENT
   -- REPRODUCTION RESEARCH BRANCH

NATIONAL INSTITUTES OF HEALTH, BETHESDA, MD

    BY   DAVID RODBARD

FORTRAN VERSION BY LOEL GRABER



TEMPERATURE =  0.0      TIME =   360.
SPECIES A: Y ZERO = 1.908236   KR =0.08871   RADIUS = 3.11100
SPECIES B: Y ZERO = 2.094193   KR =0.10887   RADIUS = 3.64000
V1 = 0.0          V2 = 0.0

GEL CONCENTRATION WHEN MOBILITY OF A = MOBILITY OF B   2.003214
GEL CONCENTRATION FOR MAXIMAL SEPARATION    6.414705
GEL CONCENTRATION FOR OPTIMAL RESOLUTION   10.826197 ←————————
GEL CONCENTRATION WHEN SPECIES A UNSTACKS   3.163477
GEL CONCENTRATION WHEN SPECIES B UNSTACKS   2.948625

ANOTHER T-MAX AND T-OPT OCCURS WHEN T = ZERO

SIZE AND CHARGE SEPARATION ANTAGONISTIC; WE SUGGEST A CHANGE OF
PH, ISOELECTRIC FOCUSING, OR ISOTACHOPHORESIS UNLESS YOU ARE
DEALING WITH DIMERS OR OLIGOMERS.

GEL CONCENTRATION   SEPARATION   RESOLUTION   MOBILITY OF A   MOBILITY OF B

    0.0            -.185958     -1.096735      1.90824         2.09419
    6.414705        0.095316     1.152947      0.51474         0.41942
   10.826197        0.070252     1.384117      0.20905         0.13879
    3.163477        0.052435     0.441305      1.00000         0.94757
    2.948625        0.044863     0.368598      1.04486         1.00000
   30.000000        0.003027     0.472873      0.00416         0.00113
```

Figure 22. Output format of program *T-OPT*. The gel concentration for optimal resolution between species A and B is indicated by the arrow. The output notes that a resolution maximum exists at %T = 0, and suggests the use of electrofocusing or isotachophoresis.

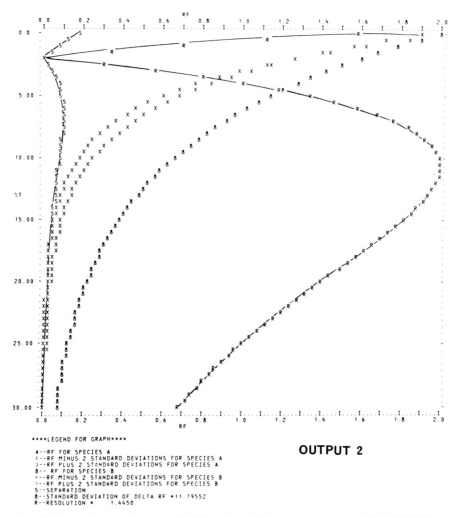

Figure 23. Output format of program *T-OPT* continued. Plots of resolution versus %T [RRRRRR] and the plot of T_{max} versus %T [SSSSSS] relevant for preparative elution-PAGE are outlined.

providing better predicted resolution than can be expected for PAGE at any %T. Such a result of the *T-OPT* computation is the necessary condition for application of one of the charge fractionation methods, isotachophoresis or electrofocusing (see Chapters 3 and 4), at least when one is dealing with the separation between two or a small number of components.

PREPARATIVE PAGE

Preparative PAGE of any kind, whether by slicing and slice extraction (40) or by successive zone elution (41), is possible only for simple mixtures of proteins. Complex mixtures have to be first resolved into relatively simple ones by preliminary chromatographic or other fractionation methods, before preparative PAGE becomes applicable. This difference in approach toward simple and complex mixtures derives from the nature of %T optimisation discussed above. Thus, in a complex mixture, when T_{opt} is computed for each of the protein zones adjacent to the band of interest, and a compromise value of T_{opt} between the two values is applied to electrophoresis, zones that had originally been distant from the protein of interest in the gel pattern now move into its proximity. This makes it impossible to separate the protein of interest sufficiently from its adjacent zones to make preparative resolution possible. This is equally true for preparative methods based on gel slicing and slice extraction . as it is for preparative elution-PAGE, although the appearance of zones in the first case may suggest separation, while the overlapping zone distributions are fully apparent in the latter method. In fact, however, bands in a stained gel pattern are merely the maxima of relatively broad zone distributions (42,43) made invisible by the insensitivity of staining methods.

Gel Slice Extraction

The isolation of proteins by excision of zones and extraction from gel slices (40) is appropriate when several zones are to be isolated simultaneously, while "elution-PAGE" (see below) appears to be labour-saving when a single component is to be prepared in milligram amounts or more. Certainly, gel slicing and slice extraction is the safer preparative method, that is, it is not as easy as in elution-PAGE to lose the entire sample by improper choice of conditions.

The optimal gel concentration for the isolation of a single component by slice extraction is a compromise between the two T_{opt} values for the resolution of the zone of interest from the adjacent zones in the gel. For simultaneous isolation of several components differing in size, some average gel concentration must be found empirically to spread out the zones as much as possible across the gel. In principle, "step-function gels" with T_{opt} values for the various components appear optimal. Pore gradient gels capable of providing these %T optima at least at one point across the gel should also improve resolution of several proteins from one another over average pore gels (44,45).

Methods for the localisation of protein zones on preparative PAGE gels have been described in Chapter 1. If transverse gel slicing is employed in the method chosen, either a diaphragm-type slicer (31) or a wire slicer with electro-vibrator (Hoefer), at the relatively high frequency of vibration of the present commercial model, appears suited for resolving gels. Reproducibility of gel slicing is far better for the resolving gels of PAGE (usually 2 to 5% crosslinked with bisacrylamide and at relatively high %T) than it is for the "non-restrictive" gels of isotachophoresis or electrofocusing (31). Furthermore, it increases in proportion to %T. Although protein may be recovered from the gel by diffusion, this is slow and only partial recovery is achieved. Instead we recommend protein elution and concentration by steady-state stacking as described for isotachophoresis in Chapter 3, except for the following considerations:

(i) The load capacity of PAGE is one to two orders of magnitude less than that of isotachophoresis. A representative value is 0.1 mg/cm^2 of gel for narrowly adjacent zones in a typical separation experiment. This raises a problem of losses by adsorption to polyacrylamide (19) if this is the matrix used for extraction and concentration by steady-state stacking. Thus, a concentration gel of a small enough gel volume must be selected to satisfy the requirement for a ratio of sample to stacking gel volume of 1:2, the further requirement for sufficient gel length to allow for quantitative slice extraction in the electric field, and the desirability of no less than 0.2-0.3 mg of protein per ml of gel to minimise the effect of protein adsorption onto polyacrylamide (19). Thus, when only a few milligrams of protein are available, the tube part of the gel funnel with collection cup (40,46; *Figure 4* of Chapter 3) should be scaled down to a 0.6 cm diameter gel tube.

(ii) As in isotachophoresis, but in contrast to electrofocusing, protein in the bands formed during PAGE is soluble. Thus, gel slices can be directly extracted by steady-state stacking without a prior solubilisation step. However, since gels in PAGE are far more restrictive to protein diffusion and migration than in isotachophoresis, the duration of electrophoretic extraction is higher, depending on gel concentration of the slices and on protein asymmetry, and stacking gel volumes should correspondingly be higher. (For the same reasons, slice extraction by diffusion is relatively inefficient in PAGE as compared to isotachophoresis. It rarely exceeds 70% even after protracted diffusion times which, depending on the factors stated above, may extend from a few hours to several days.) Subsequent to slice extraction and concentration by steady-state stacking, the electrophoresis buffer is exchanged for volatile buffer and the isolated protein separated from non-proteinaceous impurities by gel filtration, followed by lyophilisation (ref.40; Chapter 3, p.151).

Successive Zone Elution

When the isolation of a single protein from a mixture in amounts of a few milligrams or more is required, preparative elution-PAGE is applicable. Many forms of preparative elution-PAGE apparatus have been described in the literature (41). Many of them are workable if due consideration is given to the physical principles involved and to buffer characteristics. Here, however, a procedure can only be provided for the one type of elution-PAGE apparatus (*Figure 24*) that has been applied in this laboratory (47,48) and that is commercially available (Polyprep 200, 20 cm^2 cross-sectional area of gel, Buchler Scientific Instruments). In view of the low load capacity of PAGE (of the order of 0.1 mg/cm^2 of gel), even a relatively large apparatus with approximately 20 cm^2 of gel, as the one for which a detailed procedure has been provided here, is barely able to purify milligram amounts of protein. Thus, preparative elution-PAGE should employ the largest gel surface area over which a constant temperature can be maintained, that is, the largest possible apparatus and the most efficient mode of electrical heat dissipation.

The procedure of elution-PAGE has previously been described elsewhere (ref.2; Appendix III of ref.19).

POLYMERIZATION   ELECTROPHORESIS

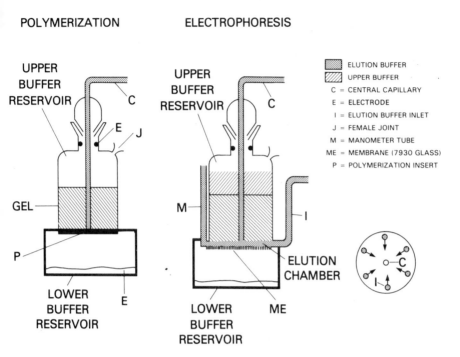

Figure 24. Diagrammatic representation of an elution-PAGE apparatus of the Polyprep-200 type (Buchler Instruments). For a detailed discussion of its design, see refs. 2, 41, 19 (Appendix III). Polymerisation insert (P) and membrane (ME) are successively attached to the gel column. The capillary, gel column, and upper buffer reservoir are water jacketed.

STEP 1: *Preparation of the Apparatus*

The bottom of the gel column is sealed with a threaded polymerisation insert (P) which provides a very smooth polypropylene surface on which the lower gel surface forms. The gel column is placed onto the lower buffer reservoir, which rests on a drip pan. Air inlets into the elution chamber from the central capillary (C) and the manometer tube (M) are closed. To ensure that polymerisation mixture does not enter the capillary, it is filled with resolving gel buffer, the elution line above the capillary is clamped by a hemostat, and the capillary tube is lightly pressed against the polymerisation insert and fixed in position by tightening the threaded joint in which the capillary tube rests. The sealing of the capillary is then tested by releasing the hemostat. If buffer flows onto the polymerisation insert the tightening procedure is repeated with slightly increased pressure of the capillary against the insert, or after lightly greasing the ground bottom surface of the capillary tube (with care not to block the capillary). The apparatus is connected to the coolant, but coolant flow is not started. The apparatus is carefully aligned vertically, to ensure parallelism between the zones and the elution chamber into which they eventually migrate electrophoretically.

137

STEP 2: Gel Polymerisation

A resolving gel of the length of an analytical-scale gel, or less, is polymerised in the gel column. The reason for minimising gel length is that passage of every protein zone to the bottom of the gel takes a long time, while the mechanical stability of the required large gel cylinders may become precarious after 10h to 15h. Crosslinking with 2 %C_{Bis} is preferable in view of the good wall adherence properties of this gel type. Strengthening of that adherence by coating of the inner apparatus walls with Gelamide-250 (Polysciences) is advisable. Evidently the need for strong wall adherence, as well as for maintenance of hydrostatic equilibration of the gel at all times during polymerisation and electrophoresis, increases with decreasing length of gel and decreasing gel concentration. The gel concentration for maximum separation (T_{max}) rather than that for maximum resolution (T_{opt}) should be considered (as an approximation until such time that an exhaustive optimisation of the conditions for preparative elution-PAGE becomes feasible) (39). Thus, the plot of T_{max} as a function of gel concentration (designated by ..S-S-S-S-.. in *Figure 23*) for each of the adjacent contaminating zones in the gel pattern allows one to find the best practical compromise of %T for elution-PAGE.

The optimal resolving gel height still needs to be found empirically. Experience has shown that a duration of 5-6 h for elution of the zone of interest subsequent to elution of the front moving boundary ($R_f = 1.0$) is desirable; the gel length is adjusted to provide such fast elution rate, within the limits compatible with mechanical stability of the gel.

Another consideration in the design of resolving gels for elution-PAGE is gel swelling as a function of electrophoresis time. Since the elution chamber is usually 1 mm high, even a slight amount of gel swelling can occlude it. To counteract swelling, the %C can be increased, but only at the price of loss of wall adherence. Alternatively, replacing the crosslinking agent by another, for example bisacrylamide by DATD, may result in decreased swelling of the gel. Allowing polymerisation of the resolving gel to proceed overnight may also swell the gel to a maximum value prior to experiment. Swelling is not a practical problem, however, at the relatively rapid elution time of 5 h suggested, and at %T values of less than 12. At higher %T it deserves attention.

The polymerisation mixture is poured into the standard tapered female joint (J) opening in the upper buffer reservoir and overlayered with water using a rigid, thin plastic tube (overlayering device) passing freely through a rubber stopper seated in the corresponding male joint, and connected to a proportioning pump delivering water at 0.3-1 ml/min. The flow of water should be as high as possible without disturbing the surface of the polymerisation mixture. Immediately after pouring the polymerisation mixture, coolant flow is initiated at 1-3.5 l/min. The three concentric polymerisation lamp units (specified above) are turned on after 2 mm of water have accumulated above the surface of the polymerisation mixture and overlayering has been stopped. Photopolymerisation is allowed to proceed for 1h. The stacking gel is then polymerised, in the same fashion, on top of the resolving gel.

STEP 3: Removal of the Polymerisation Insert

The manometer tube (M) and the central capillary (C) are opened, and the apparatus, clamped to a ringstand mounted on a lab-jack, is raised above the bench level. The polymerisation insert (P) is slowly withdrawn (unscrewed) with care to avoid exerting an abrupt pressure on the gel which could not be released rapidly enough through the narrow air openings I and C.

STEP 4: Formation of the Elution Chamber

The position of the glass membrane within the lower buffer reservoir should be set prior to the experiment to give an elution chamber of 1mm height. This is done by bringing the glass membrane to the level of the surrounding Perspex edge on which the glass column will rest, and retracting the glass membrane by a half turn in a counterclockwise direction. The top of the glass membrane is covered with elution buffer I to a height of approximately 1cm. The apparatus is lowered into the lower buffer reservoir (filled with lower buffer) by means of the lab-jack. On lowering the gel column into the lower buffer reservoir, care is taken not to entrap air under the gel. Excess elution buffer is allowed to overflow into the drip pan while the elution chamber is being formed. An air-tight connection is made between the gel column and the lower buffer reservoir by the tightening of the four wingscrews located on the top face of the lower buffer reservoir at 90° from one another. To achieve an even tightening, opposite screws are tightened consecutively and gradually. The clamp holding the gel column on the ringstand is released.

STEP 5: Filling of the Elution Chamber

The central capillary (C) is closed, and the apparatus is tilted on the bench so that the manometer tube bottom is higher than the elution buffer inlet ports into the elution chamber. Elution buffer inlet (I) is opened to elution buffer I (*PHASE BETA*) contained in a Mariotte bottle suspended from the same lab-jack-ringstand to which the gel column had been clamped. The level of the Mariotte bottle is raised sufficiently to start elution buffer flow into the elution chamber by gravity. Any remaining air in the elution chamber is progressively displaced through the manometer tube (M) or through an adjacent overflow opening of the elution chamber by the buffer. When the manometer tube is filled with elution buffer, the tilted apparatus is replaced into its vertical position, and the level of the Mariotte bottle is adjusted with the lab-jack until the level of the elution buffer in the manometer tube (M) is at the gel surface. The elution chamber has now been filled, is free of air bubbles, and the gel is hydrostatically equilibrated. The central capillary is then opened.

STEP 6: Filling of the Upper Buffer Reservoir

Upper buffer is pumped or poured (with care not to disrupt the gel surface) into joint (J) while the Mariotte bottle containing elution buffer I is raised to maintain equal levels between the buffer in manometer tube (M) and upper buffer, until the upper buffer reservoir is nearly filled. Recirculation by pump of the upper buffer between reservoir and a 2 l reservoir bottle is not started at this stage.

STEP 7: Circulation of Lower Buffer

Lower buffer is continuously pumped into the lower buffer reservoir and allowed to drain by syphoning. Since the lower buffer in the apparatus model under consideration is in contact with the glass membrane (ME), it should be as highly concentrated as practical (usually three- to ten-fold the lower buffer concentration used at the analytical scale) to obviate pH changes in the elution chamber due to the relative impermeability of the "ion permeable" Corning 7930 glass membrane.

STEP 8: Eluate Flow

The manometer tube is closed, and the tip of the central capillary (C) is connected via tubing to the elution pump. Elution buffer is drawn at 1.2-1.5 ml/min from the elution chamber into the capillary, the line connecting capillary and pump, and then into the pump. When the buffer has reached the pump, the manometer tube is opened and the Mariotte bottle is readjusted in height until the upper buffer and manometer buffer levels are equal while the elution buffer is being pumped.

STEP 9: Sample Application

The sample is underlayered onto the gel surface with a proportioning pump if it is relatively small (10-20 ml). Larger samples are delivered gently onto the gel surface and are subsequently overlayered with upper buffer, using first a proportioning pump with care not to disturb the sample-buffer interface, then a syringe (50 ml). Samples of up to 200 ml can be accommodated. Recirculation of the upper buffer is begun.

STEP 10: Electrophoresis

Because the heat dissipation from gel annuli of 1.5 cm thickness is somewhat less efficient than that from 0.6 cm gels, a regulated current of 40-50 mA/20 cm^2 should be applied instead of the one which would be strictly proportional to gel surface area (74 mA) (19). Since under analytical or preparative conditions voltage should always be as high as possible without exceeding the heat dissipation capacity of the apparatus, the value of the selected current depends on coolant flow rate, choice of operative temperature and apparatus design. The best way to test for excessive electrical heating is to remove and stain a preparative gel for protein. Annular bands should be strictly parallel to the gel surface throughout the gel if heat dissipation is adequate. If not, migration is faster at the centre of the gel than at the cooled periphery. Use of a coloured protein such as haemoglobin for this purpose may be misleading in view of its extraordinary solubility and heat resistance.

When the front edge of the stack, marked by tracking dye, reaches the elution chamber, the elution buffer I (*PHASE BETA*) is replaced by elution buffer II (*PHASE ZETA*). For convenience, and to eliminate the chance of exposing the gel to pressure changes, the two elution buffers are clamped side by side at the same level on the ringstand-lab-jack and are interconnected by a Y-tube so that by merely changing a clamp position one of the elution buffers can be replaced by the other.

140

The preparation of elution buffers I and II requires further discussion. The pH of the elution buffers is set to that of the gel buffer phase they make contact with in the elution chamber, that is initially *PHASE BETA*, until the stack reaches the elution chamber, and subsequently *PHASE ZETA*. However, at *pH (BETA)* or *(ZETA)*, the elution buffer only needs to contain the common constituent, *C6*, not any specific counterion of *C6*, since the counterion does not migrate into the gel. As in the lower buffer, chloride or potassium ions are usually used as the counterions of *C6* in the elution buffers. A second concern in making the elution buffer is to prevent rapid migration of the protein against the glass membrane before it is swept from the elution chamber by the flow of elution buffer. The necessary deceleration of the protein in the elution chamber is achieved by increasing the ionic strength of the elution buffers. Usually, they are made three to five times more concentrated than the gel phase with which they are in contact.

STEP 11: Eluate Detection and Recovery

In view of the load capacity of PAGE, of the order of 0.1 mg/cm^2 of gel, eluates at elution flow rates of 1 ml/min cannot be detected spectrophotometrically in continuous fashion as is customary in chromatography. Decreasing the elution flow rate from 1.0 to 0.1 ml/min in parallel with the decrease of R_f of the eluate helps somewhat to fight dilution (48) but is not sufficient for continuous optical detection of eluate. Re-electrophoresis of 0.1 ml aliquots of eluate is the preferred way to detect the protein of interest in the eluate, unless it can be detected by assay or isotope analysis (2). A large-scale concentration device for preparative elution-PAGE capable of concentrating the protein from 30-100 ml eluate into a 1 ml cup by steady-state stacking (49) has been devised. However, a collection cup design replacing its glass membrane by dialysis membrane (46) appears preferable.

ACKNOWLEDGEMENTS

We thank Drs Larry Hjelmeland and Geeta Kapadia for a critical review of the three manuscripts (Chapters 2-4). Mr A. Godwin (MAPS, DRS, NIH) contributed to the illustrations by expert photographic work.

REFERENCES

1. Chrambach,A. and Rodbard,D. (1971) Science **172**, 440.

2. Chrambach,A., Jovin,T.M., Svendsen,P.J. and Rodbard,D. (1976) *in* Methods of Protein Separation (Catsimpoolas,N., ed.), Plenum Press, New York, Vol.2, p. 27.[3]

3. Chrambach,A. (1978) *in* Electrokinetic Separation Methods (Righetti,P.J., van Oss,C.J. and Vanderhoff,J.W., eds), Elsevier/North Holland Biomedical Press, Amsterdam, p. 275.

[3]Available on request.

4. Ornstein,L. (1964) Ann. N.Y. Acad. Sci. **121**, 321.

5. Chrambach,A. and Rodbard,D. (1972) Sep. Sci. **7**, 663.[3]

6. Newby,A.C., Matthews,G. and Chrambach,A. (1978) Anal. Biochem. **91**, 473.

7. Chrambach,A. (1980) J. Mol. Cell. Biochem. **29**, 23.

8. Muniz,N., Rodbard,D. and Chrambach,A. (1977) Anal. Biochem. **83**, 724.

9. Rodbard,D., Levitov,C. and Chrambach,A. (1972) Sep. Sci. **7**, 705.[3]

10. Baumann,G. and Chrambach,A. (1976) Anal. Biochem. **70**, 32.

11. Wyckoff,M., Rodbard,D. and Chrambach,A. (1977) Anal. Biochem. **78**, 459.

12. Chen,B., Griffith,A., Catsimpoolas,N., Chrambach,A. and Rodbard,D. (1978) Anal. Biochem. **89**, 609.

13. Jovin,T.M. (1973) Biochemistry **12**, 871,879,890.

14. Everaerts,F.M., Beckers,J.L. and Verheggen,T.P.E.M. (1976) Isotachophoresis: Theory, Instrumentation and Applications, Elsevier, Amsterdam.

15. Svendsen,P.J. and Schaefer-Nielsen,C. (1979) *in* Electrophoresis '79 (Radola,B.J., ed.), Walter de Gruyter, Berlin-New York, p. 265.

16. Bui,C., Galea,V. and Chrambach,A. (1977) Anal. Biochem. **81**, 108.

17. Jovin,T.M. (1973) Ann. N.Y. Acad. Sci. **209**, 477.

18. Chidakel,B.E., Ellwein,L.E. and Chrambach,A. (1978) Anal. Biochem. **85**, 316.

19. Kapadia,G. and Chrambach,A. (1972) Anal. Biochem. **48**, 90.

20. Diezel,W., Kopperschläger,G. and Hofmann,E. (1972) Anal. Biochem. **48**, 617.

21. Baumann,G. and Chrambach,A. (1976) Proc. Nat. Acad. Sci. USA **73**, 732.

22. Ben-Or,S. and Chrambach,A. (1981) Arch. Biochem. Biophys. **206**, 308.

23. Dirksen,M.L. and Chrambach,A. (1972) Sep. Sci. **7**, 747.[3]

24. Hjelmeland,L.M. (1980) Proc. Nat. Acad. Sci. USA **77**, 6368.

25. Newby,A.C. and Chrambach,A. (1978) Biochem. J. **177**, 623.

26. Rodbard,D. and Chrambach,A. (1974) *in* Electrophoresis and Isoelectric Focusing on Polyacrylamide Gel (Allen,R.C. and Maurer,H.R., eds), Walter de Gruyter, Berlin, p. 28.[3]

27. Rodbard,D. and Chrambach,A. (1971) Anal. Biochem. **40**, 95.

28. Lang,U., Kahn,C.R. and Chrambach,A. (1980) Endocrinology **106**, 40.

29. Chidakel,B.E., Baumann,G., Rodbard,D. and Chrambach,A. (1975) Anal. Biochem. **66**, 540.

30. Chrambach,A., Reisfeld,R.A., Wyckoff,M. and Zaccari,J. (1967) Anal. Biochem. **20**, 150.

31. a)Tipton,H., Rumen,N.M. and Chrambach,A. (1975) Anal. Biochem. **69**, 323.
    b) Peterson,J.I., Tipton,H.W. and Chrambach,A. (1974) Anal. Biochem. **62**, 274.

32. Kohler,P.O., Bridson,W.E. and Chrambach,A. (1971) J. Clin. Endocrinol. **32**, 70.

33. Magnusson,R.P. and Jackiw,A. (1979) J. Biochem. Biophys. Methods **1**, 65.

34. Rodbard,D. (1976) *in* Methods of Protein Separation (Catsimpoolas,N., ed.), Plenum Press, New York, Vol.2, p. 145.[3]

35. Muniz,N., Rodbard,D. and Chrambach,A. (1977) Anal. Biochem. **83**, 724.

36. Hedrick,J.L. and Smith,A.J. (1968) Arch. Biochem. Biophys. **126**, 154.

37. Rodbard,D. and Chrambach,A. (1970) Proc. Nat. Acad. Sci. USA **65**, 970.

38. Rogol,A.D., Ben-David,M., Sheats,R., Rodbard,D. and Chrambach,A. (1975) Endocr. Res. Comm. **2**, 379.[3]

39. Rodbard,D., Chrambach,A. and Weiss,G.H. (1974) *in* Electrophoresis and Isoelectric Focusing on Polyacrylamide Gel (Allen,R.C. and Maurer,H.R., eds.) Walter de Gruyter, Berlin, p. 62.[3]

40. Nguyen,N.Y. and Chrambach,A. (1979) J. Biochem. Biophys. Methods **1**, 171.

41. Chrambach,A. and Nguyen,N.Y. (1978) *in* Electrokinetic Separation Methods (Righetti,P.J., van Oss,C.J. and Vanderhoff,J.W., eds), Elsevier/North Holland Biomedical Press, Amsterdam, p. 337.[3]

42. Chen,B., Griffith,A., Catsimpoolas,N., Chrambach,A. and Rodbard,D. (1978) Anal. Biochem. **89**, 609.

43. Chen,B., Chrambach,A. and Rodbard,D. (1979) Anal. Biochem. **97**, 120.

44. Rodbard,D., Kapadia,G. and Chrambach,A. (1971) Anal. Biochem. **40**, 135.

45. Kapadia,G., Chrambach,A. and Rodbard,D. (1974) *in* Electrophoresis and Isoelectric Focusing on Polyacrylamide Gel (Allen,R.C. and Maurer,H.R., eds), Walter de Gruyter, Berlin, p. 115.[3]

46. Nguyen,N.Y., DiFonzo,J. and Chrambach,A. (1980) Anal. Biochem. **106**, 78.

47. Jovin,T. Chrambach,A. and Naughton,M.A. (1969) Anal. Biochem. **9**, 351.

48. Ehwein,L.B., Huff,R.W. and Chrambach,A. (1977) Anal. Biochem. **82**, 46.

49. Wachslicht,H. and Chrambach,A. (1978) Anal. Biochem. **84**, 533.

CHAPTER 3
Gel Isotachophoresis

NGA Y.NGUYEN and ANDREAS CHRAMBACH

INTRODUCTION

Isotachophoresis, like electrofocusing, is a "charge fractionation" method, that is, it is applicable to separation problems among molecules which differ predominantly by net charge, not by size. Thus, the justification for applying isotachophoresis to a particular problem rests on the determination of an optimal pore size by Ferguson plot analysis in PAGE (Chapter 2, p.114) and specifically upon a finding of an optimal pore size at 0 %T (p. 132).

The term isotachophoresis (1) is synonymous with steady-state stacking (2) or multiphasic zone electrophoresis (MZE) (ref. 3; Chapter 2). In practice, however, it is convenient to use these terms distinctively: steady-state stacking where fractionation is achieved by confining either the protein of interest in the moving boundary ("selectively stacked") whilst other ("unstacked") proteins remain *outside* the stack, or where the protein of interest is selectively excluded from the stack ("selectively unstacked") whilst the contaminants are confined to the moving boundary: isotachophoresis where separation between the protein of interest and its contaminants occurs *inside* the stack.

To separate two proteins by isotachophoresis it is sufficient simply to stack them. However, since the stack is occupied by proteins in a contiguous fashion, it is not possible to demonstrate separation at the analytical level characteristic of PAGE, that is, in the microgram range. This is true unless one can find small-molecular weight molecules with mobilities intermediate between the two proteins ("spacers"). Since usually one cannot find such individual spacers and since multicomponent spacers in sufficient amounts to furnish the needed individual spacers cannot be separated at the steady-state in the time and migration path length available to experimentation (4,5), we will disregard the use of spacers in this discussion of separation practice. Thus to demonstrate separation in gel isotachophoresis, milligram quantities of protein are required even when standard analytical-size PAGE apparatus is employed. Therefore, even in this format, the isotachophoretic separation of proteins is preparative and limited to those problems where milligrams are available. Where they are not, one must scale down to capillary apparatus in the absence of a gel (1). However, such application presently appears difficult, in view of the lack of specific spacers, of detecting the leading and trailing edges of the stack, and thus of protein zones specifically migrating within the stack, so that we will not discuss the microgram-analytical capillary isotachophoresis of proteins (1) within the practical context of this review.

Steady-state stacking, in the narrow sense defined above, applied preparatively as a charge fractionation method, is also described here.

PROCEDURES

Optimisation of Conditions for Steady-state Stacking

The first step in protein isotachophoresis is to stack the proteins one wants to separate. The rationales and procedures described in Chapter 2 for the optimisation of stacking pH, stacking limits, and other conditions important to maintain native protein stability, for choice of apparatus, and for choice of gel, apply here with only a few additional considerations:

(i)   the pH optimisation in application to isotachophoresis is more demanding than in steady-state stacking, that is it should be pursued to an individual subsystem of the Jovin output (Chapter 2) of minimal pH (in the case of negative systems polarity) or maximal pH (in the case of positive polarity).

(ii) attention should be paid to the steepness of the pH gradient inside the stack, that is, the difference between the stacking limits *RM(2,BETA)* and *RM(1,ZETA)*. For isotachophoretic separations only a small difference in these parameters, and therefore a flat pH gradient, appears optimal in analogy to electrofocusing (6). Buffering capacity within the stack is also best under those circumstances.

(iii) since no supporting resolving gel is present, unlike PAGE, the "non-restrictive" stacking gels in isotachophoresis should possess maximal mechanical stability like electrofocusing gels. Thus, the gels should be crosslinked with DATD (at 0-4°C) (7), cast in Pyrex tubes cleaned with methanolic KOH, coated with Gelamide 250 (Polysciences), and supported by Nylon mesh exactly as for gels in electrofocusing (ref. 7 and Chapter 4). Usually 5 %T, 15 %C_{DATD} gels which are "non-restrictive" for molecules with molecular weights up to 5 x 10^5 (*Figure 1*) are applied to isotachophoresis.

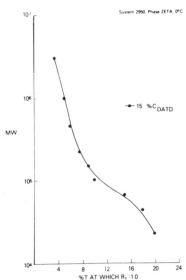

Figure 1. 'Non-restrictiveness' of 15% DATD-crosslinked polyacrylamide gel for proteins of different molecular weight (MW) considering the maximum gel concentration (%T) compatible with an R_f = 1.0 as 'non-restrictive' (Buffer system 2950, *PHASE ZETA*, pH 10.45, *RM(1,ZETA)* = 0.064, 0°C) (7).

Test of Separation within the Stack

Once the proteins to be separated are stacked at optimal pH, in a stack of minimal range of stacking limits, with addition of all the factors required for the maintenance of activity, the physical length of the stack is extended by two measures: an increase in protein load and a decrease in ionic strength. The stack is then excised, sliced and the extent of separation is determined by slice analysis.

(i) *Protein load.* Up to 15 mg protein per 0.6 cm diameter gel is desirable. An increase in the protein load leads to stack extension because protein concentration within the stack is regulated (3,4). Since the duration of electrophoresis and path length to achieve a steady-state separation both increase with protein load (and also with the degree of multiplicity of protein components) long gel tubes are required. In practice, 50 cm long Pyrex tubes of 0.6 cm I.D. and an apparatus of the type described in Chapter 2, *Figure 1,* with a 60 cm tall lower buffer reservoir (*ibid.*, Part A) are optimal.

The high protein load also affects the voltage chosen. At voltages above 100 V across the system (gel and electrolyte buffers) the protein within the stack easily heat-denatures and begins to flocculate. This is presumably due to the binding of water and of salts to the protein at the high protein concentrations (20-100 mg/ml) within the stack (4). If observed immediately, the flocculation frequently appears to be reversible upon lowering the voltage.

(ii) *Ionic strength of the operative stacking gel (PHASE ZETA).* The ionic strength is minimised to a value of 0.0020 to maximally extend the length of the stack (4). This is achieved by diluting the stacking gel buffer (*PHASE BETA*) by a factor required to reduce *ION.STR. (ZETA) (Figure 4* of Chapter 2) to the desired value of 0.0020. Reduction of ionic strength below that value appears to lead to a disruption of the steady-state (e.g. sharp zone boundaries), presumably since the buffering and current-carrying capacity of the protein then start to become significant relative to that of the buffer constituents (4).

(iii) *Electrophoresis.* Electrophoresis is conducted as in steady-state stacking except that the end point at the steady-state is determined by intermittent measurement of stack length in the tube while electrophoresis proceeds. Stack length is readily visible because at the high protein concentrations within the stack the protein itself appears yellowish, because of the razor-sharp edges of the "extended stack", and because tracking dye components frequently mark these edges, in particular the leading edge. When the extended stack has reached constant length, it is assumed that the steady-state is reached and electrophoresis is discontinued.

(iv) *Slicing of the extended stack.* The gel tube is cut a few centimetres above and below the extended stack, and the gel is rimmed out of the tube in the manner described in Chapters 1 and 2. The gel is then sliced into 1 mm slice. The wire slicer with electric vibrator (Hoefer Scientific) with a low vibration frequency appears optimal for the slicing of DATD-crosslinked gels (8). The slices are analysed for protein patterns by applying either fragments of the slices, or aliquots of slice eluates, or both, to analytical PAGE. A vertical slab gel apparatus appears optimal for that purpose. A representative analysis of an extended stack (0.6 cm gel) containing various charge isomeric isohormones of human growth hormone (4,9) is shown in *Figure 2*. Separation between species appears gradual, indicating that components within the stack are separated from one another as overlapping distributions (10). Nonetheless, the major species B is extremely sharply separated from C between sections 15 and 14, thus providing a homogeneous pool of B between sections 15 and 24. Similarly, fraction C is clearly separated from D between gel slices 10 and 9. However, C is not quantitatively separated from B in any fraction. Component B being the predominant species of the mixture, the width of its distribution overlaps with that of nearly all the other isohormones except species E (sections 2-5, not identified in *Figure 2*).

Preparative Isotachophoresis by Gel Slice Extraction

The point has already been made that the test of separation within the stack must be made at the milligram level of proteins, that is, on a preparative scale. However, isotachophoresis can cope with even larger preparative loads by using larger diameter gels. In practice, the limit of gel diameter still compatible with reasonable rates of electrophoresis and within the heat dissipation capacity of the apparatus is 1.8 cm (9).

Electrophoresis Separation and Gel Slicing

The procedure of electrophoresis of 1.8 cm diameter stacking gels is exactly the same as described in the section above for 0.6 cm diameter gels, except that usually they have to be run at somewhat lower voltage than the narrower gels to prevent the flocculation of proteins within the extended stack. The protein load is proportional to the surface area of gel; thus a 1.8 cm gel may be loaded with one order of magnitude more protein than a 0.6 cm gel, typically about 150 mg (9). Since the upper buffer reservoir described in Chapter 2 (*Figure 1* Part C) can hold six 1.8 cm diameter tubes, the procedure is capable of fractionating lg of protein per run and hence a nearly

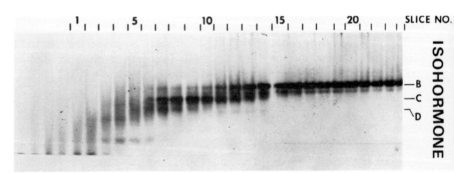

Figure 2. Analysis of protein separation achieved by isotachophoresis by re-electrophoresis according to PAGE. Isotachophoresis of 35 mg human growth hormone was carried out on a 50 cm long gel, 0.6 cm diameter, (5 %T, 15 %C$_{DATD}$) with MZE buffer system 1954.4 at $0-4°C$ (4).

industrial-scale separation can be carried out using a small, ordinary PAGE apparatus!

The sectioning of the wide diameter gel tubes and of the gel requires special procedures and apparatus. To cut the tube a few centimetres above and below the stack, a scratch is made around the circumference of the glass tube using a triangular file. Then a glowing Pyrex rod, preheated in a gas/oxygen flame, is inserted into the scratch to crack the tube. Slicing of the gel containing the stack into 1-2 mm slices is best carried out with a wire slicer (*Figure 3*) which is a scaled-up version of one described previously (11).

The only item in the procedure different from that used to test separation within the stack (p.147) concerns the recovery of the protein.

Procedure for Protein Recovery

STEP 1: Slice analysis and pooling. Gel slices are suspended in buffer and 5 μl aliquots of the diffusates are analysed by either gel electrophoresis or electrofocusing while the remaining diffusates and slices are stored at 4°C. Slices are pooled accor-

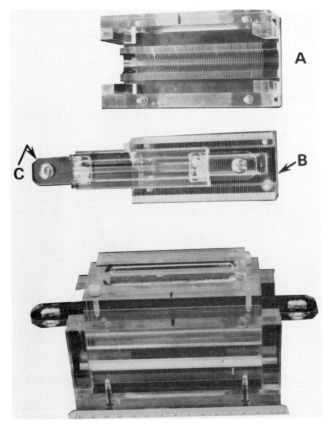

Figure 3. Wire slicer for isotachophoresis gels of 1.8 cm diameter. The design is that reported in ref. 11 (presently manufactured by MRA Corporation) adapted for gels of the wider diameter. A) slotted gel holder; B) wire grid; C) slice holding tray.

ding to the analytical gel patterns. For electrophoretic extraction of protein from the gel slices, and to simultaneously concentrate the extract, the pooled gel slices are applied to funnel-shaped steady-state stacking gels in an apparatus designed to collect the stacked proteins in a buffer-containing cup (ref. 12; see *STEP 3* below).

STEP 2: Choice of the steady-state stacking gel for extraction and concentration of protein from gel slices. A steady-state stacking system is selected as described earlier and in ref. 13 to stack the protein of interest at a pH and ionic strength at which it is stable. Since no purification is intended at this stage, stacking limits are selected by a different criterion: to provide a rapid boundary displacement rate, that is the highest values of *RM(2,BETA)* and *RM(1,ZETA)* compatible with the stacking of the protein, at the maximum pH at which it is stable.

STEP 3: Assembly of the apparatus (12). First, concentration funnels are tested to ensure that they do not leak. Circular dialysis membrane (0.002 inch thick, 1.5 inch diameter) is wetted and stretched across the bottom of the collection cup using the plastic ring as a sleeve *(Figure 4)*. The funnel is then inserted vertically into the collection cup which rests inverted on a piece of filter paper. The funnel is filled with water and allowed to stand for 1h. If after that time the filter paper has remained dry, the collection cup is considered to be watertight. The cup and funnel are then gently separated.

Next, the concentration funnels are coated with agarose. A 1% (w/v) solution of agarose (with zero electroendosmosis, e.g. Iso-Gel, Marine Colloids) is prepared in a flask fitted with an air condenser and brought to 80°C in a boiling water bath. The

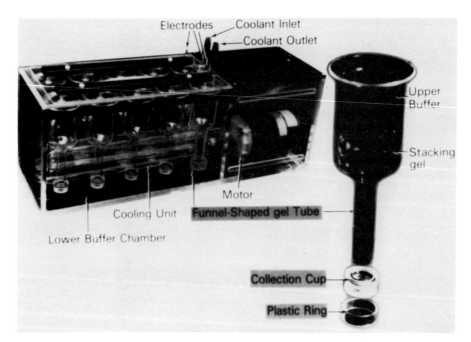

Figure 4. Apparatus for the extraction and concentration of protein from gel slices (12).

solution is dispensed into Pyrex concentration funnels previously heated in an oven at 60°C, and then allowed to drain and dry onto the walls of the funnels at 60°C for at least 3h. A stacking gel of 0.75% agarose in the selected gel buffer is then formed in the appropriate number of concentration funnels of the apparatus depicted in *Figure 4* as follows. A 1% agarose solution, prepared as described above, is mixed in a ratio of 3:1 by volume with stacking gel buffer preheated to 60°C. The mixture is poured into coated concentration funnels (sealed at the bottom with Parafilm) at room temperature and allowed to gel for 1-2h. The gel volume should be at least three times the sample volume. In practice, it is usually sufficient to form the gel up to a level 2 cm above the stem of the funnel. The agarose gel is subsequently overlayered with stacking gel buffer and allowed to stand overnight.

Recent experience has shown that agarose may interfere with the operation of some MZE buffer systems and so it may be necessary to carry out the slice extraction and protein concentration by steady-state stacking using a polyacrylamide matrix instead. The coating of the concentration funnels and the choice of gel concentration then follows the rationales and procedures given under section (iii) in the procedure for Optimisation of Conditions for Steady-state Stacking on p. 146.

After formation of either agarose or polyacrylamide gels, the Parafilm is removed and the funnels containing the gels are then suspended in the lower buffer (0-4°C) of the apparatus (*Figure 4*).

STEP 4: Electrophoretic extraction and concentration of protein from gel slices by steady-state stacking (9). Tracking dye is added to the sample consisting of pooled gel slices and buffer diffusate in 10% sucrose. The platinum wires of the upper electrodes (*Figure 4*) are bent so as to terminate about 1 cm above the samples. The samples are then overlayered carefully without disturbing the interface between upper buffer and sample. This can be done either manually with a syringe, or using a multichannel proportioning pump delivering buffer simultaneously to several funnels. When the upper buffer level reaches the electrode, the power is turned on whilst overlayering is still in progress. The funnel-shaped reservoirs should eventually be filled with upper buffer. Without attaching the collection cup, electrophoresis is started at 6-7 mA/cm^2 of gel. The collection cup is attached when the stack of extracted and concentrated protein, marked by the tracking dye, reaches 0.5 cm from the bottom of the funnel. Care must be taken to push the stem of the inverted funnel into the cup at a right angle. Electrophoresis is then continued at a reduced rate, usually at a regulated current of 3.0-3.5 mA/cm^2 of gel, giving a voltage of 50V across the system. After the stack has entered the elution cup, electrophoresis is continued for 2-3 h. The cup contents (1 ml volume with 1.3 cm I.D. 1.5 cm O.D. stems and correspondingly less with 0.6 cm stems) are collected using a Pasteur pipette and subjected to purification by gel filtration. To collect a second batch, electrophoresis may be continued overnight at the low regulated current.

STEP 5: Purification of the concentrate from impurities and exchange to volatile buffer. Sephadex G-50 (medium) is freed of soluble dextran by extraction in boiling water, or, preferably, in continuous fashion with a Soxhlet apparatus at 100°C for 24 h (6). It is then suspended in 0.02 M NH_4HCO_3 (pH 8.0) made from NH_4OH and CO_2 gas (to be completely lyophilizable). Next 60 cm x 0.9 cm columns are poured, equilibrated, and the void volumes are determined using Dextran Blue. The concentrates, not exceeding 1-2 ml in volume, are applied to these columns. The protein is eluted within the void volume and is lyophilized.

The procedure of isolation from gel slices after isotachophoresis differs from that applied to gel slices after PAGE or electrofocusing in two respects:

(i) load capacity in isotachophoresis is so high that slice extraction and concentration by steady-state stacking can be carried out without concern about protein adsorption onto the gel, in spite of the large gel volume (10-30 ml). Losses at the level of steady-state stacking due to adsorption increase to 20-30% only when the load per funnel of 1.32 cm^2 surface area (in the stem) is reduced to a few milligrams of protein (14). These adsorption losses can of course be remedied by reduction in gel diameter and volume.

(ii) zones in isotachophoresis are highly concentrated but are not precipitated, in contrast to the isoelectric zones in electrofocusing. Thus they do not have to be solubilised prior to extraction.

This procedure of protein isolation from isotachophoresis gel slices has yielded, in a representative case (9), 70% recovery and 90% purity of product.

Preparative Isotachophoresis by Successive Zone Elution

Apparatus

Many forms of preparative elution PAGE apparatus have been described in the literature (15). Many of them are workable if consideration is given to the physical principles involved and to buffer characteristics. The apparatus which has been used in this laboratory is the Buchler Polyprep 200 (20 cm^2 cross-sectional area of gel, Buchler Scientific Instruments).

In view of the high protein concentration within the stack (20-100 mg/ml) this apparatus (*Figure 24,* Chapter 2) which has a milligram-preparative scale in PAGE, has a gram-preparative, that is pilot-plant industrial-scale, for isotachophoresis (4). Thus, unless very large amounts of the protein of interest are available, the cross-sectional area of the gel must be reduced. A gel column of only 3 mm diameter has been used to separate milligram amounts of haemoglobins by isotachophoresis (16). Although apparatus of such dimensions is not widely available, it is easily assembled using a crossflow-cell attached to the bottom of an analytical-scale gel tube (Hoefer Instruments or Savant Instruments).

Methodology

The procedure of elution isotachophoresis follows essentially that for preparative PAGE by successive zone elution previously described (Chapter 2, p. 136; Appendix III of ref. 14), except that only one gel, the stacking gel, is used. Reference 4 reports the application of this procedure to the isotachophoretic separation of model proteins and of human growth hormone isohormones.

A gel of greatest possible length (200 ml in the Polyprep 200 apparatus) is prepared at the optimum pH, optimum stacking limits, an ionic strength of 0.0020, and optimum concentrations of cofactors (see p 146). The gel is prepared by the same procedure detailed for analytical-scale stacking gels, except that the temperature of the three stock solutions (monomers, stacking gel buffer and initiators; see *Figure 8* of Chapter 2) is more rigidly controlled at the temperature of polymerisation, usually 0-2°C. Protein concentrations of 20-100 mg/ml in the stack are characterised by swelling in the region of the stack, presumably through osmotic water uptake. Therefore an elastic DATD-crosslinked gel is used (usually 5 %T, 15 %C_{DATD}) which is capable of maintaining wall adherence during and after passage of the swollen stack through the gel.

Electrophoresis is allowed to proceed at 100-150 V (regulated), whilst inspecting the stack for protein flocculation. If a heat precipitate should appear anywhere in the stack, the voltage is lowered sufficiently for it to disappear.

In the case of preparative elution isotachophoresis, the analysis of the eluate is only possible by either analytical PAGE analysis of eluate fractions or by specific assay of these fractions. However, in contrast to elution-PAGE, the gel patterns of the eluate analyses do not represent the separated homogeneous charge isomeric proteins of the preparative step directly, but also the various size isomeric forms within each of the separated isotachophoretic zones. Thus, the PAGE pattern of a homogeneous charge isomer eluted from a preparative isotachophoresis gel appears heterogeneous due to the various oligomeric forms of a protein. This is illustrated in the elution pattern of a mixture of 2g each of ovalbumin and bovine serum albumin (4) (*Figure 5*). These two proteins, each visualised by PAGE, are found to separate not across a razor-sharp zone as one would predict from the appearance of their isotachophoretic patterns or from theory (3) but across a relatively narrow distribution overlap, in agreement with analytical studies of isotachophoresis (10). Nonetheless, this mixed zone represents not more than about 10% of the zone length, which is negligible in relation to the enormous protein load of the system.

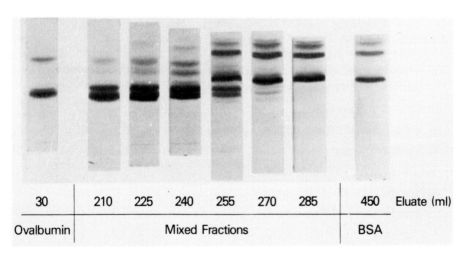

30	210	225	240	255	270	285	450	Eluate (ml)
Ovalbumin			Mixed Fractions				BSA	

Figure 5. Analysis of the eluate from a preparative elution isotachophoresis experiment by PAGE. A mixture of 2g BSA and 2g ovalbumin was loaded onto a 200 ml gel column (5 %T, 15 %C_{DATD}) and separated using MZE buffer system 2365 at $0-4$°C with an elution rate of 1.5 ml/min (4).

Preparative Steady-state Stacking by Successive Zone Elution

Steady-state stacking may be applied preparatively by selective stacking of contaminants more mobile than the protein of interest, elution of the stack containing the contaminants in an elution PAGE apparatus (see above) and re-stacking of the unstacked protein of interest into a concentrated zone (a second stack), which is eluted after the first stack (17). The procedure of preparative re-stacking is identical to that of preparative PAGE described in Chapter 2, except that two upper buffers are used in succession (*PHASES ZETA* and *TAU*), and that the corresponding elution buffers consist of *PHASES PI* and *PSI* (*Figure 7*, Chapter 2).

The choice of a secondary moving boundary, destined to catch up with the migration of the unstacked protein, is made from the table of *RESTACKING PARAMETERS* on page 3 of the Jovin output (*Figure 7*, Chapter 2). The lower stacking limit here is designated as *RM(7, PSI)* and selected to allow the protein of interest to re-stack in this boundary. Experimentally, this selection follows the procedure given above for the selective stacking of a protein. The protein is allowed to unstack in the buffer system under consideration, either by choice of stacking limits or by molecular sieving or both. After some time of electrophoresis (determined experimentally), the upper buffer is replaced by one of the composition of *PHASE TAU* corresponding to the selected restacking system. This new upper buffer will set a secondary moving boundary in motion to restack the protein and elute it in concentrated form as a stack.

It should be noted that the above procedure is applicable only when no unstacked contaminant proteins are present which would restack with the protein of interest (17). Steady-state stacking could also be applied preparatively in the opposite sense by stacking the protein of interest selectively while unstacking contaminants, but such a case has not yet been reported.

REFERENCES

1. Everaerts,F.M., Beckers,J.L. and Verheggen,T.P.E.M. (1976) Isotachophoresis: Theory, Instrumentation and Applications, Elsevier, Amsterdam.

2. Ornstein,L. (1964) Ann. N.Y. Acad. Sci. **121**, 321.

3. Jovin,T.M. (1973) Biochemistry **12**, 871, 879, 890.

4. Baumann,G. and Chrambach,A. (1976) Proc. Nat. Acad. Sci. U.S.A. **73**, 732.

5. Nguyen,N.Y. and Chrambach,A. (1978) Anal. Biochem. **94**, 202.

6. Nguyen,N.Y. and Chrambach,A. (1980) Electrophoresis **1**, 14.

7. Baumann,G. and Chrambach,A. (1976) Anal. Biochem. **70**, 32.

8. Chrambach,A., Hjelmeland,L., Nguyen,N.Y. and An der Lan,B. (1980) *in* Electrophoresis '79 (Radola, B.J., ed.), Walter de Gruyter, Berlin-New York, p . 3 .[1]

[1]Available on request.

9. Nguyen,N.Y., Baumann,G., Arbegast,D., Grindeland,R. and Chrambach,A. (1981) Prep. Biochem., in press.[1]

10. Chen,B., Rodbard,D. and Chrambach,A. (1978) Anal. Biochem. **891**, 596.

11. Chrambach,A. (1966) Anal. Biochem. **15**, 544.

12. Nguyen,N.Y., DiFonzo,J. and Chrambach,A. (1980) Anal. Biochem. **106**, 78.

13. Chrambach,A. (1980) J. Mol. and Cell. Biochem. **29**, 23.

14. Kapadia,G. and Chrambach,A. (1972) Anal. Biochem. **48**, 90.

15. Chrambach,A. and Nguyen,N.Y. (1978) *in* Electrokinetic Separation Methods (Righetti,P.J., Van Oss,C.J. and Vanderhoff,J.W., eds), Elsevier, Amsterdam, p. 337.[1]

16. Houghten,R. and Chrambach,A. (1977) Anal. Biochem. **77**, 303.

17. Kapadia,G., Vaitukaitis,J. and Chrambach,A. (1981) Prep. Biochem. **11**, No. 1.[1]

CHAPTER 4

Analytical and Preparative Gel Electrofocusing

BIRGIT AN DER LAN and ANDREAS CHRAMBACH

WHEN TO USE GEL ELECTROFOCUSING

Electrofocusing, just as isotachophoresis (Chapter 3), is a method for the resolution of molecules which differ preponderantly in their net charge ("charge isomers"): the proteins are allowed to migrate in a pH gradient and are separated at their isoelectric points (pI's). As described in Chapter 2, there are also cases other than pure "charge isomerism" where separation solely by molecular net charge is again the most efficient method. In such cases, the proteins to be separated differ in both size and net charge but in such a fashion that the resolution at %T = 0 is better than at a finite gel concentration (*Figure 22* of Chapter 2).

There are four arguments for using electrofocusing rather than isotachophoresis in such cases:

(i) in contrast to isotachophoresis, separated protein bands are not contiguous in electrofocusing, and can therefore be visualised by staining without a second dimension or fractionation step.

(ii) using isotachophoresis, the more components the system contains, the longer the time and migration path have to be to attain the steady-state. In electrofocusing, the pathlength required to reach the steady-state is independent of, and the time required to reach the steady-state is less sharply dependent on, the number and amount of proteins applied to the gel. Presumably this is because focusing allows the use of higher voltage gradients than in isotachophoresis, once carrier constituents have reached their steady-state. This again is a consequence of the low current at the steady-state, and resulting low electrical heating at equivalent voltages.

(iii) no prior knowledge nor systematic determination of the charge properties of the proteins being studied are required.

(iv) the practice of electrofocusing requires no specialised knowledge of the underlying physico-chemical principles.

However, electrofocusing also possesses inherent problems of which isotachophoresis is free:

(i) proteins tend to precipitate at or close to their isoelectric points. In liquid media this causes sedimentation, and in gels must at least retard the approach toward the isoelectric position. The insolubility of proteins at their isoelectric points also makes isolation from gels more difficult than in isotachophoresis.

(ii) uneven distribution of carrier constituents gives rise to regions of increased electrical heating. Variations in the voltage gradient from one point to another in the pH gradient also preclude the use of electrofocusing data for physico-chemical analyses of proteins.

(iii) in electrofocusing, at the "steady-state" the ionic strength is generally at least one order of magnitude lower than in isotachophoresis. This lower ionic strength favours artifactual binding by the formation of ionic bonds either between the proteins and carrier constituents, between proteins, or between proteins and other macromolecules such as nucleic acids. Such interactions can modify the apparent pI of proteins.

(iv) probably because higher voltage gradients are involved, the "steady-state" of electrofocusing decays faster than that of isotachophoresis.

(v) the rate at which proteins approach their isoelectric state depends on the shape of their titration curves. To the degree that proteins exhibit flat titration curves close to their pI, they are not able to reach their isoelectric positions on the pH gradient.

However, in general, the advantages of electrofocusing as compared to isotachophoresis outweigh the disadvantages, in particular for analytical purposes.

APPARATUS

Gel Tube Versus Horizontal Slab Apparatus

Monitoring electrofocusing as a function of time. Electrofocusing is electrophoresis into a "natural" pH gradient, that is one that is generated in the electric field rather than by a gradient maker. Such pH gradients are not static; as soon as they are generated they begin to decay without a steady-state being established. In view of such pH gradient "dynamics" (see p.170), separation between proteins also changes continuously during the time of electrofocusing. Consequently, optimal separation can only be determined by monitoring gels at various time intervals. A multiple gel tube apparatus allows protein patterns and pH gradients of individual gels to be analysed at different times without disturbing the other tubes. In contrast, the dynamics of both pH gradients and protein patterns cannot be readily monitored using either a horizontal or a vertical gel slab apparatus (1). However, once the optimum time has been determined for a particular pH gradient and protein separation, the gel slab apparatus is convenient for the purpose of comparing protein patterns amongst different samples, assuming that sufficiently concentrated protein solutions are available to allow one to use small sample volumes (less than 100μl). To date, horizontal slabs are nearly always used for this purpose; vertical slabs still present difficulties in electrofocusing with regard to the uniformity and stability of pH gradients.

Apparatus materials. Until recently, a problem of tube apparatus was the instability of plastics and plastic bonding in the presence of the strongly acidic anolytes and strongly basic catholytes commonly used for electrofocusing, forcing one to build such apparatus entirely of glass (see Chapter 2 and ref. 1). Since the advent of weakly acidic and basic anolytes and catholytes, this is a moot point, although the stability of glass, its freedom from bonding problems, leakages, and from clouding or cracking upon prolonged use, makes glass construction desirable even now. A suitable gel tube apparatus for electrofocusing is shown in *Figure 1* (see also *Figure 1* of Chapter 2).

Figure 1. Gel tube apparatus. For details, see *Figure 1* of Chapter 2. Technical drawings (A to D in *Figure 1* of Chapter 2) are available on request.

Mechanical stability of the gels. In electrofocusing, gels swell in a pH dependent fashion, and thus the wall adherence problems are exacerbated as compared to isotachophoresis. Although this does not present a problem in the horizontal slab apparatus, it does in the tube apparatus. However, it can be remedied by the same manipulations used for stacking gels (Chapters 2,3), namely, hydrostatic equilibration of the gels, equalisation of polymerisation temperatures with those of electrophoresis, mechanical support of the gels, maximising the polarity of the glass surface with a strong alkali wash, coating gel tubes with linear polyacrylamide, and the use of gels with good elastic and wall-adherence properties. If these measures are used, wall adherence is adequate for electrofocusing separations. Wall adherence is best tested by reversing the positions of anolyte and catholyte in the apparatus; if wall adherence is maintained, the shape and span of the gradient are not affected (see below and ref.2).

Heat dissipation. In electrofocusing, resolution improves as the voltage is increased. However, the price of increasing the voltage is to cause more electrical heating. Thus, the efficiency of resolution depends on the capacity of the apparatus to dissipate heat. This is of particular importance when one considers uneven conductance across the gradient. Although the overall current may be low once the gradient has been formed, local heating must occur in the regions of conductance maxima (see p. 174) which have been shown to form in every type of pH gradient after an electrofocusing time characteristic of a given set of carrier constituents (3,4). Therefore, the better the heat transfer capacity of an apparatus the less this presents a problem. In this regard, the tube apparatus is more efficient than the available types of horizontal apparatus because the gel tubes are uniformly surrounded by agitated liquid at a controlled temperature, and the gel is separated from the liquid by only 1 mm of glass. Heat dissipation from horizontal slabs should be better than from gel tubes because very thin gels can be used and thus the ratio of surface area to mass of gel can be increased. However, heat can only be dissipated across one surface of most horizontal slabs. In such an apparatus the gel is always separated from the cooled surface by several layers of material with different heat transfer characteristics, such as aluminium/water/glass, or even plastic/air/plastic in some badly designed models. The most efficient way to cool horizontal slabs is with Peltier cooling coupled to the recently developed beryllium oxide plates (MRA Corporation), separated from the gel by a thin layer of plastic ("Gel bond" from Marine Colloids Div. FMC Corporation).

Miscellaneous problems of slab apparatus. Some of the problems inherent in horizontal slabs do not exist for gel tubes:

(i) CO_2 is absorbed across the surface of the slab open to air, which may affect pH;

(ii)   moisture is lost from the surface because it is difficult to maintain a water-saturated atmosphere over the gel at all times; partial dehydration increases the voltage locally, causing band distortion; it also alters the gel structure, leading to either channelling or loss of permeability;

(iii) electrode and wick contact with the gel may be uneven, resulting in an uneven field strength across the gel; in addition, some areas of the wicks may dry out;

(iv) sample volume capacity is rather limited, compared to gel tubes, since the sample is applied either in narrow slits in the gel or by adsorption onto strips of paper, whereas a 0.6 cm diameter gel tube can be readily loaded with 0.5-1.0 ml of sample;

(v)   unless the horizontal alignment is perfect, syphoning occurs across the gel between the anolyte and catholyte reservoirs;

(vi) in many models the electrolyte reservoirs are too small to be used with relatively insoluble and/or weak electrolytes (see below).

With skill and ingenuity these difficulties with the horizontal slab technique can probably be overcome, but they are reflected in most patterns obtained in horizontal gels by uneven band migration distances in the various channels across the slab and by band asymmetry.

Voltage Control Devices

Voltage control across the gel itself, rather than across electrolytes and gel, becomes important as soon as one uses the terminal carrier constituents of the electrofocusing gel, that is weak electrolytes, as anolyte and catholyte (for reasons described below). Providing one uses several identical gels, the voltage can be monitored using two shielded platinum wires inserted into the upper buffer reservoir and making contact with the top and bottom respectively of one gel (*Figure 2*). The lower wire is inserted into the neighbouring grommet of this gel tube and is angled so as to make contact with the bottom of the gel; the top wire extends into the tube itself through a hole in the electrode holder and is positioned vertically with a Nylon screw on this holder (5). A power supply which has been modified so as to regulate the voltage delivered across the gel, and can thus compensate automatically for changes in the conductance of the reservoirs during electrofocusing, has been designed in the authors' laboratory (An der Lan,B., Chidakel,B.E. and Chrambach,A., in preparation[1]).

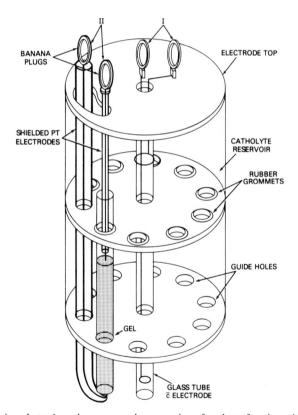

Figure 2. Monitoring electrode and power supply connections for electrofocusing with weakly acidic anolytes and weakly basic catholytes. Electrodes I and II are connected to a power supply modified to provide voltage regulation across the gels independently of the voltage drop across the electrolyte reservoirs. Only electrodes II are connected to the power supply if it has not been modified. In that case, monitoring electrodes I are connected to an external voltmeter.

[1]Available on request.

Gradient Monitoring Devices

Determination of pH gradients requires no special apparatus. One can slice the gel with a razor blade into 8 or 16 sections either by eye, or using a ruler as a guide. After soaking the sections for 30 min in degassed dilute (e.g. 0.025 M) KCl solution, one can take manual pH measurements. Corrections of pH for carrier ampholyte concentration, temperature, viscosity, and CO_2 absorption are usually neglected, although these factors may cause errors of hundredths of a pH unit. For most purposes, such neglect is justified since, in practice, the aims of electrofocusing are that proteins attain a constant pH, and are optimally resolved, rather than an accurate physicochemical determination of their isoelectric points.

When it is important to determine pH gradients of gels which are later to be stained or assayed, or when several pH gradients have to be determined at frequent time intervals, it is convenient to measure the pH gradients directly on the gels using a contact electrode (DESAGA No. 122060). This is best done manually, allowing enough time at each position for the pH to assume a steady value. A semi-automated pH gradient measuring device (*Figure 3*; also see ref. 6) is commercially available (Hoefer Scientific) but should not be used without attention to the following:

(i) the response time of the Ingold contact electrode used in the device is slow, requiring a minimum of 9 min scanning time per 4.5 cm gel (at present, the Hoefer in-

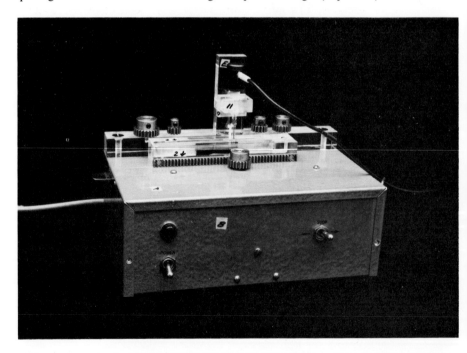

Figure 3. pH gradient measuring device for cylindrical electrofocusing gels. The gel is placed into the carriage (Part 2). A contact combination pH electrode (Ingold Electrodes Inc.) is fastened into electrode holder 11. As the carriage moves past the stationary electrode, the pH gradient on the gel is automatically recorded. The scanning speed should be no more than 0.5 cm/min. For details and definition of other part numbers see reference 6.

strument scans much faster); the time has to be even longer with old electrodes or electrodes clogged with polyacrylamide or protein at the membrane surface. Contact electrodes should therefore be cleaned after each use. A cleaning solution containing NaOCl is available from Ingold Instrumentation Laboratory, but is not always effective. In fact, no single cleaning procedure (e.g. wiping, 3 M KCl, 1 M HCl) has proved effective in prolonged use of this electrode.

(ii) the pH values obtained by the measuring device for the beginning and end of the gradient may be inaccurate, also because of the slow electrode response. To overcome this, the electrode should be positioned, before scanning, over one end of the gel and allowed to equilibrate. At the end of the scan, the electrode should again be allowed to remain in position over the other end of the gel until the pH reading is steady.

(iii) when pH-dependent gel swelling introduces significant differences in diameter along the gel, the contact electrode has to be guided into the troughs along the gel surface by manual adjustment of its height.

To date, the performance of the instrument has shown it to be labour-saving and useful for approximate pH determinations, accurate to within a few tenths of a pH unit. For exact determinations, the values derived from automated scans should be checked by manual pH measurement.

In contrast to pH gradient determination, the determination of protein activity profiles along the pH gradient does require a special apparatus, namely a transverse gel slicer capable of providing sequential 1 mm thick gel slices reproducibly. Slicing of electrofocusing gels is more difficult than slicing most gels (Chapters 1 and 2) because of the need in electrofocusing for elastic, open-pore gels typically of the 5 %T, 15 %C_{DATD} type (see the section below and ref. 7). At this time, no commercially available transverse gel slicer is able to section these gels with low enough variance (8,9). However, two recent modifications of a marketed vibrating wire slicer (Hoefer Instruments or Bio-Rad Laboratories Ltd.) allow one to reproducibly section 6 %T, and, with some care, even 5 %T, 15 %C_{DATD} gels: one decreases the vibration frequency of the sectioning wires with a rheostat, and uses a slotted grid on which the gel rests and which allows the wires to pass below the plane of the gel, thus ensuring that these elastic gels are completely sectioned (5). We have also found that a 0.5 x 10cm bed of 1-2% agarose, a few millimetres thick, is an effective alternative to the slotted grid.

GEL

Electrofocusing, like isotachophoresis (Chapter 3), requires a "non-restrictive", anticonvective gel. The selection of a suitable gel is, however, more demanding than in isotachophoresis for three reasons:

(i) migration velocities decrease as proteins approach their pI; thus any molecular sieving effect will not only retard the migration of the protein but may appear to halt it altogether, giving rise to a fallacious pI.

(ii) gels swell to a varying extent along the pH gradient, making the wall adherence problem more severe.

(iii) isoelectric precipitation at the steady-state requires maximum pore sizes to allow for preparative solubilisation and extraction from the gel in the least possible time.

Polyacrylamide Gel

The most successful compromise solution for the three conflicting requirements described above is a 5 %T, 15 %C$_{DATD}$ gel (7), or, in case exact transverse gel slicing is required to locate a protein activity, a 6 %T gel with the same crosslinking. These gels are more elastic and have better wall-adherence properties than any other presently known gel type. As shown in *Figure 1* of Chapter 3, these gels are not as "non-restrictive" as one may wish; a 6 %T gel will retard molecules with molecular weights in excess of 100,000 and a 5 %T those in excess of 500,000. The criteria of stacking on the basis of which this evaluation of "non-restrictiveness" was made [pH 10.5, *RM(1,9)* = 0.064; see Chapter 2] are, however, probably not stringent enough for electrofocusing. As pointed out above, retardation appears to become more effective as the migration rates decrease on the approach towards the pI, whereas it remains constant at the constant migration velocity in isotachophoresis.

Wall adherence can be easily tested using a tube apparatus as described on p.159 that is by a polarity reversal of anolyte and catholyte; if wall adherence is inadequate, the pH gradient shifts towards the pH of the lower electrolyte in which the gel is immersed (2).

Sephadex Gel

In cases where polyacrylamide is noxious to protein activities or where very large proteins are being fractionated, electrofocusing can be carried out on Sephadex G-75 (Ultrodex, LKB) gel cylinders. The Sephadex is packed into glass tubes between plugs of polyacrylamide (ref. 10 and An der Lan,B., Allenmark,S., Fitze,P. and Jackiw,A., in preparation[1]). However, these gels have the disadvantage, compared to polyacrylamide gels, of being unsuitable for gel slicing by the usual procedures. At present, such gels have to be sliced by hand on a cold plate after they have been frozen and extruded from the gel tube. An automatic guillotine-type of gel slicer (e.g. a Mickle gel slicer from Joyce-Loebl Ltd.) could also be used but it would be necessary to use a refrigerant other than solid carbon dioxide, since absorption of this gas into the gel can alter the pH of the slices.

Agarose Gel

In spite of the popularity of agarose for horizontal slab electrofocusing, and its usefulness for isotachophoresis, at least in some buffer systems (Chapter 3), its use for electrofocusing in gel tubes presents some problems, even when wall adherence is increased by coating the tubes with agarose prior to gelation. After some hours of electrofocusing at 10 - 20V/cm, the wall adherence starts to deteriorate, ultimately to the extent that the gels fall out of the tubes. This occurs even when using agarose which contains no measurable charged (sulphate) groups ("IsoGel", Marine Colloids Div. FMC Corporation), indicating that this problem does not derive from charge effects. Another possible cause of the mechanical instability of agarose in electrofocusing is pH-dependent swelling and/or shrinkage.

An additional problem is that pH gradients formed in agarose gels are less stable than those formed in polyacrylamide, particularly at higher %T. Gradient instability may therefore be related to the larger pore size of agarose. Alternatively, the postulated isotachophoretic nature of pH gradient formation (15) may be the cause of this relative instability. Some support for the latter idea comes from the observation that it is not possible to stack proteins in some buffer systems if agarose is used as a matrix (Chapter 3).

FORMATION OF pH GRADIENTS

Originally, electrofocusing studies were carried out using synthetic carrier ampholytes to generate the pH gradient. The finding that natural pH gradients can be generated from mixtures of ordinary buffers with different pK's has considerably increased the freedom in designing pH gradients for achieving optimal resolution (5). Several options are now available:

(i) flat pH gradients, narrowly encompassing the pI's of the species to be resolved, can be prepared by "buffer electrofocusing" (11).

(ii) carrier constituents with pI's outside the desired pH range can be displaced into the catholyte/anolyte by choosing suitable pH's for these electrolytes, that is by "constituent displacement" (12).

(iii) pH gradients can be shifted by controlling the pH of the anolyte (13).

"Ampholine" Versus Buffers

At the start of an electrofocusing fractionation, it is always advantageous to select carrier ampholytes providing a pH gradient across a wide pH range, thus avoiding any assumptions as to the number of proteins, or their apparent pI values (pI'). Steep pH gradients are best generated from suitable synthetic carrier ampholyte mixtures, here referred to collectively as "Ampholine". These are available from commercial sources (pI range 2-11 Servalyt, 3-10 Pharmalyte or Bio-Lyte, or 3.5-10 LKB Ampholine; see *Table 1*) or can be synthesised in the laboratory (14). Since the pH gradient is steep, protein migration velocities remain relatively high until the proteins reach the vicinity of their isoelectric positions. Wide-range pH gradients of this type are, however, the most difficult to stabilise. Their main function is to define the required range of the pH gradient. As expected, resolution appears less than that achieved by flatter pH gradients (5,11). Flat pH gradients in the range close to the pI's of the proteins of interest can be prepared either by using narrow pI range "Ampholine" (*Table 1*), or by mixing suitable buffers (11). *Table 2* gives a selection of buffers that have proved suitable for generating gradients for gel electrofocusing. Alternatively, flat pH gradients can be generated by "constituent displacement" by choosing suitable pH's for the reservoir electrolytes (12).

At present, the manipulation of pH gradients has to be conducted in a systematic but empirical manner, because our understanding of the mechanism of pH gradient formation is still insufficient to predict the pH gradient from the pK's of carrier constituents, anolyte, and catholyte. Also, the number and chemical nature of buffers with useful pK's is quite limited, particularly when one chooses amphoteric buffers which provide better gradient stability than non-amphoteric buffers (presumably to the degree that their pK's approach their pI's). Furthermore, the labour involved in weighing more than a few buffer constituents is a serious limitation in designing buffer pH gradients. With present knowledge it is only possible to make qualitative changes in the pH gradient by modifying the carrier constituent buffer mixture; that is, addition of acids will make the pH gradient more acidic, addition of bases will make it more basic (11). Another method for shifting the gradient toward a higher pH is to use anolytes with higher pK's, until the pH of the anolyte approximates that of the anodic terminus of the electrofocusing gel (13). Increasing the anolyte pH further would result in carrier constituent displacement from the gel into the anolyte (12).

Amphoteric Versus Non-amphoteric Carrier Constituents

The rates at which natural pH gradients form do not depend on whether all carrier constituents are amphoteric, non-amphoteric, or whether both types are mixed; in other words pH gradients may form by a non-isoelectric mechanism.[2] The practical question is which carrier constituents to select to obtain a desired pH gradient. The theories of neither isotachophoresis nor electrofocusing consider such cases. We learn from the classical theory of electrofocusing (17), which is concerned only with amphoteric carrier constituents, that a good constituent should provide the maximum conductance in its zero-mobility condition (the pI). This condition is only fulfilled to the degree that the pK's of acidic and basic functional groups on the carrier constituent are close to one another. This restriction of carrier constituents to ampholytes with close pK's rules out practically all available amphoteric buffers, which usually have widely divergent acidic and basic pK values. Non-amphoteric buffers at zero mobility, that is two pH units from their pK, exhibit even less conductance, and also no buffering capacity.

Nonetheless, gradients do form from every form of carrier constituent: non-amphoteric buffers in their zero-mobility condition, that is with minimum conductance and buffering capacity, produce useful pH gradients, either by themselves (6,15,18) or mixed with amphoteric buffers, even with those whose pK's are not close to their pI's (11).

---

[2]Indirect evidence exists that pH gradients may be generated by an isotachophoretic mechanism under conditions where the leading constituent is arrested at zero mobility (15). If this concept is correct, conductance would be provided throughout the train of constituents, aligned in order of constituent mobility, because the constituents are at pH's well removed from their pK's, that is they are charged. To predict, on the basis of this hypothesis, a pH gradient from the pK's of the constituents, one would have to compute the steady-state sequential moving boundaries and the operative pH of all the phases (16). Until the required computer program is designed for that purpose, we have no choice but to arrive at the desired pH gradient empirically, through systematic variation of constituents.

Table 1. pI Ranges of Commercially-available Carrier Ampholytes.

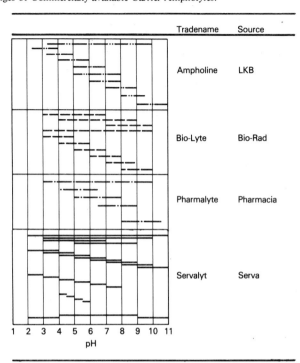

Flat Versus Steep pH Gradients

The narrower the pH range of the gradient, the more stable it seems to be, providing the pH's of anolyte and catholyte are near those of the gel termini. Also, the more neutral the pH range the more stable the pH gradient, at least if one uses the terminal carrier constituents as anolyte/catholyte (see p 169). However, from the viewpoint of separation, it is not yet clear to what degree gradient flatness is desirable. Our earlier notion of "the flatter the better" seems questionable, since proteins with flat titration curves around their pI's seem unable to migrate at appreciable rates into the gel. Also, not surprisingly, band width increases with pH gradient flatness; resolution is thus lowered to such an extent that it nearly, although not completely, offsets the improvement in separation (5,11,19). Another problem with flat pH gradients is that the focusing time markedly increases because of the low migration rates of the proteins. Thus, fractionations in "medium steep" pH gradients, such as the 2-pH unit ranges of present commercially-available carrier ampholyte preparations (*Table 1*), appear to be a satisfactory compromise. This is also the pH range of buffer focusing which has been achieved with presently available amphoteric buffers (see *Table 2*).

pH Gradient Linearity

Choice of carrier constituents is often based on their ability to provide linear gradients. Linearity is not, however, essential for effective electrofocusing separations.

Table 2. Component Buffers of Mixtures used for the Formation of Natural pH Gradients.

Constituent	pI	pK	pI-pK
MES	3.07 *	6.15	3.08
ACES	3.45 *	6.90	3.45
BES	3.57 *	7.15	3.58
MOPS	3.60 *	7.20	3.60
TES	3.75 *	7.50	3.75
HEPES	3.77 *	7.55	3.77
lactic acid		3.80	
EPPS	4.00 *	8.00	4.00
acetic acid		4.75	
TAPS	4.20 *	8.40	4.20
propionic acid		4.87	
tricine	5.07 *	8.15	3.08
bicine	5.17 *	8.35	
gly-gly	5.20 *	8.40	
pyridine	-	5.25	
asparagine	5.40 *	8.85	3.45
threonine	5.59	9.18	3.59
serine	5.68	9.15	3.47
hydroxyproline	5.82	9.73	3.91
isoleucine	6.04	9.76	3.72
glycine	6.06	9.78	3.72
Bistris		6.80 *	
Bistrispropane		6.80 *	
β-alanine	6.90	10.20	3.30
imidazole		6.95 *	
GABA	7.50	11.30	3.80
GACA	7.60	10.75	3.15
triethanolamine		8.35	
AEPD		8.80	
AMP		8.80	
ammonia		9.25	
lysine	9.47	10.53	1.06
ethanolamine		9.50	
triethylamine		11.01 **	

1. Data from a) Calbiochem Catalogue; b) Handbook of Chemistry and Physics; c) Merck Index.
2. All pK's and pI's at 25°C except (*) 20°C; (**) 18°C.
3. All pI's for sulfonic acid Good's buffers calculated as pI = $pK/2$, i.e. pK (R-SO$_3$H) = 0.00.
4. Compounds with limited solubility (e.g. glutamate and histidine) or pI-pK values larger than 4 (e.g. taurine) have been excluded.

What is important is that the pH gradient be moderately flat in the isoelectric region of the proteins to be separated. The linearity of pH gradients depends on the number of carrier constituents selected. Perfectly usable non-linear pH gradients covering 2-3 pH units have been made with only 8-10 buffer constituents (15,18).

Electrofocusing at High Ionic Strength

When proteins require high ionic strength for stability and maintenance of activities, it may be increased by adding salt to all phases (20). This of course increases the electrical heat generated, an effect that can only be overcome by using very low voltages and compensating by extending the electrophoresis time. A possible alternative to the addition of salt is to conduct electrofocusing at high concentrations of carrier constituents. However, even at concentrations of between 0.1 and 1.0M, the conductance, and thus the ionic strength, is still low at any point in the pH gradient. Nevertheless, it is possible that the native properties of proteins may be retained under these conditions. "Ampholine" (*Table 1*), at final concentrations of up to 20%, and/or buffers with solubilities in the molar range (*Table 2*) may be used for this purpose.

The distribution of carrier constituents along the pH gradient is not continuous. After some time of electrofocusing, voltage peaks appear, usually at neutrality (see below). Attempts have been made to bridge these "conductance gaps" by adding amphoteric compounds with widely divergent pK values, such as trimethylamino-propane-sulfonate to the carrier constituents (21). However, even at concentrations of more than 1M, these compounds do not abolish "conductance gaps", undoubtedly just because they fail to be conductive.

Another way to render electrofocusing gels more conductive or more uniformly conductive is to add "poor carrier ampholytes" with widely divergent pK values, such as many of the amino acids, to the system. While in the transient state, these may increase conductance, although at the expense of perturbing the steady-state and therefore, probably, resolution.

Electrofocusing in the Presence of Detergents

Ionic detergents are compatible with the formation of pH gradients if one is willing to wait. As pointed out above, the increase in conductance after such additions forces one to reduce the voltage so as not to exceed the capacity of the apparatus to dissipate heat, although the introduction of an apparatus for electrofocusing which has extremely efficient electrical cooling (see p.160) may alter this situation. In contrast, non-ionic and amphoteric detergents are perfectly compatible with electrofocusing procedures. They do not cause any apparent alteration in the pH gradient (22) or conductance, although they may inhibit interactions between proteins and carrier ampholytes (23). Suitable amphoteric detergents of the alkyl-sulfobetaine type are commercially available (e.g. Zwittergent 3-14 from Calbiochem-Behring). Similar detergents with rigid hydrocarbon backbones, which, under suitable conditions, appear not to denature proteins, have been described (24).

CHOICE OF ANOLYTE AND CATHOLYTE

Classical electrofocusing used a strongly acidic anolyte and a strongly basic catholyte. The rationale for this was that carrier constituents would protonate or deprotonate when they migrated to the ends of the gel and made contact with extremes of pH, and thus reverse their direction of migration and remain within the gel. However, pH gradients can be generated equally well if one uses the terminal carrier ampholytes of the pH gradient as anolyte and catholyte in which case there is the additional advantage that the stability of pH gradients with time is enhanced (25,26,27). It is important to avoid the presence of counterions which would carry excessive current through the pH gradient gel, causing heating. Therefore if the terminal constituent is an ampholyte, the isoelectric form should be selected. Similarly if the terminal constituent is non-amphoteric, it should be the fully protonated acid, or the fully deprotonated base, two pH units or more from the pK. In buffer electrofocusing, the "terminal" species is easy to recognise; it is simply the species in the carrier constituent mixture with the highest or lowest pK or pI. When using "Ampholine" to generate the pH gradient, the anolyte and catholyte are added to a final concentration of 0.02-0.03 M to a 1-2% carrier ampholyte mixture used for the gel. All carrier constituents in the mixture with pI's exceeding the pH of the catholyte or less than the pH of the anolyte will migrate into the electrolyte chambers ("constituent displacement", ref. 12) leaving the anolyte and catholyte species at the ends of the gel as the terminal constituents.

It appears that the pH gradient is regulated in its range by the choice of anolyte as well as that of the carrier constituents (13), though in practice the operative pH of the anodic terminus is not the same as that of the chosen anolyte. For example, using pI range 4-8 Ampholine with cacodylic acid, pH 3.3 (pK 6.21), as anolyte gives rise to pH gradients with an anodic terminus of pH 5. In the same system, dimethylacrylic acid, pH 2.8 (pK 5.12), as anolyte generates a terminus at pH 4 (13). There is no simple way of predicting operative terminal pH's to arrive at a desired anodic and cathodic terminal pH. It may thus be necessary to systematically vary anolytes in their uncharged form, two pH units or more from their pK's or at their pI, until the chosen pH gives rise to the desired operative terminal pH (13). Alternatively one may achieve this by systematic constituent displacement.

The more concentrated the anolyte and catholyte, the more stable are the pH gradients (25). From considerations of solubility and cost, 0.1 M solutions are usually used. Since both available isoelectric anolytes/catholytes and fully protonated or deprotonated acids and bases have low buffering capacity, large anolyte and catholyte reservoirs are required in order to maintain the pH in the electrolyte chambers. The recommended gel tube apparatus (see p 158) provides reservoirs of 500 ml or more, but many types of horizontal slab apparatus do not have such large reservoirs. Since electrolytic reactions may easily give rise to pH changes even in large reservoirs, the pH should be monitored periodically during electrophoresis, and electrolyte solutions should be replaced if substantial changes in pH or conductance make it necessary. Thus, the use of the terminal constituents as anolyte and catholyte introduces some experimental difficulties and expense compared to the use of strong acid and base as anolyte and catholyte. However, these appear justified because gradient stability is required, particularly for flat pH gradients, to ensure that proteins reach their isoelectric endpoints, at which resolution is reproducible.

ELECTROFOCUSING DYNAMICS

pH Gradients

Originally, it was thought that electrofocusing produces an "equilibrium state", first of the carrier constituents, giving rise to a stable pH gradient, and then of the proteins at their pI positions. The generally held notion that protein fractionations take place during the steady-state, that is before the gradient begins to decay, appears to stem from the observation that protein patterns form relatively rapidly. These patterns, however, may not be constant because they do not necessarily reflect the respective isoelectric positions. In reality, one can define the steady-state both for the carrier constituents and the proteins only in relation to particular voltage-time frames; that is the pH gradient is formed and decays (and the proteins approach their isoelectric positions) as a function of the voltage and the length of time applied. At 20-40 V/cm gel, for example, a natural pH gradient forms in a few hours, by the alignment of carrier constituents in order of pI or constituent mobility; one may consider this alignment a "steady-state". Under most conditions of gel electrofocusing, however, pH gradients appear to be in continuous motion, such that the "steady-state" in between their formation and decay is, in fact, transient.

Changes in the distribution of carrier constituents are partly responsible for and illustrate the dynamics of pH gradients (ref. 28 and *Figure 4*). Thus in the presence of strongly acidic and basic anolyte/catholyte, carrier ampholytes in a pI range 3.5-10 Ampholine system first disappear gradually from the neutral centre of pH gradients,

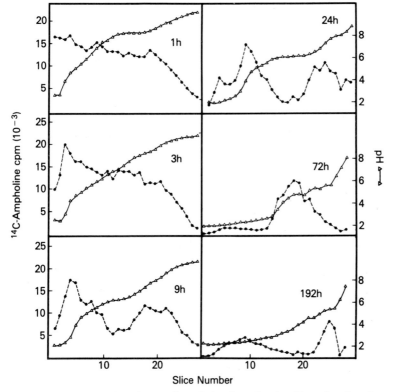

Figure 4. Kinetics of the formation of pH gradients during electrofocusing. Electrofocusing gels containing [14C]Ampholine were electrophoresed for periods of 1h to 192h. The pH gradients and the distribution of [14C]Ampholine were then analysed. (Adapted from Reference 28).

and accumulate in one acidic and one basic peak. Subsequently, the basic, neutral and acidic constituents migrate in succession into the cathodic reservoir, while acid follows progressively from the anodic reservoir, until the pH gradient has decayed and the gel has acidified. Corresponding to the movements of carrier constituents, the pH gradient itself achieves its so-called "stable" form, that is its greatest span, within a few hours of electrofocusing under representative conditions, then flattens and distorts progressively while usually undergoing a "cathodic drift". One can also cause the opposite "anodic drift" by selecting a catholyte which is a strong base compared to the acidity of the anolyte (2). Thus, no true, enduring "steady-state" of pH gradients can usually be discerned. Narrow range pH gradients between weakly acidic anolyte and weakly basic catholyte have been stabilised for several days (11,26,27), possibly because some carrier constituents in the transient state contribute to the conductance of the system (the underlying assumption being that uneven conductance contributes to pH gradient decay). Under conditions of pH gradient decay, there is no steady-state of protein *positions* either. However, within the voltage-time frame of pH gradient decay, the proteins do approach a constant pH asymptotically. For them, therefore, one can discern a steady-state with regard to pH although the location of this pH on the gradient varies.

Instability of pH gradients is important in the practice of electrofocusing because the migration of proteins towards their isoelectric pH is slow (see the following section). Furthermore, it is necessary to "reach", or closely approach, isoelectric protein positions for two reasons:

(i)　reproducibility of the relative positions of proteins would depend much more heavily on the constancy of gel concentration and voltage gradients between experiments than is usually obtainable, if one were to carry out separations during the transient state, that is before the asymptotic phase of the approach toward the pI by all proteins in a mixture, and thus before "pattern constancy". For the same reasons resolution would become less reproducible.

(ii)　characterisation of proteins on the basis of pH would be meaningless and arbitrary unless the pH at least approximately equals the pI'.

The following measures can be employed to stabilise the pH gradient, either individually or combined:

(i)　the pH range may be narrowed to, at most, 2-3 pH units, comprising the pI's of species which are to be separated. Narrowing can be achieved either by the choice of carrier constituents in buffer electrofocusing (11), or by constituent displacement using a higher anolyte pH than the anodic terminus of the selected pH range, and a lower catholyte pH than the cathodic terminus of the selected pH range (12). Alternatively, it may be achieved by decreasing the pH difference between the anodic terminus and the uncharged anolyte (e.g. a fully protonated acid; see ref. 13).

(ii)　the pH of the anodic terminal constituent can be made the same as that of the anolyte, and that of the cathodic terminal constituent the same as that of the catholyte (25).

(iii)　the anolyte and catholyte concentrations can be made as high as solubility and cost will allow (25). Usually 0.1 M solutions are used.

(iv)　wall adherence of the gel should be ensured by thorough cleaning, coating, mechanical support and hydrostatic equilibration. Improvement in wall adherence is always reflected in more stable gradients and should be tested by reversing the polarity (anolyte once at the top, then at the bottom) and demonstrating absence of any effect on the pH gradient (2).

(v) the carrier ampholyte concentration can be increased.

(vi) the viscosity can be increased. This measure, however, seems only effective in some cases (see e.g. ref. 29), not in others (see e.g. ref. 28), for reasons not yet understood.

Proteins

The migration of proteins towards their isoelectric positions on the pH gradient is usually slow (*Figure 5*). The rate is governed by the titration curve of the protein. Furthermore, the protein continuously decelerates as it approaches its isoelectric position on the pH gradient. In part, the slow migration is also due to molecular sieving by the gel. This is not only the case when obviously restrictive pore sizes have been used. Even when the pore size has been shown experimentally to be "non-restrictive" (7) it may become progressively more restrictive as the pI is approached because of the greater sensitivity of low migration velocities to any small retarding effect of the matrix.

In order to achieve reproducible separations at optimal resolution, it is necessary to determine experimentally the electrofocusing time required for migration to a constant pH. This pH, if it remains constant for at least twice the time required to reach it, is by definition the apparent pI (pI') characteristic for the protein in the particular milieu of carrier constituents, ionic strength, and temperature. Only this pH value can serve as a physical constant for identifying and characterising the protein. The lack of evidence that a protein has migrated to a constant pH invalidates most "pI" data reported to date in gel electrofocusing studies. Thus, for each separation problem it is first necessary to determine experimentally the time required to reach constant pH values for the proteins of interest. Then one should attempt to stabilise the pH gradient for the required time period at the selected (maximum) voltage as described above (p 172).

The asymptotic approach toward the pI'-position on a gel is frequently accom-

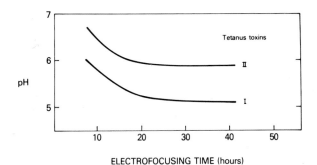

Figure 5. Effect of electrofocusing time on the apparent isoelectric points of proteins. Tetanus toxins were electrofocused using Ampholine, pH range 3.5 - 10, in polyacrylamide gel (6 %T, 15 %C$_{DATD}$) for varying periods of time and then the apparent pI of each toxin was determined (Adapted from ref. 5).

panied by isoelectric precipitation of the protein because of its relatively high concentration in the band. This precipitation is relatively harmless on an analytical scale since zones of protein precipitate are still capable of migrating, presumably through a process of re-solubilisation and re-precipitation, although it may be a further retarding factor in the approach toward the pI. In contrast, protein precipitation does appear important on the preparative scale. Protein solubilisation from the band seems to become increasingly difficult the longer isoelectric precipitation of the protein in the band is allowed to proceed. Probably for the same reason, the recovery of protein activity also seems to diminish with increasing electrofocusing time when electrofocusing times are much extended after the pI has been reached (30).

Conductance

Another dynamic element in electrofocusing is the generation after some time of voltage maxima within the gel, usually at about pH 6. The uneven distribution of conductance across the pH gradient, starting with voltage maxima at the ends of the pH gradient and changing progressively into a voltage maximum at its neutral centre, remained unrecognised until the recent development of a method for measuring conductance (3,4). The phenomenon has been observed in pH gradients formed from all carrier ampholytes so far examined, as well as those generated by lysine alone (anodic voltage maximum), glutamic acid alone (cathodic voltage maximum) and in a gel containing no explicit carrier constituents at all (neutral voltage maximum) (3,4). These data indicate that voltage maxima (conductance gaps) form in the regions of the pH gradient from which carrier constituents have migrated.

Local voltage maxima correspond to a reduction in voltage, and thus to slower migration, in other parts of the gel. This slower migration may prevent the attainment of an isoelectric position by the protein, and produce the illusion of a plateau value of constant pH, that is of a fallacious pI′. This problem may be detected by determining the pI′ after applying the sample from both ends of the gel. No remedy exists to date other than to try a very slow separation with the addition of 0.1 M KCl to all phases (20). Attempts to bridge conductance gaps with amphoteric additives have as yet been unsuccessful.

PROCEDURE FOR ANALYTICAL GEL ELECTROFOCUSING

An abbreviated procedure for gel electrofocusing has been previously reported (31). Many procedural elements are the same as for PAGE and isotachophoresis, (Chapters 2 and 3) but for ease of reading some of these are redescribed here.

STEP 1: Apparatus

Pyrex gel tubes 12 cm long (0.6 cm I.D., 0.8 cm O.D.) and lightly fire-polished at the ends are used. They are washed with methanolic KOH, rinsed to neutrality, and coated with 1% Gelamide 250 (Polysciences or Union Carbide) by filling the tubes and draining then onto tissue paper (2). After allowing them to dry for several days at room temperature, the tubes are stored dust-free until use. Squares (2 x 2 cm) of nylon mesh (88 micron, Small Parts Inc.) and 2 mm sections of 0.7 cm I.D. Tygon tubing are also prepared for use in mechanical support of the gel after polymerisation.

The gel tube apparatus is identical to that used for PAGE and isotachophoresis (see *Figure 1* of Chapter 2) and is reproduced in *Figure 1*. If voltage monitoring or voltage control across the gels is intended, and the gel composition is the same in each tube, upper and lower monitoring electrodes are installed in the upper buffer reservoir as shown in *Figure 2*. The requisite number of tubes, each sealed at one end with Parafilm, is inserted into the upper buffer reservoir where they are held vertically by rubber grommets and guide holes. The remaining grommets are sealed with rubber bungs. The upper reservoir is then seated in the lower buffer reservoir.

Since polymerisation should be conducted at the same temperature as electrofocusing, the lower buffer reservoir should be filled with electrolyte (see *STEP 3*) before the polymerisation. Usually the temperature of the apparatus is kept at 0-4°C using a coolant flow rate of 1.5-3.5 l/min and continuously stirring the lower buffer reservoir.

STEP 2A: Polymerisation of Polyacrylamide Gel

Usually gel concentrations of 5 or 6 %T, 15 %C_{DATD} are used (7); *Figure 1* of Chapter 3 gives a rough estimate of the restrictiveness of these gels to proteins of various sizes. The highest gel concentration which is "non-restrictive" should be used since it is the one with the best mechanical stability and wall-adherence properties. However, it should be kept in mind that the criteria of "non-restrictiveness" in *Figure 1* (Chapter 3) do not appear sufficiently stringent for the purposes of electrofocusing (see discussion above, p.164). Some workers continue to use bisacrylamide-crosslinked gels (see Chapter 5) although these are more restrictive than DATD-crosslinked gels.

The polymerisation mixture is prepared in a 25 ml amber bottle containing a small magnetic stirring bar and equilibrated to the same temperature as the apparatus. An acrylamide-DATD stock solution, ampholyte or buffer mixture, and initiator stock solutions are mixed in the correct volumes (*Table 3*). The gel mixture is deaerated for 5 min using a vacuum pump. For reproducible deaeration, a custom-made evacuator is available (see p.108). After deaeration, TEMED is added (5μl per 10 ml gel mixture), each tube is filled with 1.6 ml of gel mixture, and then carefully overlayered

---

Table 3. Preparation of Polyacrylamide Gel for Electrofocusing.

In a typical application, 10 ml of polymerisation mixture for an electrofocusing gel are prepared at 0-4°C as follows:

5 ml   10 %T, 15 %C_{DATD} (8.5 g acrylamide + 1.5 g DATD/100 ml)
2.5 ml   4% "Ampholine" (or 2.5 ml buffer mixture, 0.4 M for each constituent)
2.5 ml   0.06% potassium persulphate, 0.002% riboflavin.

After mixing, the polymerisation mixture is evacuated to 10 mm Hg for 5 min at 0°C and then 5 μl TEMED are added to initiate polymerisation. A polymerisation time of 30 min is allowed.

---

with a water-layer 2 mm high using a 1 ml syringe fitted with a 23-gauge needle. Alternatively a 100 μl volume Hamilton syringe can be used. In either case, great care must be taken not to disturb the gel surface. The gels are photopolymerised by illuminating them with a fluorescent lamp for 30 min. For even polymerisation a bank of lamps should be used as described on p.108.

Basic (amine) carrier constituents appear to catalyse the polymerisation of cross-linked polyacrylamide, presumably in the manner of TEMED, and this results in too fast a rate of polymerisation giving a short average chain length and poor mechanical stability of the gel. To avoid this, concentrations of the three polymerisation initiators should be lowered when basic carrier constituents are used, to obtain polymerisation within about 10 min. In practice, it is easiest to keep the concentrations of two polymerisation initiators constant and to reduce the third systematically to achieve such a rate. At high pH, where such over-catalysis occurs, systematic reduction in potassium persulphate levels is the most effective method to reduce excessive polymerisation rates (1,31).

STEP 2B: Preparation of Cylindrical Sephadex Gels (10)

For each gel tube, 0.2 g of the Sephadex G-75 (Ultrodex, LKB), or other gel filtration medium depending on the size of the proteins, are made into a slurry with about 5 ml carrier ampholyte or buffer mixture of the desired pH range and concentration and allowed to swell for a few hours according to the manufacturer's instructions. Pyrex tubes at least 20 cm in length (0.6 cm I.D., 0.8 cm O.D.), are prepared and lightly fire-polished at one end to prevent damage to gaskets. The tubes are inserted into a plastic filter holder (Part B in *Figure 6*) which holds a 1 cm diameter Nylon mesh (100 μ) filter tightly against the end of the tube. The assembly is placed on an evacuation chamber, (Part A in *Figure 6*). Each tube is filled with a vortexed suspension of gel, taking care to avoid trapping air bubbles, and this is allowed to pack under house vacuum. The long tubes allow the entire volume of the slurry to be applied at once. As soon as the tubes are packed, the vacuum is released and the upper end of the Sephadex column is sealed by polymerising 0.5 ml of acrylamide solution (6-10 %T, 2 %C_{Bis}) containing anolyte or catholyte. When the polyacrylamide gel plug has formed, the tube is inverted, about 0.5 ml of the Sephadex removed from the other end, and a similar plug, containing this time the other electrolyte, is polymerised within the tube. A dye can be added to the polymerisation mixture to mark the respective plugs. If desired, the excess glass tubing can be broken off at this stage. This prevents a large voltage drop which can be observed in the gel tube above the gel (32) and is possibly due to displacement of electrolyte by isoelectric constituents moving out of the gel in the course of the "cathodic drift" (28). If they are well sealed with Parafilm, packed gels can be stored in the refrigerator for at least a week before use.

STEP 3: Setting up the Electrofocusing Apparatus

Once the polyacrylamide gel has been prepared, the upper buffer reservoir is removed and inverted. The water overlay is gently absorbed onto paper tissue and the Parafilm seals are removed by sliding a razor blade across the end of each tube. Each is replac-

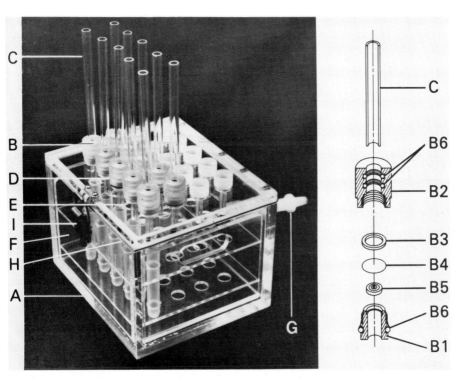

Figure 6. Vacuum chamber for packing cylindrical Sephadex gels for electrofocusing (An der Lan,B., Allenmark,S., Sullivan,J., Fitze,P. and Jackiw,A., in preparation[1]). A Perspex box (A) can hold twenty Sephadex gel tube assemblies (B), which can be used to simultaneously pack twenty 0.6 cm I.D. Pyrex tubes (C). The holes (D), in which the holders are seated, serve to collect the fluid into an elution nipple which drains into tubes (E) held in place in a tube rack (F). The chamber is evacuated via the connector (G). An airtight seal is provided with a silicon rubber gasket (H), not visible in this photograph, and a spring clip (I). An exploded view of the filter holder assembly (B) is shown on the right. It consists of a modified Millipore filter holder (B1), with its screen (B5) supporting a 1 cm diameter Nylon mesh (100 μ) filter (B4). The 0.8 cm O.D. glass tube (C) is held against the nylon mesh with a knurled plastic (Lexan) tube holder (B2) and a Teflon gasket (B3). Airtight seals are provided by two O-rings in the tube holder and one on the filter holder (B6).

ed by a Nylon mesh support (2 x 2 cm Nylon mesh, 88 micron, Small Parts Inc.), held in place by a 2 mm section of 0.7 cm I.D. standard Tygon tubing. Care must be taken not to trap air between the mesh support and the gel. The lower buffer reservoir is filled with anolyte if the sample protein is anionic, or catholyte if the protein is cationic, and the upper buffer reservoir is then filled with the other electrolyte. Electrolyte levels in both reservoirs should be the same. The concentration of each electrolyte should be at least 0.1 M and have the same composition as the terminal constituents, that is, those at the ends of the gels once the gradients have formed. In addition, both electrolytes should have no counterions other than H^+ and OH^-. The pH of the anolyte should be as close as possible to that of the anodic end of the gel after formation of the pH gradient, as determined on blank gels. The apparatus is vertically aligned by means of the level on the electrode holder (cover D in *Figure 1* of Chapter

2), and the adjustment screws on the apparatus base (*ibid*. part B). Finally the apparatus is checked for the absence of air bubbles at the tops and bottoms of the gels.

STEP 4A: Application of Samples to Polyacrylamide Gels

The sample should contain 5-2,000 μg protein per band in a gel of 0.6 cm diameter. Any non-proteinaceous contaminants which might affect either the formation of the gradient or the effective pI of the proteins should be removed prior to electrofocusing (e.g. nucleic acids; see p. 194). Experience has shown that the sample volume should be less than 1 ml, preferably less than 0.5 ml. For staining purposes, loads of 5 to 25 μg per band are best; for preparative purposes, up to 1-2 mg per band can be tolerated per gel of 0.6 cm diameter, depending on the overlap of distribution and separation between the species of interest (19,33). The load capacity for other gel diameters is proportional to gel surface area.

The protein sample is prepared either in upper reservoir electrolyte, or in a carrier constituent mixture, at such a pH that in either case the proteins are charged. For example, for anionic proteins a mixture of pI range 7-9 or 8-10 with an actual pH 8 or 9, respectively, is suitable to charge the proteins negatively. A pH close to or at the isoelectric pH of the sample should be avoided; this would frequently be obtained if the sample were prepared in a 3-10 pI range ampholyte mixture, or if ampholyte mixtures diffused from the gel into an unbuffered sample prior to turning on the power. The addition of carrier ampholytes to the sample does alter the pH gradient of the gel slightly, although this alteration is insignificant if the sample volume is small.

The density of the sample is usually increased by adding 10-30% sucrose or glycerol; higher concentrations may dehydrate the gel surface. To visualise the sample, one can add a dye, for example Evans blue (pI = 5.35) or methyl blue (pI = 3.60). These amphoteric dyes simultaneously serve as pH markers when they arrive at their isoelectric position; thereafter they diffuse rapidly.

If the sample is small, that is 200 μl or less on an analytical size gel, it is easiest to layer the sample under the upper electrolyte by directing the tip of a 100 μl Hamilton syringe to just above the surface of the gel. If the sample is larger, it is pipetted onto the gel surface before filling the upper buffer reservoir; then the sample is carefully overlayered with upper electrolyte using a Pasteur pipette until the tube is completely full. The upper reservoir is then filled to the required level. Using either procedure of sample application it is important to ensure that the interface between the sample and upper electrolyte remains undisturbed. An alternative method of sample application is to mix the sample throughout the gel by including it in the gel polymerisation mixture. This method is particularly useful when analysing complex mixtures, the components of which may have very different pI's (see p.205). It has the advantages that it is possible to use large sample volumes and also samples are not exposed to the extremes of pH. The major disadvantages are, firstly, that a proportion of the sample proteins become irreversibly occluded in the polyacrylamide gel matrix during polymerisation which increases the background staining of the gel and, secondly, the samples may be modified by the polymerisation reaction. Finally, the proteins are likely to be less soluble since the pH of the gel mixture is usually close to that of the pI's of the proteins.

STEP 4B: Application of Samples onto Cylindrical Sephadex Gels

If the sample volume is small (50 μl or less), it can be injected with a Hamilton syringe into the column through the polyacrylamide plug. If it is large, but less than the void volume of the column, it is useful to dialyse carrier constituents into the sample and then to apply the sample directly to the Sephadex column while it is still under vacuum, before the polyacrylamide plugs are made. To monitor sample application, that is to prevent the sample from being sucked through the column, a few microlitres of Blue Dextran (Pharmacia) can be applied immediately before the sample. Unfortunately the second method of sample application may not allow the proteins to be fully charged, and therefore soluble, before electrofocusing, since they will be applied at the average pH of the constituent mixture. Larger sample volumes can only be applied in the same way as for polyacrylamide gels, that is over the polyacrylamide plug in either anolyte or catholyte, as appropriate. It is obvious that, in this case, the plug should be non-restrictive for the protein, and so the use of 15 %C_{DATD} gels of the lowest possible %T (5% or less) is recommended for the upper gel plug.

STEP 5: Electrofocusing

The sample may be applied just prior to electrofocusing if the pH of the ampholyte mixture, or the transient state pH in the gel segment traversed by the protein, are not noxious to the protein or make it insoluble. In that case, the power is turned to a regulated current of 1 mA/tube of 0.6 cm diameter as soon as the sample has been loaded. When the voltage has reached the desired maximum value compatible with the heat dissipation capacity of the apparatus, usually 20-40 V/cm of gel, the power supply should be switched to voltage control.

Power supplies regulating wattage at below 1 watt are not yet available. Thus, wattage control, from the start of electrofocusing, can still only be used with at least 5 to 10 gel tubes. Also, as the current decreases asymptotically during the later stages of electrofocusing, the voltage limit of these power supplies is usually reached, and the power supply automatically switches to voltage control.

When a strongly acidic anolyte and a strongly basic catholyte are used, the voltage drop across the reservoirs is negligible and 200 V can be applied across the entire system, which approximately equals the voltage across the gel. However, when weakly acidic and basic, uncharged anolyte and catholyte are used, the voltage drop across the reservoirs is far larger and cannot be neglected. It is then necessary to place platinum electrodes in contact with the top and bottom of one representative gel in the upper reservoir (*Figure 2*) and to measure the voltage across the gel with an external voltmeter, and to regulate the voltage across the system on the basis of that measurement. Unfortunately, in view of voltage changes during electrophoresis both in the gel and the reservoirs, the regulated voltage across the system cannot precisely maintain the desired voltage of 20 or 40 V/cm of gel, so that re-adjustment during the course of electrofocusing is required. This problem has recently been solved by modifying a power supply so as to allow one to regulate the voltage delivered to the ends of the gel tube itself (An der Lan,B., Chidakel,B.E. and Chrambach,A., in preparation[1]).

The protein load is also of consequence for the initial power settings. When the protein load exceeds 100-200 μg/gel, the starting zone is sensitive to electrical heating for the same reasons that extended stacks in preparative isotachophoresis are (see Chapter 3), that is, there is a large potential drop across the protein zone, presumably because the protein immobilises water and salt. The effect is manifested by the formation of a white zone of protein precipitate above the gel. If it is detected immediately, it can usually be reversed by regulating the voltage across the gel to 20 V/cm or less.

Frequently it is advantageous to allow the pH gradient to form before applying the protein. Under these conditions, the pH range to which the protein is exposed is exactly known, denaturation and precipitation at unfavourable pH can be averted, and the maximum migration rate toward the pI-position of the protein is achieved. The procedure consists of applying a voltage across the gel, and waiting for the current to decline exponentially to a steady low value before applying the sample. Usually, at 20-40 V/cm of gel, a suitable current level is reached in approximately 1 h. Thus prefocusing of the gels does not appreciably decrease the total electrofocusing time available for fractionation prior to gradient decay.

Prefocusing in the presence of charged reducing agents can also be used to reduce or abolish the oxidising conditions in polyacrylamide gel (34). This can be done by loading 5-20 μmoles of thioglycolic acid, brought to the pH of the cathodic terminus by titrating with the catholyte, at the cathodic end of the gel in the same manner as a protein sample, and allowing it to sweep the gel before applying the protein. Alternatively, it may be loaded simultaneously with the protein by underlayering in a denser solution under the protein sample. (Since reduction of the gel is pH dependent, uniform reduction would require that the thioglycolate be loaded before the pH gradient has formed, that is before a voltage is applied. However, in polyacrylamide gels, such a uniform reduction cannot be achieved prior to polymerisation because polymerisation has to take place under oxidative conditions.)

STEP 6A: *Removal of Polyacrylamide Gels and pH Gradient Determination*

Gel tubes are withdrawn successively at time intervals such as 4, 8, 16, 24, 48 h of electrofocusing. In each case, a small rubber stopper is placed into the top of the tube. The tube is then withdrawn from underneath the reservoir until it has slipped out of the grommet and the stopper placed in the top has sealed the hole. The stopper is then tightened by pulling from underneath, and the bottom surface of the upper buffer reservoir around the stopper is rinsed with a washbottle to avoid any cross-contamination between anolyte and catholyte. The reservoir is then replaced and electrofocusing is continued for the remaining gels. Some workers find this technique difficult to use and prefer to empty the upper reservoir prior to tube removal. Care must again be taken to avoid contamination of upper reservoir electrolyte with lower reservoir electrolyte.

The anodic end of the gel can be marked by inserting a hypodermic needle to form an air-bubble; to make the mark more visible, the needle may be first coated with a water-insoluble marker such as Indian ink. Gels are removed from their tubes by rimming with water using a 23 gauge 6 cm long hypodermic needle connected to a syringe, siphon or water tap. Coating the tube with Gelamide 250 makes it somewhat more difficult to remove the gel than from uncoated tubes. Furthermore, the need for mechanically labile "non-restrictive" gels in electrofocusing and the differential gel swelling as a function of pH requires extreme care in gel removal. Water should emerge dropwise from the needle; care should be taken not to insert the needle into the gel during rimming; the tube should be held vertically, close to the bench surface, to allow the gel to slide onto the bench without breaking; both ends of the gel should be rimmed until it slides out of the tube freely, without need for air pressure.

The pH gradients of gels can be determined on the same gel as is used for activity determination or for staining, either manually using a contact electrode or, alternatively, using a semi-automatic pH gradient measuring apparatus (*Figure 3*), as detailed above in the section on Gradient Monitoring Devices (p. 162). The gel is placed into the groove of the apparatus and covered with 0.025 M KCl. The contact electrode is then allowed to pass over the gel at a rate of not more than 0.5 cm/min, while the pH is plotted on an automatic recorder. Alternatively, a representative gel can be sliced either with a razor blade or with a gel slicer as described in the apparatus section. Slices are placed into 0.5 ml dilute KCl, and the pH determined manually. The assumption is then made that the pH gradients on the gels analysed by staining or activity are identical to that determined on the representative gel. This assumption requires that gel composition and volumes, sample ionic strength and volume be exactly identical in all gels.

STEP 6B: *Removal of Sephadex Gels and pH Gradient Determination*

After focusing, gel tubes are removed and wrapped in Parafilm and placed in a $-20°C$ freezer until they are completely frozen. Very fast cooling is not advisable because the tubes tend to crack. A cooled cutting surface is prepared by placing a piece of plate glass on a bed of crushed dry ice. After the tube has been slightly warmed by hand, the frozen Sephadex gel can be extruded from the glass tube onto the glass plate using a close-fitting glass rod. This procedure should be practised once or twice on a blank gel in order to be able to gauge the minimum amount of thawing necessary. A razor blade kept at room temperature will cut through frozen Sephadex quite easily, whereas a cold blade will tend to break it. The pieces should be picked up with cooled forceps and placed into tubes for pH or activity measurement, in the same manner as for polyacrylamide gel slices.

STEP 7: *Staining or Activity Determination*

Staining gels after electrofocusing is more problematic than after PAGE because the ampholytes may have fixation and staining characteristics similar to those of proteins. Advantage is taken of the fact that, in general, proteins are insoluble in 3.5% perchloric acid (PCA), while ampholytes tend to be soluble (35). On the other hand, using buffer electrofocusing, proteins can be fixed in the relatively effective fixative trichloroacetic acid (TCA) (36); Coomassie brilliant blue G-250 is used as the stain in both cases.

When using TCA as a fixative, the preferred procedure is that described by Diezel *et al.* (36). Each gel is soaked in a forty-fold excess of 12.5% TCA for 5 min to fix the proteins and then sufficient 0.25% aqueous Coomassie brilliant blue G-250 is added to give a final concentration of 0.0125%. After 30 min in the staining solution the gel is transferred to a 9 ml screw-cap tube containing 5% acetic acid. The protein bands become immediately apparent without appreciable background and the pattern is stable for at least several months. A modification for particularly acid-soluble proteins is given on p.118.

In the PCA procedure of Vesterberg (procedure B of ref. 37) the gel is incubated at 60°C for 30 min in 40 ml of a solution *A* (9g sulfosalicylic acid, 30g TCA in 236 ml water). It is then placed into a 10 ml screw-cap tube filled with solution *B* (0.1g Coomassie brilliant blue G-250, 200 ml water, and 8.7 ml 61.3% PCA are mixed, made to 300 ml final volume, and filtered), and re-incubated at 60°C for 30 min. The gel is stored in solution *B* diluted a hundred-fold.

Staining of Sephadex gel columns by the usual methods is not possible. The position of protein bands can be determined by estimating the protein concentration in the eluates by the Lowry procedure, using TCA precipitation prior to protein estimation if the eluate is too dilute. Presumably the "contact print technique", commonly used with thin-layer Sephadex separations, can be adapted to Sephadex tubes, since it has been applied to cylindrical gels in PAGE.

For activity determination, polyacrylamide gels are sliced into 1 mm segments as described earlier (p 163) and Sephadex gels with a razor blade (p 181). If necessary, gel slices are incubated at an extreme acidic or basic pH for minimal times at 0°C to solubilise the isoelectric protein precipitates. The main problem with this procedure is the denaturation of proteins. Alternatively gel slices are incubated directly in the assay mixture. Any measure that suppresses isoelectric precipitation, such as the addition of 20% glycerol to the gel (28), or a decrease in electrofocusing time after the isoelectric point has been reached, appears beneficial to the recovery of activities (30). When the protein requires a solubiliser, such as Triton X-100, to prevent isoelectric precipitation, one must bear in mind that the detergent may continuously migrate out of the gel in the form of a complex with carrier ampholytes; this may cause the protein to precipitate irreversibly (23). Also, an increase in "non-restrictiveness" or a decrease in the adsorptive forces of the matrix may help in the recovery of activity. Thus, Ultrogel (an agarose-polyacrylamide mixed polymer) (30) or Sephadex G-75 (Ultrodex) (An der Lan *et al., op. cit.*) have allowed recoveries of activity which could not be achieved with polyacrylamide.

STEP 8: Protein Identification and Characterisation by pI'

The band distances relative to gel length, and the plots of pH relative to gel length, are used to assign a pH value to each band. These pH values plotted as a function of electrofocusing time yield a curve (*Figure 5*), the asymptotic "plateau" value of which is the characteristic pI' of the protein. To ascertain whether a particular band is an artifact of electrofocusing, excision of the band and re-focusing of the gel slice in an identical system is generally desirable. In some cases, such as when the physicochemical significance of pI' values is important, one can also test for the association of protein and carrier ampholyte at the pI' by carrying out the electrofocusing in dissociating media, for example concentrated urea solution (23,38), or by using buffers as carrier constituents, or by comparing the asymptotic plateau values of pH obtained when the sample is applied at either end of the gel.

PREPARATIVE GEL ELECTROFOCUSING

The principal advantages of preparative electrofocusing in gels, as compared to that in sucrose density gradients, are threefold:
(i)   at high protein loads the isoelectric precipitation of proteins does not interfere with separation;
(ii)   the gel matrix acts as an anti-convective medium, increasing band sharpness;
(iii) no mixing of bands occurs during elution.
The disadvantage of electrofocusing in gels compared to liquid media, is the need to solubilise, diffuse, or electrophorese the protein out of gel slices after the separation. In practice, gel slicing and extraction of the protein from gel slices is the only preparative gel electrofocusing method generally applicable at this time. The procedures for extracting proteins from gel slices essentially follow those described for electrofocusing on an analytical scale. Alternatively this can be achieved by steady-state stacking as described earlier but with the following modifications.

Apparatus

Gel electrofocusing apparatus. As in PAGE and isotachophoresis (Chapters 2 and 3) a preparative scale is achieved by increasing the surface area of the gel. The dimensions of the gel, and hence the load capacity, are limited by the capacity of the apparatus to dissipate heat at practical rates. Although resolution in electrofocusing increases in proportion to gel length, the long tubes and elongated lower reservoirs of isotachophoresis (Chapter 3) are only employed when necessary. Usually, preparative electrofocusing is carried out on 1.8 cm diameter gels employing a sufficiently short gel to allow one to use the same lower buffer reservoir as is used for analytical scale gels (*Figure 1*). Tubes need to be coated with Gelamide 250 as described above, and the gels should be supported mechanically with Nylon mesh, since it is even more difficult to maintain adequate wall adherence of the gel in wide diameter tubes than in 0.6 cm analytical tubes.

Slicers. The non-uniformity of the gel along the pH gradient is probably responsible for the fact that longitudinal or most transverse wire slicers are unsuitable for gel electrofocusing. For longitudinal slicing of gels, the manual use of a microtome knife blade is recommended, while for transverse slicing, the diaphragm slicer may be used (see p119) to provide 2 mm gel slices with 15% DATD crosslinked gels. The commercial electric vibrating wire slicer (see p163) may lend itself to sectioning of 1.8 cm diameter gels when the vibration frequency is decreased sufficiently using a rheostat.

pH Gradient measuring devices. The manual contact electrode technique (p. 162) can be used for wide gels. Alternatively, the apparatus depicted in *Figure 3* has the option of a 1.8 cm diameter groove.

Power supply. The greater the diameter of the gel, the less efficient is the dissipation of electrical heat. Thus electrofocusing on a preparative scale should be conducted at lower voltages than on an analytical scale. In a representative case (19) it was found necessary to lower the voltage from the 15 V/cm appropriate for analytical scale gels to 10 V/cm.

Procedure

STEPS 1-3: Gel Electrofocusing

The analytical gel electrofocusing conditions providing optimal resolution (11) are applied to one or several gels of 1.8 cm diameter, using an upper buffer reservoir with rubber grommets suitable for a gel of this size. The voltage should be reduced, and coolant flow rate and the rate of magnetic stirring in the lower buffer reservoir maximised, to compensate for the diminished rate of heat dissipation from wider diameter gels. If several gels are run, all should have exactly the same volume, ionic strength, and protein load, to prevent field strength differences between gels. The voltage, protein load, and gel length should be as large as possible for the required separation, and these parameters should be determined on an analytical scale prior to the preparative experiment. Depending on the electrofocusing conditions and on the protein, between 4 and 8 mg/cm^2 of gel protein band may be loaded (19,33); protein load per gel is proportional to gel surface area. Gel electrofocusing is started at 4 mA/cm^2 of gel, regulated current, until the voltage across the gel reaches 10-12 V/cm. Thereafter, a regulated voltage at the maximum permissible value should be maintained. At the time at which the proteins of interest reach their pI', or at any later time providing improved resolution (11), electrofocusing is terminated.

STEP 4: Gel Removal, Slicing, Protein Solubilisation, Re-analysis and Pooling

The gels are removed from the tube carefully, using the same procedure as for preparative isotachophoresis gels (Chapter 3, p. 148). The gels are sliced into 2 mm sections by means of the diaphragm slicer. Extraction of proteins from gel slices containing isoelectrically precipitated zones requires their solubilisation. This is achieved by very short exposure to an extreme of pH while attempting to reduce, as much as possible, the concomitant protein denaturation by carrying out the solubilisation in an ice-water bath for the shortest possible time. Typically, corresponding section numbers from all gels are pooled into scintillation vials, barely covered with 0.05 M NH$_4$OH, and kept in an ice-water bath for 10 min, then neutralised with the acid constituent of the multiphasic buffer system selected for concentration (see p.150) or with NaH$_2$PO$_4$. To allow for loading in the subsequent gel electrophoretic steps, sucrose is added to each slice suspension to a final concentration of 10%, and aliquots are analysed by electrofocusing or PAGE on an analytical scale, while the slice suspension is stored at 4°C. Slices are then pooled immediately according to band pattern.

STEP 5: Protein Recovery

Protein is recovered from the pooled gel slices using the steady-state stacking extraction and concentration apparatus described in Chapter 3 (p 150). The choice of isotachophoresis buffer for the extraction and concentration of protein from gel slices, the procedure of steady-state stacking by which these are accomplished, and the final removal of non-proteinaceous, polyacrylamide-like contaminants and non-volatile buffers are the same as described in Chapter 3 (see p.151).

Until now the yields of protein from electrofocusing gel slices have been of the order of 70% (19,39). The purity of protein samples of a few milligrams has been as low as 50% when gel slices were extracted by steady-state stacking on polyacrylamide gels, whereas it became better than 90% when agarose was used for steady-state stacking (40).

The bulk of carrier ampholytes can be removed from the protein by a gel filtration step. Quantitative removal by dialysis, or, if the bonding is sufficiently strong, electrodialysis, is applicable to buffer carrier constituents but not to commercial grade synthetic carrier ampholyte mixtures since these contain, at least in some cases, large molecular weight non-dialysable compounds; the latter are also removed to a lesser degree during the gel filtration step compared to those species within the nominal molecular weight distribution of the particular synthetic carrier ampholyte mixture (41). This problem can be avoided by passing the commercial grade synthetic carrier ampholyte mixture through an Amicon UM-2 ultrafiltration membrane prior to use in electrofocusing. Alternatively, one can remove residual carrier ampholyte mixtures from proteins by an additional filtration step such as ion exchange chromatography (42), although at the risk of losing some or even all of the isolated protein by adsorption to the particular fractionation matrix used or to the walls of the containers. Weighed against this risk, it is often preferable to tolerate a small amount of contamination of the protein with synthetic carrier ampholyte mixture.

REFERENCES

1. Chrambach,A., Jovin,T.M., Svendsen,P.J. and Rodbard,D. (1976) *in* Methods of Protein Separation (Catsimpoolas,N., ed.), Plenum Press, New York, Vol.2, p. 27.[1]

2. Nguyen,N.Y., McCormick,A.G. and Chrambach,A. (1978) Anal. Biochem. **88**, 186.

3. Jackiw,B.A. and Brown,R.K. (1980) Electrophoresis 1, 107.

4. Jackiw,B.A., Chidakel,B.E., Chrambach,A. and Brown,R.K. (1980) Electrophoresis 1, 102.

5. Chrambach,A., Hjelmeland,L., Nguyen,N.Y. and An der Lan,B. (1980) *in* Electrophoresis '79 (Radola,B.J., ed.), Walter de Gruyter, Berlin-New York, p. 3.[1]

6. Chidakel,B.E., Nguyen,N.Y. and Chrambach,A. (1977) Anal. Biochem. **77**, 216.

7. Baumann,G. and Chrambach,A. (1976) Anal. Biochem. **70**, 32.

8. Tipton,H., Rumen,N.M. and Chrambach,A. (1975) Anal. Biochem. **69**, 323.

9. Peterson,J.I., Tipton,H.W. and Chrambach,A. (1974) Anal. Biochem. **62**, 274.

10. Jackiw,A. (1979) XI International Congress Biochem. Montreal, Abstract p.716.

11. Nguyen,N.Y. and Chrambach,A. (1980) Electrophoresis **1**, 14.

12. McCormick,A., Miles,L.E.M. and Chrambach,A. (1976) Anal. Biochem. **75**, 314.

13. An der Lan,B. and Chrambach,A. (1980) Electrophoresis **1**, 23.

14. Vinogradov,S.N., Lowenkron,S., Andonian,M.R., Bagshaw,J., Felgenhauer, K. and Pak,S.J. (1973) Biochem. Biophys. Res. Commun. **54**, 501.

15. Chrambach,A. and Nguyen,N.Y. (1977) *in* Electrofocusing and Isotachophoresis (Radola,B.J. and Graesslin,D., eds), Walter de Gruyter, Berlin-New York, p. 51.

16. Nguyen,N.Y., Rodbard,D., Svendsen,P.J. and Chrambach,A. (1977) Anal. Biochem. **77**, 39.

17. Svensson,H. (1962) Acta Chem. Scand. **16**, 456.

18. Nguyen,N.Y. and Chrambach,A. (1976) Anal. Biochem. **74**, 145.

19. Nguyen,N.Y., Grindeland,R.E. and Chrambach,A. (1981) Prep. Biochem., **11**, 173.

20. Righetti,P.G. and Chrambach,A. (1978) Anal. Biochem. **90**, 633.

21. Hjelmeland,L.M., Allenmark,S., An der Lan,B., Jackiw,B.A., Nguyen,N.Y. and Chrambach,A. (1981) Electrophoresis, in press.[1]

22. Hjelmeland,L.M., Nebert,D.W. and Chrambach,A. (1979) Anal. Biochem. **95**, 201.

23. Gianazza,E., Astorri,C. and Righetti,P.G. (1979) J. Chromatogr. **171**, 161.

24. Hjelmeland,L.M. (1980) Proc. Nat. Acad. Sci. USA **77**, 6368.

25. Nguyen,N.Y. and Chrambach,A. (1977) Anal. Biochem. **79**, 462.

26. Nguyen,N.Y. and Chrambach,A. (1977) Anal. Biochem. **82**, 54.

27. Nguyen,N.Y. and Chrambach,A. (1977) Anal. Biochem. **82**, 226.

28. Baumann,G. and Chrambach,A. (1975) *in* Progress in Isoelectric Focusing and Isotachophoresis (Righetti,P.G., ed.), Elsevier, Excerpta Medica, North Holland, Assoc. Sci. Publ., Amsterdam, p. 13.

29. Doerr,P. and Chrambach,A. (1971) Anal. Biochem. **42**, 96.

30. Salokangas,A., Eppenberger,U. and Chrambach,A. (1981) Prep. Biochem., in press.

31. Chrambach,A. (1980) Mol. and Cell. Biochem. **29**, 23.

32. Jackiw,B.A. and Chrambach,A. (1980) Electrophoresis **1**, 150.

33. Finlayson,G.R. and Chrambach,A. (1971) Anal. Biochem. **40**, 292.

34. Dirksen,M.L. and Chrambach,A. (1972) Sep. Sci. **7**, 747.[1]

35. Reisner,A.H., Nemes,P. and Bucholtz,C. (1975) Anal. Biochem. **64**, 509.

36. Diezel,W., Kopperschlaeger,G. and Hofmann,E. (1972) Anal. Biochem. **48**, 617.

37. Vesterberg,O., Hansen,L. and Sjosten,A. (1977) Biochim. Biophys. Acta **491**, 160

38. Righetti,P.G. and Gianazza,E. (1978) Biochim. Biophys. Acta **532**, 137.

39. Nguyen,N.Y. and Chrambach,A. (1979) J. Biochem. Biophys. Methods **1**, 971.

40. Nguyen,N.Y. Baumann,G., Arbegast,D., Grindeland,R.E. and Chrambach,A. (1981) Prep. Biochem., **11**, 139.

41. Baumann,G. and Chrambach,A. (1975) Anal. Biochem. **64**, 530.

42. Baumann,G. and Chrambach,A. (1975) Anal. Biochem. **69**, 649.

CHAPTER 5

Two-Dimensional Gel Electrophoresis

JOHN SINCLAIR and DAVID RICKWOOD

INTRODUCTION

The electrophoretic separation of proteins has become one of the main methods for the fractionation and characterisation of all types of proteins at both the analytical and preparative levels. This form of separation depends not only on the molecular weight of polypeptides but also on their overall charge which in turn depends on the amino acid composition of the polypeptides, the presence or absence of detergents and the pH of the solution.

As described in detail elsewhere in this book there are two types of separation based on either the mobility of polypeptides related to their size and net charge, or equilibrium separation methods, such as electrofocusing, which separate polypeptides only on the basis of charge independently of their molecular weights. Irrespective of the method chosen, there is a real possibility that two or more polypeptides may comigrate as a single band particularly if the original sample is a complex mixture of proteins. The comigration of polypeptides cannot only mask the true complexity of the proteins but also it can mask any changes in the amount of the components present in any band. It is for these reasons that two-dimensional separations have been developed incorporating separation by electrophoresis in one direction followed by a second separation at right angles to the first.

Two-dimensional separations should be designed so that the polypeptides are separated on different bases in each direction. Systems that do separate polypeptides on a similar basis in each direction give an essentially diagonal array of spots, the resolution of which may not be much greater than that obtained using a single dimension. Besides improved separations, additional information about the basic parameters of polypeptides can be obtained by careful choice of the type of separation used in each dimension. Thus electrophoresis in the presence of SDS gives an estimate of molecular weight. On the other hand electrofocusing gives a separation based on the isoelectric point of polypeptides which reflects the amino acid composition of polypeptides and in some cases their degree of post-translational modification. Hence using these two parameters one can obtain rapid estimates of the molecular weight and isoelectric point of a protein.

However, the nature of the sample proteins themselves may preclude some types of separation. For example, some proteins may be extremely basic or acidic such that their isoelectric points lie outside the range of the pH gradient of the gel. In addition, two-dimensional separations do not guarantee that individual spots represent single polypeptides, especially if the original sample contains many proteins (e.g. total cellular proteins); rather the use of two-dimensional separations minimises the possibility of comigration of polypeptides. Another factor to be remembered is that comigration of polypeptides may arise not because they are of similar size or charge, but rather because they are interacting with each other and hence they are migrating as an aggregate rather than individually. Proteins from nuclei and membranes frequently form very stable aggregates and so the separation of such samples must be carried out in media which ensure their complete dissociation. In this chapter some of the different techniques used for two-dimensional separations are described together with the variations used for particular types of samples.

APPARATUS

In most cases the amount of specialised apparatus required for two-dimensional gels is minimal though, as discussed at the end of this section, where large numbers of gels are to be run on a routine basis it may well be worth investing in one of the more specialised types of apparatus for running several gels at once.

First-dimensional Gels

The first-dimensional runs are usually carried out in gel rods using a simple apparatus which can take 8 or 12 gels (see p. 19). In some cases it is feasible to run the samples on slab gels and then slice the slab gel into strips ready for the second-dimensional separation. The advantage of running the samples on a slab gel is that all are separated under exactly the same conditions and hence the separations of different samples are more similar than when running a number of different rod gels. On the other hand the limited physical strength of thin strips of slab gels may prove a real difficulty during the extensive handling of gels during equilibration with buffers. In addition, when the first-dimensional separation is electrofocusing the use of gel rods facilitates the mixing of the sample throughout the gel prior to electrophoresis as is required by some methods.

Second-dimensional Gels

The second-dimensional separations are always carried out in slab gels. In some early papers quite thick slab gels, 0.35 cm thick, were used in order to accommodate relatively thick first-dimensional gels. The problem with such gels is that they cannot be dried down successfully; generally gels can only be dried down if they are 0.15 cm or less thick. The material used for the slab gel plates is also important. Glass is generally used although the glass plates do tend to be fragile. While plastic plates are more durable, their low thermal conductivity as compared with glass means that they are less suitable in that the gel can overheat more easily resulting in distortion of the protein bands.

The necessity to apply the first-dimensional gel to the slab gel also affects the design of the plates. The actual method of application is variable and in some cases

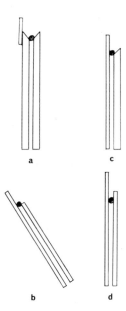

Figure 1. Methods of applying the first-dimensional gel to the second-dimensional slab gel. (a) use of bevelled plates (0.6 cm thick), (b) use of normal plates (0.3 cm thick), (c) a combination of bevelled and normal plates, (d) first-dimensional gel squashed between 0.3 cm thick plates.

this means that the plates must be modified. An example of this is the use of bevelled plates (*Figure 1a*) which, because they are thicker than normal, form a 'V' at the top in which the first-dimensional gel can be positioned and sealed in place. These bevelled plates also require an additional backing plate when the gels are run. In some cases the use of thick gel plates may reduce the cooling of the gels significantly so that it is necessary to use a lower current in order to keep the gels cool. The normal thin gel plates can also be used. In this case a notch into which the gels fit can be formed by slanting the plates (*Figure 1b*) and the gels are sealed in position. A combination of an ordinary backing plate and one bevelled plate can also be used effectively (*Figure 1c*) and it has the advantage of being simpler and of giving a greater degree of cooling than if both of the plates are of thick glass.

However, one problem in all of the cases described above is that the rod gel is very close to the top of the plates. If the rod gel is to be sealed in place using an acrylamide stacking gel with the systems described, usually the rod gel is never quite enclosed by the stacking gel as the mixture at the top never polymerises because the air inhibits polymerisation. An alternative is to seal the gel in position with agarose although the exposure of the gel to hot agarose may adversely affect the proteins being separated. This problem can be overcome by squashing the first-dimensional gel between the gel plates during assembly (*Figure 1d*). Rod gels originally much thicker than the slab gels can be squashed satisfactorily and neither the gels nor the quality of separation are affected by this procedure. However, one minor disadvantage of this system is that the rod gel must be equilibrated with the correct buffers before preparing the second-dimensional gel.

The types of apparatus described above can only take a single gel. In the authors' laboratory up to four sets of apparatus are run in parallel and this type of set-up will suit many potential users. However, in some cases where it is necessary to run large numbers of gels it is worthwhile considering the use of an apparatus that can run several gels simultaneously. The other advantage of running gels on such sets of apparatus is that the gels are likely to be more comparable because they are run under identical conditions. Extra wide gels, up to 80 cm wide, have been described (1) but most of the other types of apparatus devised are designed to run a number of standard size gels. An example of a multi-gel apparatus is shown in *Figure 2,* and *Table 1* is a list of some references which will be found useful if the reader wishes to construct such an apparatus. However, it is important to realise that there is not only the problem of running a large number of gels but also of processing them through staining, destaining, photography and in some cases autoradiography. Frequently, there are more problems in processing large numbers of gels than in running them.

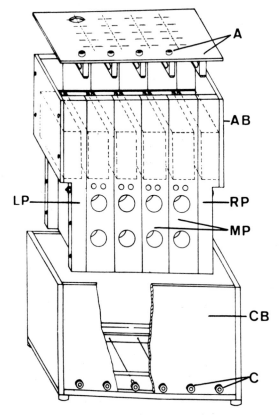

Figure 2. Apparatus for running multiple gel slabs. The perspective view of the apparatus is shown. A, Perspex plate with anode terminals; MP, middle part; RP and LP, right and left side parts, respectively; AB, anode buffer chamber; CB, cathode buffer chamber; C, cathode terminals. (Reproduced from ref. 13 with permission.)

Table 1. Apparatus for Running Multiple Gel Slabs.

Authors	Number of gels	Gel dimensions	Reference
E.Kaltschmidt and H.G.Wittmann	5	19 x 20 cm	13
B.Dean	8	13 x 16 cm	21
N.G.Anderson and N.L.Anderson	10	17 x 17 cm	20
J.I.Garrels	4	20 x 17 cm	1
M.I.Jones, W.E.Massingham and S.P.Spragg	10	17 x 17 cm	25

GENERAL TECHNIQUES FOR TWO-DIMENSIONAL GELS

The art of running two-dimensional gels can only be learned by the experience gained in the laboratory and the actual sequence of operations will depend on the exact two-dimensional technique used. This section deals specifically with experimental procedures and general techniques common to most two-dimensional gel electrophoretic separations. In addition the reader should read the appropriate sections of Chapter 1 which deal with the general techniques as applied to one-dimensional gels.

Solutions and Apparatus

As stated earlier, acrylamide solutions are very toxic and must be handled with the greatest of care. The solutions used for two-dimensional gel electrophoresis are similar to those used for one-dimensional separations. However, the overall greater complexity of two-dimensional separations makes it important that, whenever possible, all solutions required for the preparation and running of gels are prepared in advance as stock solutions. Most solutions can be stored at $-20°C$ or at $5°C$ in the presence of a bacterial inhibitor such as chloroform. There are some limitations to this approach. For example, ammonium persulphate solutions should be prepared on the same day as the gels. In addition the presence of urea in solutions can lead to the formation of cyanate ions which can modify and degrade polypeptides. The rate of formation of cyanate ions increases with temperature but they can be neutralised by the addition of compounds containing a free amino group. Thus urea solutions should be freshly prepared, deionised before use by passage through a mixed-bed resin, kept cool and, where possible, buffered with Tris buffer.

Originally in situations where a reducing agent was required, 2-mercaptoethanol was used. However, at concentrations at which it is effective (approx. 10 mM), 2-mercaptoethanol inhibits the polymerisation of polyacrylamide gels. Hence reducing agents effective at lower concentrations, such as dithiothreitol, should be used whenever possible. In the following protocols dithiothreitol is used throughout as the reducing agent in all cases where the use of 2-mercaptoethanol might adversely affect the gels. Where this differs from the original protocol it is noted in the text. Another inhibitor of polymerisation is oxygen and hence the rate of polymerisation can be enhanced by degassing the gel solutions under vacuum before use; this also minimises the formation of air bubbles during polymerisation of the gel.

As noted in the preceding section the apparatus used for two-dimensional gels is fairly conventional. The most important points are to ensure that the gel tubes and plates are cleaned to ensure even adhesion of the gel to the glass. This is particularly true when working with slab gels where the gel may have a tendency to tear; gel plates must be washed in chromic acid, washed with distilled water and then ethanol and dried. Some methods also recommend that gel tubes and plates are siliconised before use (see p. 28) though this does introduce the problem that gels, particularly the low-percentage acrylamide tube gels used in the first dimension, will slip out of the tube. Thus when using siliconised gel tubes it is advisable to cover the bottom of the gel tube with Nylon gauze or perforated Parafilm before running the gel.

Preparation of the Sample

The sample and the solution that the sample is loaded in vary greatly from one type of separation to another. Hence there is a similar diversity in the exact way in which the sample is prepared and so here only general hints are given. The reader should also consult the appropriate earlier sections of this book.

Perhaps the most important rule in the preparation of samples is to ensure that there is no particulate material arising from protein aggregation in the sample solution, otherwise severe streaking can occur resulting from the slow dissolution of the aggregates in the sample buffer. As a routine precaution samples should be centrifuged for 5 min in a microfuge at about 10,000xg before applying them to the first-dimensional gel.

Another problem can be the presence of nucleic acids in the sample which, in interacting with the polypeptides, can adversely affect the resolution obtained on gels. This problem can be especially serious in separations which employ electrofocusing in one of the dimensions. Hence, particularly when preparing protein samples from nucleoprotein complexes, steps must be taken to minimise the contamination by nucleic acids.

In some cases nucleic acids can be removed by selective extraction or precipitation procedures, for example the acid-extraction of histones from chromatin (2) and the LiCl precipitation of rRNA for the isolation of ribosomal proteins from ribosomes (3). High molecular weight nucleic acids can be removed either by rate-zonal or iso-pycnic centrifugation (4). Isopycnic separation on CsCl or Cs_2SO_4 gradients is usually considered to be better because, not only is there no possibility of large protein aggregates being pelleted, but also isopycnic centrifugation can be used to concentrate the protein sample. Some proteins form a pellicle at the top of the gradient after isopycnic centrifugation which is very difficult to dissolve in urea or even SDS and in such cases it is better to lower either the sample loading or alternatively the concentration of gradient solute to ensure that the proteins form a band within the gradient. Neither of these centrifugation methods removes low molecular weight nucleic acids.

Nucleic acids can also be removed from the sample by passing it over a column of ion-exchange material. One such material that can be used is hydroxyapatite which preferentially binds nucleic acids even in solutions of high ionic strength containing high concentrations of other dissociants such as urea or guanidine hydrochloride (4). The problem of all such column methods is that losses may occur as the result of non-specific adsorption onto the ion-exchange medium. In addition the conditions for the adsorption of nucleic acids may lead to some selective losses of very acidic proteins.

The final method of removing nucleic acids is by enzymatic digestion using specific nucleases. Such procedures do, however, add additional proteins to the sample unless insolubilised enzymes attached to Sepharose are used. It is particularly important to ensure that both the sample and the nucleases are free of any proteolytic activities. Finally the activity of nucleases may be partially inhibited by the binding of sample proteins to the nucleic acids.

In separations involving rate-zonal electrophoresis it is advantageous to add a dye which will mark the ion front to ensure that no proteins are lost off of the bottom of the gel. In the case of equilibrium methods, for example electrofocusing, either gels should be run for different lengths of time or samples should be loaded at both ends to ensure that polypeptides reach their equilibrium positions.

Preparation of the First-dimensional Gel

This is the simplest part of the two-dimensional separation. The gels are prepared in tubes using the techniques described in detail in Chapter 1 and the proteins are fractionated using one of the techniques described in the later sections of this chapter. It should be emphasised that the final quality of the two-dimensional separation is very dependent on the degree of resolution obtained in the first dimension. Hence the reader should be prepared to try variations of the techniques described in this chapter in order to optimise the separation obtained. After electrophoresis the gels (still in their tubes) can be stored in the deep freeze at $-20°C$ as long as the orientation of the gel is clearly marked together with its identity.

Equilibration of the First-dimensional Gel for the Second Dimension

The gel is removed from the tube by cracking the tube or by extruding it under gentle pressure after rimming with a needle as necessary (see p. 43). Polyacrylamide gels that have been frozen and thawed are generally easier to remove from tubes. It is very important that proteins from the skin are not transferred to the gel and so plastic disposable gloves should be worn at all times during these manipulations. In addition do not allow the gel to come into contact with any paper surface since gels, particularly low-percentage polyacrylamide gels, readily stick to such surfaces. If it is necessary to put the gel down, put it into a trough of aluminium foil. Otherwise keep it in the palm of the glove.

The first priority is to mark which end of the gel is which and it is of course very important that any such orientation mark remains throughout the equilibration of the gel in the various buffers. In the authors' laboratory one end of the gel is marked by the insertion of a small piece of fine gauge wire (approx. 0.1 mm diameter). In rate-zonal runs the wire marker can have a dual purpose; if the wire is inserted at the ion front, which is usually shown by the position of a marker dye, then by cutting of the gels at this point before the second-dimensional separation one can normalise the position of spots in the gel even if the gels have run different distances. In the case of equilibrium systems there is no ion front and it is necessary simply to adopt a fixed convention, for example, to always mark, say, the bottom end of the gel. Identification of the two ends of low-percentage polyacrylamide electrofocusing gels is aided by the fact that the acid end swells after removal of the gel from the tube. Besides using wire to mark one end of the gel it is also possible to inject one end with a small amount of India ink.

At this stage the gels can also be stored in suitably labelled aluminium-foil troughs in the deep freeze at $-20°C$. There appears to be no degradation of the polypeptides nor diffusion of the protein bands under these conditions.

The next stage is to equilibrate the gel rod ready for the second dimension. It should be stressed that all such equilibration procedures in terms of times, temperatures and buffers tend to be somewhat empirical. The total equilibration time recommended in various methods varies from 5 min to 2 h. Longer times and higher temperatures will increase the rate of diffusion of molecules and hence may adversely affect the final resolution obtained. The published procedures should be satisfactory provided that the types of gel run are exactly as given in the published method. However the reader may wish to optimise the equilibration conditions for his own particular type of separation.

One problem of the equilibration process is that it leads not only to the equilibration of pH and ionic content of the gel but also to the loss of macromolecular material from the gel. This effect can be helpful in electrofocusing gels in removing low molecular weight ampholytes. However, less desirable is the potential loss of polypeptides, particularly low molecular weight species, from the gel. The loss of polypeptides during equilibration can be as high as 25% and hence for low loadings of protein it can be particularly serious. The loss of protein can be greatly reduced by putting the gel through the staining procedure (see p 197) before the equilibration is done. This fixes the polypeptides and minimises their diffusion during the equilibration. This procedure does not appear to result in a higher retention of the polypeptides in the first-dimensional gel during electrophoresis in the second dimension. One advantage of this approach is that it is possible to analyse distribution of polypeptides before running the second dimension.

Preparation of the Second-dimensional Gel

It is advisable to set the apparatus up early and to start preparing the second-dimensional gel during the equilibration of the first-dimensional gel. The polyacrylamide slab gel is set up in the standard way (see p. 35). The resolving gel can either be uniform or, for very heterogeneous samples, can consist of a gradient. The gradient can be either linear or exponential and prepared using the appropriate type of gradient maker (see p. 73). When preparing such gradients there is always a possibility of variations of the gradient profile from one gel to another and so it is advisable to run appropriate standard marker polypeptides in the second dimension. If the sample mixture contains a discrete mixture of high and low mobility components a discontinuous gradient can be used, though frequently separations on such gels lead to artifactual bands at the interface between the two gel concentrations. If the polypeptides of the sample have similar mobilities then frequently an optimal separation can be obtained using a single gel concentration which is not only simpler but also gives the most reproducible results.

The first-dimensional gel is usually placed in the top of the slab gel plates and sealed in place with a stacking gel of low-percentage polyacrylamide. It is important that the percentage of polyacrylamide of the stacking gel is less than that of the first-dimensional gel that it surrounds, otherwise a significant proportion of the protein will remain in the first-dimensional gel. When sealing the first-dimensional gel with the polyacrylamide stacking gel, the gel closest to the air frequently does not polymerise leaving the top of the gel exposed and this can adversely affect the separation in the second dimension. The problem can be overcome by casting the stacking gel on top of the resolving gel as usual and then sealing the first-dimensional gel in position with 1% agarose in the appropriate buffer; this method has been widely adopted. When using agarose for sealing the gel in position, resolution can be improved by blotting the top of the stacking gel and draining all the excess liquid from the first-dimensional gel after equilibration. Whichever method is used it is important to ensure that the first-dimensional gel is as straight as possible and that there are no air bubbles between it and the surrounding gel. Before the sealing gel sets, a sample well should be made at one end of the slab gel which can be used for calibrating the gel with standard marker polypeptides, usually in the presence of a marker dye. The gels are then placed into a vertical slab gel apparatus and electrophoresed until the marker dye reaches the bottom of the gel.

Analysis of the Distribution of Polypeptides

The methods for detecting polypeptides fall into two main groups; staining methods, and for radioactively labelled polypeptides, autoradiographic methods. Most staining procedures are fairly similar and do not differ greatly from those described for one-dimensional gels in Chapter 1.

Detection of Polypeptides by Staining Procedures

Most variations relate to the concentration of the stain, composition of the stain solvent, temperature and length of time that the gel is stained for. Again, frequently the conditions used are empirical in many cases though some studies have indicated that some stains, notably Coomassie blue, are more sensitive than others, being able to detect microgram amounts of polypeptides in gels.

The staining methods given in *Table 2* have been used for gel separations described later in this chapter and are typical examples of staining procedures. Many staining techniques are similar and a full bibliography can be found at the end of this volume.

One problem associated with staining gels which contain ampholytes or which have been run as the second dimension to a gel containing ampholytes is that the ampholytes themselves may behave as small polypeptides. Consequently they can bind SDS, migrate into the gel, running in a position just behind the dye front, and bind most stains used for detecting proteins. If the slab gel contains a high percentage of acrylamide or is a gradient gel with a high percentage acrylamide at the bottom, it is possible to let the marker dye run off the end of the gel together with the ampholytes. The high concentration of acrylamide will tend to slow down the migration of the sample

Table 2. A Summary of Staining Techniques for Polyacrylamide Gels.

	Staining procedure	Destaining procedure	Additional information	Reference
A.	3-4h in 0.1% Coomassie blue in methanol:water:acetic acid (5:5:1)	Overnight by diffusion against methanol:water: acetic acid (5:5:1)		19
B.	20 min in 0.1% Coomassie blue in 50% TCA	Several changes of 7% acetic acid	Removal of ampholytes	10
C.	15 min in 0.55% amido black in 50% acetic acid	40h in 1% acetic acid		13
D.	3h in 0.25% Coomassie blue in methanol:water:acetic acid (5:5:1)	Several changes of 5% methanol, 10% acetic acid		18
E.	Overnight in 25% isopropyl alcohol, 10% acetic acid, 0.025-0.05% Coomassie blue, followed by 6-9 h in 10% isopropyl alcohol, 10% acetic acid, 0.0025-0.005% Coomassie blue	Several changes of 10% acetic acid	An additional optional staining step overnight in 10% acetic acid containing 0.0025% Coomassie blue helps intensify the gel pattern	12
F.	1-4 h in 0.1% Coomassie brilliant blue R250 in 7.5% acetic acid, 50% methanol in water	Overnight in 7.5% acetic acid, 50% methanol in water		14
G.	3h at 80°C, or overnight at room temperature in 0.1% amido black in 0.7% acetic acid, 30% ethanol in water	Several changes of 7% acetic acid, 20% ethanol in water		15

polypeptides preventing their loss from the bottom of the gel. A number of other various methods have been devised to minimise the interference of ampholytes. For example, soaking the gel in 50% ethanol, 7% (v/v) glacial acetic acid and 0.005% Coomassie blue (to prevent destaining of the polypeptides), as described by O'Farrell (10) removes most of the ampholytes except for a very fine band at the dye front. In addition there is usually a difference in the hue of the stain when bound to polypeptides as compared with ampholytes and this makes each clearly distinguishable. Problems will arise though if the stained ampholytes mask the presence of any low molecular weight polypeptides. In such cases extraction of the ampholytes is essential.

Another method of visualisation of polypeptides, which is used to a lesser extent than commonly used polypeptide stains but requires no staining or destaining, involves the precipitation of unbound SDS in the gel by soaking it for 2h in 10 volumes of 4M potassium acetate at 25°C. This results in the gel turning opaque except where SDS is already bound to polypeptides (5). Special types of polypeptides can also frequently be detected on the basis of their specific properties. For example, on non-denaturing gels, enzymes which have a spectrophotometric assay can be specifically stained by incubating the gel in the appropriate enzyme assay mixture while glycoproteins can be detected by using the periodate-Schiff reaction (6). Appendix I provides a reference list of the methods which have been devised for detecting polypeptides in gels.

Detection of Labelled Polypeptides by Autoradiography

The use of radioisotopes, as previously mentioned, provides a much more sensitive method for locating polypeptides, being as much as four orders of magnitude more sensitive than the sensitivity obtained by conventional staining methods. The lower loadings of sample possible greatly enhance the resolution in both quantitative and qualitative terms.

Proteins can be labelled *in vivo* by exposing the cells to radioactive amino acids. However, frequently such labelling procedures are hampered by the endogenous pools of amino acids, and metabolism of the precursors can lead to non-specific incorporation. Proteins can usually be labelled to much higher specific activities by labelling them *in vitro* either by labelling them during synthesis *in vitro* or by chemical modification of the proteins, for example, by iodination (7) or by reductive methylation (8), the latter of which involves reacting some of the amino groups of the protein with formaldehyde and sodium borohydride; either reagent can be isotopically labelled. A full bibliography of the methods available for labelling proteins *in vitro* is given in Appendix II. A detailed description of the autoradiography and fluorography procedures for detecting radiolabelled proteins after electrophoresis is given in Chapter 1.

As mentioned above, labelling proteins with radioactive amino acids *in vivo* can lead to randomisation of the label into other molecules as a result of metabolic processes. Similarly, studies of post-translational modifications of proteins *in vivo* such as phosphorylation and acetylation can also be affected by the randomisation of label. Thus phosphate is readily incorporated into lipids and nucleic acids as well as proteins and contamination of the latter in particular may cause problems when analysing phosphoproteins. Even when labelling proteins *in vitro* many of the procedures will also label contaminants such as nucleic acids and lipids and subsequently they may mask the distribution of polypeptides in the gel. A typical method used for removing nucleic acid contamination from a polyacrylamide gel which is to be autoradiographed is to run the gels and then hydrolyse the nucleic acids by soaking the gel in 7% TCA for 2h followed by incubation in 7% TCA at 90°C for 30 min. The gel is washed in more 7% TCA to remove the labelled nucleotides, shrunk and dried as required and then autoradiographed. However, in some cases, for example electrofocusing, it is necessary to remove the nucleic acids before running the sample in order to optimise the separation; this is discussed elsewhere in this chapter (see p.194).

Comparative Analysis of Two-dimensional Gels

Usually gel electrophoresis is undertaken on a comparative basis between one, two or many different samples. The use of a one-dimensional separation using a slab gel allows all the samples to be electrophoresed under the same conditions, and any minor variations in running conditions will affect all the samples identically.

However usually two-dimensional gels can be used for only one sample unless special sets of apparatus are used. Due to variabilities such as minor fluctuations in the shape and range of the pH gradient of electrofocusing gels, the distance of run on non-equilibrium gels, and stretching of the first-dimensional gel when it is applied to the slab gel, gel patterns are very rarely completely superimposable. Consequently, it is difficult if not impossible to analyse proteins by their absolute migration distances.

Although microdensitometric tracings linked to computer analysis have been used (9), most analyses are done visually using the ability to superimpose certain areas of gels which then allows the presence, absence or reduced intensity of certain spots to be noted. On most occasions major protein spots can be used as internal markers or reference points from which minor proteins can be analysed from their relative positions to these markers. Whenever possible, standard proteins should be run, especially in the second dimension, to aid the comparison of gels run at different times. When relating protein synthesis to the relative intensity of autoradiogram spots using labelled amino acids, it must be realised that the percentage of any one amino acid can vary widely between different polypeptides. For example, if [^{35}S]-methionine is used, some correction must be made to account for the fact that proteins, such as actin, have between two and four times the methionine content of most other proteins. The same argument also holds true for any one labelled amino acid and no correction can be made unless the amino acid composition of the proteins to be compared is known. Also when labelling proteins *in vivo* the intracellular amino acid pools can vary significantly not only between different cell types but also in different physiological states of a single cell type. Difficulties can also arise because the size of individual spots is related to both the exposure time and the radioactivity of a protein spot. Thus two or three closely positioned protein spots can appear as one autoradiogram spot. Reducing the autoradiographic exposure is sometimes able to resolve this problem. Similarly, when using isotopic labelling to analyse protein modification, an increase in spot intensity can be due to either an increase in the amount of modified protein or an increase in modification of that protein. Coomassie blue staining may be able to help distinguish between these alternatives.

TWO-DIMENSIONAL SEPARATIONS OF PROTEINS ON THE BASIS OF ISOELECTRIC POINTS AND MOLECULAR WEIGHTS

While there are many different types of two-dimensional separation for proteins the authors have chosen this method and its variations as the main method, not only because it has been shown to be suitable for a wide variety of proteins but also because it gives both increased resolution of polypeptides and additional information on their molecular weights, amino acid composition and in some cases their post-translational modifications. There are a large number of variations in terms of this single method and here only the most established methods will be described which involve electrofocusing in the first dimension and SDS gel electrophoresis in the second.

The technique of electrofocusing separates polypeptides on the basis of their isoelectric points. Consequently, information about the amino acid composition of the proteins can be obtained. The molecular weight of the protein only affects the rate at which the protein reaches its isoelectric point and, because the separation is an equilibrium one, the length of the run does not affect the separation obtained as long as the pH gradient remains stable. This form of separation, followed by a second separation on the basis of molecular weight, usually SDS-PAGE facilitates enormously the task of characterising polypeptide species. The ability of this type of two-dimensional separation to reveal additional information about polypeptides as well as its inherent sensitivity and high resolving power have made it an invaluable tool in the separation and analysis of a wide variety of proteins.

First-dimensional Separation

The use of electrofocusing for separating proteins has been described in detail in Chapter 4 and the reader is advised to consult this chapter for background information on this technique. The pH gradients for focusing are usually formed by subjecting commercially prepared ampholytes to an electric potential across the ends of the gel (see Chapter 4). The pH gradient is usually formed within the first hour of electrophoresis, the polypeptides move more slowly and eventually reach their isoelectric position in the gradient where they have no net charge and hence cease to migrate. As long as sufficient time is given for focusing, the proteins will take up positions along the gel based solely on their isoelectric point and no other factor, though when using denaturing conditions it should be remembered that the isoelectric point of the denatured polypeptides may differ from that of the native proteins. Although this is an equilibrium method, a number of factors can affect the separation obtained. Different brands and even different batches of the same ampholyte can vary and hence produce slightly different gradients. In addition, the gradient may tend to drift towards the basic end. Both of these difficulties can be overcome by checking the pH gradient of a blank gel prepared and run at the same time as the other gels. The other problem that must always be considered is that the isoelectric points of some of the sample polypeptides may lie outside the range of the gradient and as such they may migrate off of the gel. Also because electrofocusing is a very sensitive technique and can resolve proteins with only a single charge difference, it is important to avoid any procedures which might result in chemical modification of polypeptides, hence leading to the formation of artifactual microheterogeneity.

A number of workers have developed two-dimensional separations involving the combination of electrofocusing and SDS-PAGE. However, it was O'Farrell (10) who first pointed out that using isotopically labelled proteins it was possible to resolve several thousand different polypeptides. Using the less sensitive conventional staining procedures far fewer proteins can be resolved since more sample must be loaded onto the first-dimensional gel. The method devised by O'Farrell has been widely used for separating the proteins of both prokaryotic and eukaryotic organisms and can be used for the separation of many types of cellular proteins. However, it is important not to apply the method blindly because the method is not applicable to all proteins.

As previously mentioned, because this separation is very sensitive to the charge of polypeptides any procedures which might cause artifactual changes in the isoelectric point of samples must be avoided (see General Techniques). Samples which contain high amounts of nucleic acids must also be treated to remove the nucleic acids before electrofocusing, otherwise they may interact with basic proteins and ampholytes. The result of such interactions are the smearing or streaking of the polypeptide spots in the first dimension. The removal of nucleic acids can be carried out by any of the conventional means as described in the General Techniques section (p.194).

The gels and buffer systems used are given in *Table 3*. The gel mix is degassed and then polymerisation is initiated by the addition of 10 μl freshly prepared 10% ammonium persulphate. The gel solution is pipetted into gel tubes (0.25 cm internal diameter and 13 cm long) up to within 0.5 cm of the top, overlayered with water and allowed to set for approximately 60 min. After aspiration of the overlay the upper surface of the gel is allowed to equilibrate for a further hour with 25 μl of sample buffer:

9.5 M urea
5% 2-mercaptoethanol
2% Nonidet P-40
1.6% Ampholines (pH 5-7)
0.4% Ampholines (pH 3.5-10)

After the sample buffer has been removed the gels are placed in the electrophoresis apparatus containing 10 mM H_3PO_4 in the lower reservoir which is connected to the positive terminal and 20 mM NaOH in the upper reservoir which is connected to the negative terminal. About 25 μl of fresh sample buffer is layered onto each gel and the gels are prerun first for 15 min at 200 V then for 30 min at 300 V and finally for 30 min at 400 V. After prerunning the gel the NaOH is removed from the upper reservoir and the liquid above the gels is gently removed by aspiration. Usually about 25 μl of sample containing approximately 20 μg of radioactively labelled protein is loaded onto the gel in the sample buffer but if the rod gel is made 1 cm shorter a maximum of 100 μl of sample can be applied. If the sample is not a complex mixture of polypeptides less can be loaded onto the gel. When polypeptides are to be detected by Coomassie blue staining the loading should be increased to between 50 μg and 250 μg protein depending on the complexity of the sample. The sample is then overlayered with 10 μl of overlay buffer [9 M urea, 0.8% Ampholine (pH 5-7), 0.2% Ampholine (pH 3.5-10)]. The upper chamber is refilled with fresh 20 mM NaOH and electrophoresis is continued for 13.5 h at 400 V or 18 h at 300 V. In both cases for the final hour the voltage is increased to 800 V to sharpen the bands.

Table 3. Preparation of the First-dimensional Electrofocusing Gel (O'Farrell Method).[a]

To make 10 ml of the first-dimensional gel, mix:
    5.5 g of urea (ultrapure)
    1.33 ml 28.38% acrylamide, 1.62% bisacrylamide
    2 ml 10% Nonidet P-40 (NP 40)
    0.4 ml 40% Ampholines (pH 5-7)
    0.1 ml 40% Ampholines (pH 3.5-10)
    1.95 ml water
    5 μl TEMED
    10 μl 10% ammonium persulphate

[a]An alternative recipe for DATD-crosslinked gels is given in Chapter 4.

The rod gels are extruded from the tubes under gentle pressure and equilibrated for two periods of 60 min in 5 ml of SDS sample buffer comprising 2.3% SDS, 5% 2-mercaptoethanol, 10% glycerol, 62.5 mM Tris-HCl (pH 6.8).

Second-dimensional Separation

An exponential gradient slab gel of 10-16% acrylamide is prepared 16.4 cm wide by 14.6 cm high and 0.8 mm thick with a 4% acrylamide stacking gel. The resolving gel is prepared using a gradient maker with two gradient solutions prepared from stock acrylamide solutions as shown in *Table 4*. The ammonium persulphate is added just prior to pouring the gel solutions into the gradient maker. The degassed light solution (16 ml) is poured into the reservoir chamber of a standard exponential gradient maker (see p. 76) and 5 ml of degassed dense solution is poured into the mixing chamber which is kept well stirred using a magnetic stirrer. The tap between the chambers is opened and the gel is poured using a flow rate of about 5 ml/min by either gravity or using a peristaltic pump until the gel is about 2.5 cm from the top of the plates. The mixing chamber of the gel pourer will still contain some residual acrylamide solution and it is very important that it is discarded and the apparatus washed out before the gel solution polymerises. The gel is overlayered with water and allowed to set for about one hour. After removal of the overlay by aspiration and its replacement with a four-fold dilution of the lower gel buffer the gel is left overnight.

While the first-dimensional rod gels are being equilibrated, the stacking gel is prepared as given in *Table 4*, and polymerisation is initiated by the addition of 10% ammonium persulphate. The buffer overlay on the gradient resolving gel is removed and the stacking gel is poured up to the top of the plates. Air is excluded from the stacking gel either by overlayering with water or by placing a Teflon strip over the gel

Table 4. Preparation of the Gel for the Second-dimensional Separation (O'Farrell Method).

Resolving gel

Light solution:
    4 ml 0.4% SDS, 1.5M Tris-HCl (pH 8.8)
    5.3 ml 29.2% acrylamide, 0.8% bisacrylamide
    6.7 ml water
    8 μl TEMED
    25 μl 10% ammonium persulphate

Dense solution:
    2 ml 0.4% SDS, 1.5M Tris-HCl (pH 8.8)
    4.3 ml 29.2% acrylamide, 0.8% bisacrylamide
    1.7 ml 75% glycerol
    4 μl TEMED
    10 μl 10% ammonium persulphate

Stacking gel
    1.25 ml 0.4% SDS, 0.5M Tris-HCl (pH 6.8)
    0.75 ml 29.2% acrylamide, 0.8% bisacrylamide
    3.0 ml water
    5 μl TEMED
    15 μl 10% ammonium persulphate

to ensure an airtight seal. After polymerisation the overlay or Teflon strip is removed, the top of the gel is washed with water and the excess liquid is removed by blotting or aspiration. Depending on which sort of two-dimensional gel plates are used the rod gel is sealed to the stacking gel by pipetting a small amount of hot agarose sealing gel, containing 1% agarose in SDS sample buffer, into the bevel of the gel plates or into the groove formed by leaning the plates at an angle as described earlier in this chapter. Care should be taken to ensure that no air bubbles are trapped around the rod gel. The agarose should seal the rod gel to the stacking gel in about 5 min. The gel is electrophoresed at a constant current of 20 mA for about 4 h or until the marker dye runs off (see below) using 0.192 M glycine, 25 mM Tris base and 0.1% SDS as the electrode buffer containing 2 drops of 1% bromophenol blue as a marker dye. After electrophoresis the gel is carefully removed from between the gel plates and the polypeptide spots located by autoradiography (see p 199) or Coomassie blue staining (see *Table 2,* procedure B). A typical separation using this technique is shown in *Figure 3.*

The pH gradient formed in the first dimension using the recipe given above does not tend to be stable above about pH 7. Also, as mentioned previously, varying results may be obtained with different types of ampholytes and hence for comparative studies the same ampholytes should be used. The limitation of this system lies in the pH range of the gradient. Although the pH 3.5-10 Ampholines give a wide range it does not encompass all types of proteins. For example very basic or acidic polypeptides will either fail to enter the gel or migrate off of the end of the gel. Quantitatively similar results should be seen if the electrophoretic buffers in the first dimension are interchanged and the sample loaded at the acidic end of the rod gel and it may be found that basic polypeptides will tend to focus better if this procedure is used (see later). Whereas a broad range pH gradient will give the best separation for a

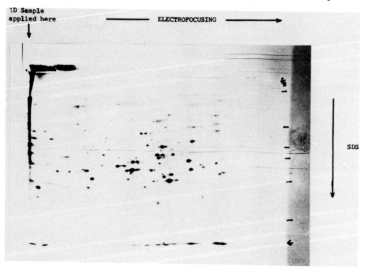

Figure 3.   Two-dimensional separation of proteins using the O'Farrell method. Total cellular proteins were labelled with [³⁵S]-methionine, separated using the O'Farrell technique described in the text, and the polypeptides were located by autoradiography.

complex mixture of proteins, a narrow range pH gradient can be effectively substituted for less complex mixtures or samples of known isoelectric points. If narrow range gradients are to be used it may be necessary to change the first-dimensional buffer systems. By using radioactively labelled proteins and detection by autoradiography, sensitivity can be increased more than four-fold over ordinary staining procedures. Consequently, a protein which constitutes 10^{-4}-10^{-5}% of the total protein can be visualised. Such high sensitivity and resolution can prove to be invaluable for many different types of protein analysis.

Modifications to the Basic Technique

Samples which have been lysed in the presence of SDS can still be electrophoresed in the first dimension by simply adding solid urea to a final concentration of 9.5 M. The SDS does not appear to affect the isoelectric point of the protein in this environment since the interaction of SDS with proteins in the presence of high concentrations of urea and non-ionic detergent is minimal. One disadvantage of the technique described above is the restriction on the volume of the sample which is essential if one is to get good resolution. A similar system routinely used in the authors' laboratory which enables much larger sample volumes to be used consists of setting the sample in the first-dimensional electrofocusing gel. This method also avoids the possibility that electrofocusing artifacts may arise as a result of the sample being applied in very acid or alkaline environments. Although some retention of the sample will occur, the possibility of increasing the sample volume more than compensates for this. Also the pH gradient does tend to be more stable over the higher pH ranges. The gels for the first dimension are made individually according to the recipe given in *Table 5*. The gel mixes are degassed and 10% ammonium persulphate is added to initiate the polymerisation. The gels are poured in gel tubes 14 cm long and 0.25 cm internal diameter to give a gel 10 cm long, overlayered with water and allowed to set for 60 min. The gels are transferred to the electrophoresis apparatus and electrophoresed at a constant voltage of 150 V for 24 h using 5% 1,2-diaminoethane in the upper (cathode) chamber and 5% orthophosphoric acid in the lower (anode) chamber. After electrophoresis the gels are extruded under gentle pressure and equilibrated for the second-dimensional separation. The gels are incubated separately at 37°C with shaking for 30 min in 50 ml each of a series of buffers. These are:

(i)   8 M urea, 0.1 M sodium phosphate (pH 7.0), 1% SDS,
      1 mM dithiothreitol, 5 mM Tris base.
(ii) 8 M urea, 0.01 M sodium phosphate (pH 7.0), 1% SDS,
      1 mM dithiothreitol, 5 mM Tris base.
(iii) 8 M urea, 0.01 M sodium phosphate (pH 7.0), 0.1% SDS,
      1 mM dithiothreitol, 5 mM Tris base.

The second-dimensional gel is similar to the gel system of Laemmli (see p. 26) but consists of a linear 5-20% acrylamide gradient gel containing 4 M urea with a 4% acrylamide stacking gel and is prepared as given in *Table 6*.

The bottom of the gel plates is sealed with 10 ml of the dense solution polymerised by the addition of a further 5 μl of TEMED and 0.2 ml of fresh 10% ammonium

Two-Dimensional Gel Electrophoresis

Table 5. Preparation of the First-dimensional Electrofocusing Gel.

0.5 ml of sample in 8M urea containing up to 800 μg of protein
0.133 ml 28.38% acrylamide, 1.62% bisacrylamide
0.3 g ultrapure urea
0.2 ml 10% Nonidet P-40
0.05 ml 40% Ampholines (pH 3.5-10)
2 μl TEMED
5 μl 10% ammonium persulphate

Table 6. Preparation of the Gel for the Second-dimensional Separation.

Resolving gel

Light solution (30 ml):
    1.5 g acrylamide
    0.04 g bisacrylamide
    3.75 ml 3M Tris-HCl (pH 8.8)
    0.3 ml 10% SDS
    9 ml water
    10 μl TEMED
    15 ml 8M urea
    1.5 ml 8 mg/ml riboflavin

Dense solution (30 ml):
    6.0 g acrylamide
    0.16 g bisacrylamide
    3.75 ml 3M Tris-HCl (pH 8.8)
    0.3 ml 10% SDS
    4.5 g sucrose
    3 ml water
    10 μl TEMED
    15 ml 8M urea
    1.5 ml 8 mg/ml riboflavin

Stacking gel
    2.5 ml 1M Tris-HCl (pH 6.8)
    1.75 ml 40% acrylamide, 1.06% bisacrylamide
    5.15 ml water
    10 ml 8M urea
    0.2 ml 10% SDS
    7.5 μl TEMED
    300 μl 5% ammonium persulphate

persulphate. Riboflavin solution (*Table 6*) is used for polymerisation; this is added to the light and dense solutions just before they are poured into the gradient maker. The gel is overlayered with 5 ml of overlay buffer prepared by mixing 2.54 ml 3 M Tris-HCl (pH 8.8), 17.26 ml water, 0.2 ml 10% SDS and left to polymerise with the aid of a fluorescent light source. At this stage the gel can be left to stand overnight before addition of the stacking gel.

The stacking gel is sealed with a Teflon strip and allowed to polymerise for 60 min. After setting, the Teflon strip is removed and the top of the gel is blotted dry. The rod gel is applied to the top of the gel as in O'Farrell's method except substituting 1 mM dithiothreitol for 5% 2-mercaptoethanol. A sample well is made at one end of the stacking gel for a one-dimensional separation of marker proteins made up in 8 M urea, 50 mM Tris-HCl (pH 8.0), 10% SDS, 1% 2-mercaptoethanol, 10% glycerol, and 2 drops of 1% bromophenol blue. The bromophenol blue acts as the tracking dye. The slab gel is electrophoresed at 150 V constant voltage overnight or until the tracking dye runs off, using 0.192 M glycine, 25 mM Tris and 0.1% SDS as the running buffer. After electrophoresis the gel is carefully removed from the gel plates and stained for 3-4 h in 0.1% Coomassie blue in water:methanol:acetic acid (5:5:1 by volume) and destained overnight with several changes in the same solution without the Coomassie blue stain. A typical separation is shown in *Figure 4*.

Factors Affecting the Separation

If this method is used it is essential to reduce high ionic concentrations of the sample by extensive dialysis of the sample before incorporating it into the electrofocusing gel. It has been found that SDS can interfere with the first-dimensional separation; although it does not affect the isoelectric point of the proteins the SDS does adversely affect the linearity of the pH gradient. One problem associated with all electrofocusing separations is the manifestation of a phenomenon known as gradient drift which is discussed in greater detail in Chapter 4. Fortunately this effect only becomes really significant if runs are extended longer than 24 h. Recently attempts to use electrofocusing for the separation of basic proteins have been made. This method,

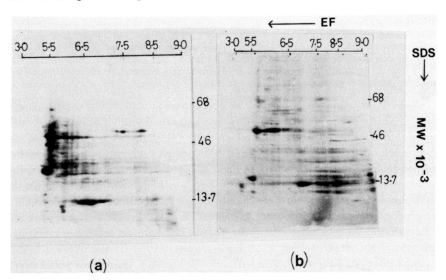

Figure 4. Two-dimensional separation of proteins using a modified O'Farrell system. Proteins of (a) cytoplasm and (b) nuclei of *Dictyostelium discoideum* were depleted of nucleic acid contamination by isopycnic centrifugation and then separated by two-dimensional gel electrophoresis using electrofocusing (EF) in the first dimension and SDS-PAGE in the second dimension as described in the text.

termed non-equilibrium pH gradient electrophoresis, involves running the sample towards the cathode for short time periods (11). This separation uses the same gel system as that originally devised by O'Farrell, the only differences being that the gels are electrophoresed without a prerunning, and for time periods of only 4-5 h at 400 V. Because this system is no longer an equilibrium system it is important to, once again, standardise the running conditions for comparative electrophoresis. A real disadvantage of this modification is that it gives little or no information on the isoelectric points of polypeptides.

Modifications for Membrane Proteins

The basic two-dimensional separation techniques described in the previous section have been used for a wide variety of proteins of widely differing properties. However such separations of membrane proteins are difficult because of their tendency to aggregate. However in this case the use of ionic and non-ionic detergents has helped to obtain high resolution separations of both prokaryotic and eukaryotic membrane proteins. A successful separation of bacterial cell envelope proteins, mammalian membrane proteins and viral protein preparations has been obtained by Ames and Nikaido (12). The membrane, or envelope proteins, are solubilised in 2% SDS, 0.5 mM $MgCl_2$ and 50 mM Tris-HCl (pH 6.8) by heating the proteins in this solution at 70°C for 30 min. The supernatant is cooled, centrifuged and diluted with two volumes of sample dilution buffer (9.5 M urea; 2% Ampholines consisting of 0.4% pH 3.5-10, 0.8% pH 5-6, 0.8% pH 6-8; 5% 2-mercaptoethanol; 8% Nonidet P-40). It is important to use a final percentage ratio of Nonidet P-40 to SDS of 8 to prevent streaking of the spots in the first dimension.

Electrofocusing is carried out as described in the previous section except that the total concentration of Ampholines in the rod gel is 2% comprised of the following ranges, pH 4-6, pH 6-8 and pH 3.5-10 in the ratio of 2:2:1 by volume, and that polymerisation is initiated by the addition of 0.4 ml of 0.14 mg/ml riboflavin and 1% (v/v) TEMED to 15 ml of gel solution. The gels are poured and overlayered with 8 M urea. Polymerisation is allowed to continue for 30 min using a fluorescent lamp. A maximum of 100 μl (approx. 100 μg) of protein is then loaded onto each gel (see previous section) and without prerunning the gels are electrophoresed for 18 h at 300 V followed by 1.5 h at 400 V to sharpen the bands. The second dimension involves SDS-PAGE using a 10% acrylamide resolving gel with a 5% acrylamide stacking gel made from a stock solution of 30% acrylamide, 0.8% bisacrylamide using the discontinuous buffer system of Laemmli (p. 26). The second-dimension gel is prepared as given in *Table 7*. The gel mix is degassed and polymerised by the addition of fresh 10% ammonium persulphate solution. Also prepare 20 ml of 5% acrylamide stacking gel as described in *Table 7*. The stacking gel mix is degassed and polymerised by addition of fresh 10% ammonium persulphate. The resolving gel (about 1 mm thick) is poured with a 2.5 cm deep stacking gel on top. The electrofocusing gel is equilibrated and sealed to the stacking gel by hot agarose. Electrophoresis in the second dimension is carried out at an initial constant current of 15 mA which is raised to 30 mA and electrophoresis is continued until the bromophenol blue marker dye has run off the gel. After electrophoresis the gel is carefully removed from the gel

Table 7. Preparation of the Second-dimensional Gel for Membrane Proteins.

Resolving gel

For 40 ml of 10% acrylamide resolving gel use:
    13.3 ml 30% acrylamide, 0.8% bisacrylamide
    0.4 ml 10% SDS
    7.5 ml 3M Tris-HCl (pH 8.8)
    10 μl TEMED
    18.7 ml water
    0.1 ml 10% ammonium persulphate

Stacking gel

    3.33 ml 30% acrylamide, 0.8% bisacrylamide
    0.2 ml 10% SDS
    2.5 ml 1M Tris-HCl (pH 6.8)
    5 μl TEMED
    14 ml water
    50 μl 10% ammonium persulphate

---

plates, then stained and destained to visualise the polypeptides (see *Table 2*, procedure E).

SEPARATION OF SPECIAL CLASSES OF PROTEINS

Most proteins can be fractionated using the combination of electrofocusing and SDS gel electrophoresis. However some types of proteins are routinely separated using other two-dimensional separation systems either because the samples are unsuited for the more usual method or because the other methods have become established for historical reasons. As an example of the latter case, ribosomal proteins are always separated using a standard system which was one of the first two-dimensional techniques devised and now, universally accepted, allows the comparison of the results of all workers using many different types of ribosomes.

While the methods described in this section have been applied to particular types of proteins this does not mean that such techniques will not be suitable for other types of protein. Thus, should the frequently used combination of electrofocusing and SDS gel electrophoresis prove unsuitable, one of these other methods may give the degree of resolution required.

Ribosomal Proteins

The method of Kaltschmidt and Wittmann (13) with modifications by Howard and Traut (14) is routinely used to separate eukaryotic and prokaryotic ribosomal proteins. It consists of a discontinuous gel in the first dimension followed by a continuous one in the second, both dimensions containing high concentrations of urea. The first-dimensional rod gel consists of the sample set in an agarose large pore gel in the middle of a smaller pore acrylamide resolving gel. Originally Kaltschmidt and Wittmann used a large pore acrylamide sample gel but loss of sample due to the immobilisation of protein in the acrylamide sample gel gave rise to the use of the modification devised by Howard and Traut of using an agarose sample gel. Under these conditions proteins in the sample gel having overall positive or negative charges will stack and migrate in opposite directions.

The composition of gels and buffers for the first dimension are prepared as given in *Table 8*. The gel solution is degassed and polymerised by the addition of 1.5 ml of fresh 7% ammonium persulphate. The solution is sufficient for 20 rod gels. Glass tubes 18 cm long and 0.5 cm inside diameter are sealed at one end with either Parafilm or similar material or with a rubber bung (see Chapter 1). Half of the tube is filled with the resolving gel, overlayered with water, and left to polymerise for 40 min. After polymerisation the overlay is removed. The protein sample containing 50-100 μg of protein in 0.1 ml to 0.15 ml, is mixed at 65°C with an equal volume of 1% agarose in running buffer (pH 8.6). The running buffer (pH 8.6) is prepared from:

360 g urea
2.4 g EDTA-Na$_2$
9.6 g boric acid
14.53 g Tris base
and water to make 1 litre

The sample in agarose is layered onto the resolving gel, overlayered with water and left to set. After removal of the overlay the tube is filled to within 0.3 cm of the top with more resolving gel and this is again overlayered with water and left to polymerise. Electrophoresis in the first dimension is carried out using the running buffer (pH 8.6) with the cathode in the top reservoir the buffer of which also contains 0.5% pyronine G as a tracking dye. Electrophoresis is initially carried out at 3 mA/ gel tube for 30 min to allow the proteins to stack, then the current is increased to 6 mA/gel and electrophoresis is continued for 5-6 h until the tracking dye reaches the end of the gel. After electrophoresis the rod gels are carefully extruded and each equilibrated in 150 ml of equilibration buffer (pH 5.2) containing:

8 M urea
0.074% glacial acetic acid
12 mM KOH
for a total of 60 min with at least two changes of buffer.

The rod gel is then transferred to the slab gel apparatus and sealed in place with the second-dimensional resolving gel (pH 4.5) which is given in *Table 8*.

The solution is degassed prior to use and polymerised by the addition of 33 μl of fresh 10% ammonium persulphate for 10 ml of gel solution. A tracking dye of 0.1% pyronine G in 20% glycerol is layered over the top of the rod gel under the running buffer, which contains 14 g glycine and 1.5 ml glacial acetic acid per litre (pH 4.0). The slab gel is electrophoresed with the anode at the top for 30-60 min at 40 V to allow the proteins to stack then the voltage is increased to 80-150 V for 6-12 h. The actual voltage used makes little difference to the separation of the proteins unless it results in a very high current, hence the main criterion used for picking a particular voltage is usually convenience. After running, the gel is removed from the gel plates

Table 8. Preparation of Gels for the Separation of Ribosomal Proteins (13,14).

First-dimensional gel (pH 8.6)

  54.0 g urea
  6.0 g acrylamide
  0.2 g bisacrylamide
  1.2 g EDTA Na_2
  4.8 g boric acid
  7.3 g Tris base
  0.45 ml TEMED
  and water to make 148.5 ml

Second-dimensional gel (pH 4.5)

  36 g urea
  18 g acrylamide
  0.5 g bisacrylamide
  0.96 ml 5M KOH
  0.58 ml TEMED
  5.3 ml glacial acetic acid
  and water to make 96.7 ml

---

and stained (see *Table 2*, procedure C or F). A typical separation is shown in *Figure 5*.

An alternative to placing the sample in an agarose sample gel is to load equal amounts of protein onto two shorter resolving gels (each 9 cm long). Electrophoresis of one of the gels is then undertaken from the anode to the cathode, with 0.5% pyronine G as tracking dye, and the other from cathode to anode with 0.1% bromophenol blue as tracking dye, under the same conditions as previously described. After equilibrating the gels as before, they are placed in the slab gel apparatus with the two sample origins adjacent to each other in the centre of the slab gel and electrophoresis in the second dimension is carried out as previously described.

Because this separation utilises a continuous buffer system in the second dimension it is essential to reduce the ionic content of the rod gel using the sequential equilibration in buffer prior to the second-dimensional separation. Otherwise, zone sharpening, which is seen when proteins migrate from a region of low to high conductivity, will not occur.

Certain adaptations of this, including variations of pH and the composition of the resolving gel in the first dimension and the inclusion of SDS in the second dimension, have been used successfully to separate ribosomal proteins. Details will not be given here but brief descriptions of, and references to, them can be found in *Table 9*.

Histones

The difficulties of using electrofocusing as a separation method for histones has led to the adoption of alternative techniques for two-dimensional analysis. Separation at low pH in low acrylamide concentrations will separate histones on the basis of charge which is a reflection of their amino acid composition. Separations in the presence of SDS will separate histones on the basis of size; although compared with other proteins histones do migrate anomalously. Other separations depend on the fact that

211

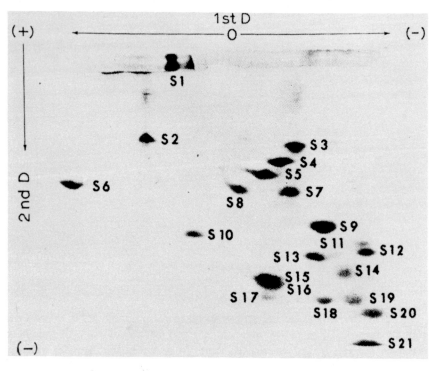

Figure 5. Two-dimensional separation of ribosomal proteins. *E. coli* MRE 600 30S ribosomal subunit proteins were extracted with 3M LiCl, 4M urea and separated by two-dimensional gel electrophoresis as described in the text. (Reproduced from ref. 14 with permission.)

Table 9. Other Two-dimensional Gel Methods for the Separation of Ribosomal Proteins.

Authors	Gel	Reference
L.J.Mets and L.Borograd	4% acrylamide rod gel in 8M urea at pH 5.0 in the first dimension followed by separation in a 10% acrylamide slab gel containing SDS.	22
O.H.W.Martini	4% acrylamide rod gel in urea-acetate in the first dimension followed by separation in a 12% acrylamide slab gel containing SDS.	23
W.L.Hoffman and R.M.Dowben	8% acrylamide rod gel at pH 4.5 in urea-acetate containing Triton X-100 followed by separation in a 15% acrylamide slab gel containing SDS.	24

non-ionic detergents such as Triton X-100, Triton DF-16 or Lubrol-WX will bind to proteins in proportion to the degree of hydrophobicity of the proteins. This binding greatly reduces protein mobility and separates sub-species of the various histones differently to that obtained in acid-urea or SDS-PAGE.

Histone proteins have been separated very successfully on two-dimensional systems incorporating acid-urea gels in the first dimension with Triton-acid-urea gels in the second as described by Hoffmann and Chalkley (15). Because the basis of separation of the histones in these gel systems is different, good resolution is usually obtained. One problem encountered when using Triton gels is the possibility of artifactual electrophoresis patterns caused by oxidation particularly of methionine residues in the histones. Triton has less affinity for oxidised histones and this may lead to artifactual heterogeneity. Therefore it is advantageous to keep histones in a reducing environment, such as 1 mM dithiothreitol, to prevent this. The buffers and gel systems for this separation are described in *Table 10*.

The gel solution has a final composition of 0.9 M acetic acid, 2.5 M urea, 15% acrylamide and 0.1% bisacrylamide. The gel mix is degassed and poured into cylindrical gel tubes 0.3 cm x 7 cm, and overlayered with 0.5 ml of cold 0.9 M acetic acid. After polymerisation the rod gels are placed in the electrophoresis apparatus and pre-electrophoresed for 4 h at 130 V using 0.9 M acetic acid as electrophoresis buffer. The samples are then layered onto the top of the rod gels and electrophoresed from the anode to the cathode at 1.5 mA/gel for approximately 2h at room temperature. A tracking dye of methyl green can be used, the blue component of which moves just ahead of histone H4 in this system.

The second-dimensional gel is as given in *Table 10*. The solutions are degassed and mixed in the same proportions as above. The gel mix is then poured to form slab gels of dimensions 0.3 cm x 14 cm x 10 cm overlayered with 0.9 M acetic acid, 1% Triton X-100 and allowed to polymerise. After polymerisation the slab gel is pre-electrophoresed for 12h at 20 mA constant current using 0.9 M acetic acid, 1% Triton X-100 as the electrophoresis buffer after which the electrophoresis buffer is poured off and the first-dimensional gel is sealed to the slab gel using 2 ml of Triton gel mix. This is in turn overlayered with 0.9 M acetic acid, 1% Triton X-100 and allowed to polymerise for approximately 1h. The second-dimensional gel is then electrophoresed from positive to negative using 0.9 M acetic acid, 1% Triton X-100 as electrode buffer for 15 h at 20 mA constant current. After electrophoresis the gel is carefully removed from the gel plates and stained in 0.1% amido black in 0.7% acetic acid, 30% ethanol for 3 h at 80°C or overnight at room temperature and then destained in 7% acetic acid, 20% ethanol.

Two-dimensional separations of histones and 'histone-like' proteins from lower eukaryotes are routinely carried out in the authors' laboratory by the combination of

Table 10. Preparation of Gels for the Two-dimensional Separation of Histones.

First-dimensional gel

    3.0 ml 40% acrylamide, 0.267% bisacrylamide
    1.0 ml 43.2% glacial acetic acid, 4% TEMED
    4.0 ml 5M urea, 0.2% ammonium persulphate

Second-dimensional gel

    3.0 ml 40% acrylamide, 0.267% bisacrylamide
    1.0 ml 43.2% glacial acetic acid, 4% TEMED
    4.0 ml 2% Triton X-100, 5M urea, 0.2% ammonium persulphate

Two-Dimensional Gel Electrophoresis

an acid-urea gel in the first dimension followed by an SDS separation at high pH in the second. The acid-urea system is a modification of Reisfeld *et al.* (16) and the SDS slab gel based on that of Thomas and Kornberg (17).

All solutions with the exception of the 9 M urea can be stored for several weeks frozen in the dark. The resolving gel (pH 4.7) is prepared according to the protocol given in *Table 11*. After mixing the solutions in the proportions given the gel mixture is adjusted to pH 4.7 by the addition of acetic acid prior to the addition of the ammonium persulphate. After degassing the gel mix, polymerisation is initiated by addition of fresh 15% ammonium persulphate and the gel mixture is then poured into cylindrical gel tubes 14 cm x 0.5 cm to the 10 cm mark. The gels are overlayered with water and left to polymerise for about 1 h. After polymerisation the overlay is removed and is replaced with approximately 150 µl of stacking gel made up according to the recipe given in *Table 11*. This is again overlayered with water and allowed to polymerise with the aid of a fluorescent light source for 30 min. After polymerisation, 100-200 µl of histone sample containing 200-300 µg of protein in 7 M urea is layered onto the stacking gel and electrophoresed from positive to negative at 150 V for 3 h, using 31.2 g β-alanine and 8.0 ml glacial acetic acid per litre as electrophoresis buffer. Pyronine G in urea can be used as a marker dye. After electrophoresis the rod gels are gently extruded under pressure and equilibrated in the following buffers for 1 h in each buffer at 37°C,

Table 11. Preparation of Gels for the Two-dimensional Separation of Histones.

First-dimensional gel

Resolving gel:
  4.0 ml 60% acrylamide, 0.4% bisacrylamide
  2.0 ml 0.48 M KOH, 17.2% glacial acetic acid, 4% TEMED
  9.0 ml 9M urea
  50 µl 15% ammonium persulphate

Stacking gel:
  2.0 ml 10% acrylamide, 2.5% bisacrylamide
  1.0 ml 0.048 M KOH, 2.87% glacial acetic acid, 0.46% TEMED
  4.0 ml 9M urea
  1.0 ml 0.4 mg/ml riboflavin

Second-dimensional gel

Resolving gel (50 ml):
  30 ml 30% acrylamide, 0.15% bisacrylamide
  0.5 ml 10% SDS
  12.5 ml 3M Tris-HCl (pH 8.8)
  12.5 µl TEMED
  water to 50 ml final volume

Stacking gel (20 ml):
  3.33 ml 30% acrylamide, 0.8% bisacrylamide
  0.2 ml 10% SDS
  2.5 ml 1M Tris-HCl (pH 6.8)
  10 µl TEMED
  water to 20 ml final volume

(i) 1% SDS, 1 mM dithiothreitol, 0.1 M phosphate, 10 mM Tris-HCl (pH 7.0);
(ii) 1% SDS, 1 mM dithiothreitol, 0.01 M phosphate, 10 mM Tris-HCl (pH 7.0);
(iii) 0.1% SDS, 1 mM dithiothreitol, 0.01 M phosphate buffer, 10 mM Tris-HCl (pH 7.0).

After equilibration they are transferred to the second-dimensional gel.

The gel mixture is degassed and polymerisation is initiated by addition of 125 μl of 10% fresh ammonium persulphate. After pouring the gel is overlayered with 0.75 M Tris-HCl (pH 8.8), 0.1% SDS and allowed to polymerise for 1 h. After polymerisation, the overlay is poured off and the top of the resolving gel is wetted with some of the stacking gel solution without ammonium persulphate. The stacking gel is prepared as given in *Table 11*.

The stacking gel is polymerised by the addition of 100 μl of 10% fresh ammonium persulphate and overlayered with water. After polymerisation the overlay is removed and the rod gel is sealed to the stacker using 1% agarose in 62.5 mM Tris-HCl (pH 6.8), 2.3% SDS, 1 mM dithiothreitol. Bromophenol blue can be used as a marker dye. The gel is then electrophoresed at 150 V constant voltage overnight, or until the marker dye runs off the gel, using 0.05 M Tris base, 0.38 M glycine and 0.1% SDS as the electrode buffer. After electrophoresis the gel is carefully removed from the gel plates and stained in 0.1% Coomassie blue in methanol:water:glacial acetic acid (5:5:1 by volume) for 3 h. Destaining is by diffusion overnight in the same solvent without the Coomassie blue stain. A typical separation of histones is shown in *Figure 6*.

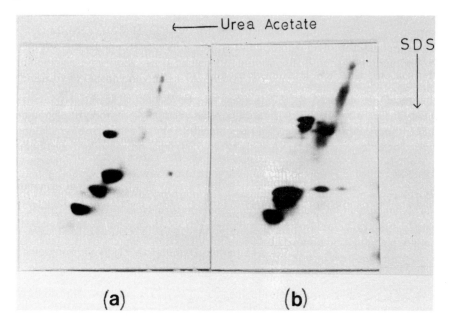

Figure 6. Two-dimensional separation of histones. Purified histones of (a) *Dictyostelium discoideum* and (b) calf thymus were separated by two-dimensional gel electrophoresis as described in the text.

Nuclear Proteins

Problems caused by protein aggregation prevent the use of ribosomal separation procedures for some proteins especially for nuclear proteins. Orrick *et al.* (18) have developed a high resolution two-dimensional separating system for extracts of both whole nuclei and nucleoli which can be used as an alternative separation method to that devised by O'Farrell (10) which is usually used. The method consists of a separation using a continuous buffer system at acid pH in the first dimension followed by a separation on the basis of molecular weights in the presence of urea and SDS in the second. The gels are prepared according to the protocol given in *Table 12.*

Gels 9.5 cm long are poured in gel tubes 12 cm long and with an internal diameter of 0.5 cm. The gels are left to polymerise for 60 min. After pre-electrophoresis for 2 h at 120 V, using 0.9 M acetic acid in 4.5 M urea as the electrode buffer, 20-50 μl of sample mixture, containing approximately 500 μg of protein in 0.9 M acetic acid, 10 M urea, 1% 2-mercaptoethanol, is then loaded onto the gel and electrophoresis is continued for a further 5 h at 120 V equivalent to 2.5 mA per gel.

After electrophoresis in the first dimension the gel may be sectioned longitudinally so that the distribution of proteins within the gel can be visualised by staining immediately. The other half is prepared for electrophoresis in the second dimension. The gels are incubated for 35 min at 45°C in each of the following solutions:

(i)   2% SDS, 0.1 M sodium phosphate, 6 M urea, 1 mM dithiothreitol[1] (pH 7.1);
(ii) 1% SDS, 10 mM sodium phosphate, 6 M urea, 1 mM dithiothreitol[1] (pH 7.1);
(iii) 0.1% SDS, 10 mM sodium phosphate, 6 M urea, 1 mM dithiothreitol[1] (pH 7.1).

The second-dimensional gel is prepared as described in *Table 12.* Using this recipe 25 ml of the gel mixture is poured to form gel slabs 10 cm x 9.5 cm x 0.3 cm and overlayered with water.

The portion of the rod gel equilibrated with SDS buffer is then gently soaked in the sealing gel mixture which is identical to the second-dimensional slab gel except that it contains a final concentration of only 10 mM phosphate (pH 7.1). The first-dimensional gel is sealed to the second-dimensional gel using about 5 ml of sealing gel. The gels are electrophoresed for 16 h at 50 mA per gel slab with 0.1% SDS, 0.1 M phosphate (pH 7.1) as the electrophoresis buffer. After electrophoresis the gel is carefully removed from the gel plates and stained.

Table 12.   Preparation of Gels for the Separation of Nuclear Proteins.

First-dimensional gel
     5.0 ml 40% acrylamide, 1.4% bisacrylamide in 4M urea
     5.0 ml 4% TEMED in 2M urea
     5.0 ml 0.21% ammonium persulphate, 21% glacial acetic acid in 6M urea

Second-dimensional gel
     18 ml 20% acrylamide, 0.52% bisacrylamide in 8M urea
     7.5 ml 0.2% TEMED, 0.4% SDS in 0.4M sodium phosphate (pH 7.1)
     4.5 ml 0.5% ammonium persulphate in 8M urea

---

[1]In the original paper 1% 2-mercaptoethanol was used as a reducing agent.

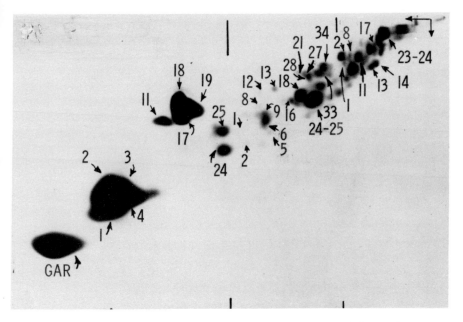

Figure 7. Two-dimensional separation of nucleolar proteins. Acid-soluble nucleolar proteins were extracted from purified rat-liver nucleoli and separated by two-dimensional gel electrophoresis as described in the text. (Reproduced from ref. 26 with permission.)

The first-dimensional gel separates the proteins on the basis of charge which is related to the amino acid composition of the polypeptides. Even so, as larger polypeptides move more slowly than smaller polypeptides of the same amino acid composition, the polypeptides are also separated on the basis of molecular weight. Consequently, the final pattern on the slab gel tends to be essentially diagonal in nature (*Figure 7*) and hence the degree of resolution may be poorer than that obtained using electrofocusing and SDS-PAGE. However, polypeptides whose amino acid composition is markedly different from the other polypeptides in the sample will be located away from the diagonal.

REFERENCES

1. Garrels,J.I. (1979) J. Biol. Chem. **254**, 7961.

2. Johns,E.W. (1976) *in* Subnuclear Components: Preparation and Fractionation (Birnie,G.D., ed.), Butterworths, London and Boston, p. 202.

3. Kruh,J., Schapira,G., Lareau,J. and Dreyfus,J.C. (1964) Biochim. Biophys. Acta **87**, 669.

4. MacGillivray,A.J. (1976) *in* Subnuclear Components: Preparation and Fractionation (Birnie,G.D., ed.), Butterworths, London and Boston.

5. Higgins,R.C. and Dahmus,M.E. (1979) Anal. Biochem. **93**, 257.

6. Gordon,A.H. (1975) Electrophoresis of Proteins in Polyacrylamide and Starch Gels (Laboratory Techniques in Biochemistry and Molecular Biology; Work,T.S. and Work,E., eds), North-Holland, Amsterdam, Oxford, Vol. 1, p. 190.

7. Hunter,W.M. and Greenwood,F.C. (1962) Nature (London) **194**, 495.

8. Rice,R.H. and Means,G.E. (1971) J. Biol. Chem. **246**, 831.

9. Garrels,J.I. (1979) J. Biol. Chem. **254**, 7961.

10. O'Farrell,P.H. (1975) J. Biol. Chem. **250**, 4007.

11. O'Farrell,P.H. (1977) Cell **12**, 1137.

12. Ames,G.F-L. and Nikaido,K. (1976) Biochemistry **15**, 616.

13. Kaltschmidt,E. and Wittmann,H.G. (1970) Anal. Biochem. **36**, 401.

14. Howard,G.A. and Traut,R.R. (1973) FEBS Lett. **29**, 177.

15. Hoffmann,P. and Chalkley,R. (1976) Anal. Biochem. **76**, 539.

16. Reisfeld,R.A. Lewis,U.J. and Williams,D.E. (1962) Nature (London) **195**, 281.

17. Thomas,J. and Kornberg,R. (1975) Proc. Nat. Acad. Sci. U.S.A. **72**, 2626.

18. Orrick,L., Olson,M. and Busch,H. (1973) Proc. Nat. Acad. Sci. U.S.A. **70**, 1316.

19. Sinclair,J.H. and Rickwood,D. (1980) Paper in preparation.

20. Anderson,N.G. and Anderson,N.L. (1978) Anal. Biochem. **85**, 341.

21. Dean,B. (1979) Anal. Biochem. **99**, 105.

22. Mets,L.J. and Borograd,L. (1974) Anal. Biochem. **57**, 200.

23. Martini,O.H.W. (1974) Ph.D.thesis (University of London).

24. Hoffman,W.L. and Dowben,R.M. (1978) Anal. Biochem. **89**, 540.

25. Jones,M.I., Massingham,W.E. and Spragg,S.P. (1980) Anal. Biochem. **106**, 446.

26. Ballal,N.R., Kang,Y.J., Olson,M.O.J. and Busch,H. (1975) J. Biol. Chem. **250**, 5921.

CHAPTER 6

Peptide Mapping by Limited Proteolysis using SDS-Polyacrylamide Gel Electrophoresis

B.DAVID HAMES

INTRODUCTION

Fractionation of proteins by polyacrylamide gel electrophoresis in SDS-containing buffers (SDS-PAGE) is a high resolution method widely used to investigate the composition of sample mixtures. However, because the fractionation is on the basis of molecular size alone, it is not possible to deduce that because two polypeptides have identical mobility in SDS-PAGE they must be related. Conversely, two proteins may be closely related but have different mobilities because of a precursor-product relationship, or as a result of artifactual degradation during preparation. In all these cases it is desirable to establish the true relationship between two or more polypeptide bands. Many different techniques may be applied to this problem (see Chapter 1), for example, analysis of amino acid composition, immunological characterisation and peptide mapping. Peptide mapping, in particular, has been taken as a stringent test of protein identity and has been recently reviewed by James (1). Modifications to the classical methodology of two-dimensional tryptic fingerprinting have allowed the technique to be used with the small amounts of protein usually available after analytical polyacrylamide gel electrophoresis (p. 81 , refs. 2,3,4). In addition, Cleveland *et al*. (5) have recently introduced a sensitive test for protein identity, also based upon proteolytic cleavage, but which takes only a few hours to perform. Furthermore, several polypeptides can be compared simultaneously and the only apparatus required is the standard slab gel electrophoresis equipment. Although this method is particularly suitable for the mapping of polypeptides fractionated in the first instance by SDS-PAGE, it is equally applicable to a pure protein from any source.

In the Cleveland method the polypeptides of interest are located after SDS-PAGE by a brief period of staining and destaining, then cut out and the gel slices placed in sample wells of a polyacrylamide gel slab prepared according to the SDS-discontinuous buffer system. These sample wells also receive known amounts of a specific protease which is active in the presence of SDS. During the second electrophoretic separation the protease digests each polypeptide and the resulting peptides are separated according to their molecular weight. The peptide banding pattern is characteristic of the protein substrate and the proteolytic enzyme used, so that relationships between polypeptides are evident simply by comparing peptide banding patterns.

It must be emphasised that peptide mapping by two-dimensional methods is still a more stringent test of polypeptide identity. The value of one-dimensional peptide mapping by the Cleveland procedure is its rapidity and simplicity in estimating the relatedness of several protein samples. Usually, if two peptide profiles are dissimilar by this method then the original proteins are not closely related. On the other hand, similar profiles, especially when using a single protease, are no proof of identity. Fortunately, at least seven proteases of different specificity can be used in this methodology. This, together with the ability to digest with one protease, reisolate a peptide fragment, and digest with a second protease, extends the stringency of the method in sequence comparison between polypeptides. This chapter describes the basic methodology of the Cleveland procedure and some modifications which may be of use to the experimentalist.

APPARATUS

The only apparatus required for peptide mapping in SDS-PAGE is the standard slab gel electrophoresis equipment described in Chapter 1 (p. 20).

METHODOLOGY

Experimental Procedure

If polypeptide bands are to be analysed without elution the initial SDS-PAGE fractionation of the protein mixture should be carried out using the slab gel format so that gel fragments will fit easily into the sample wells of the second gel in which proteolytic digestion occurs. This first slab gel is prepared and electrophoresed as described earlier using the SDS-discontinuous buffer system (Chapter 1, p.26). After electrophoresis the slab gel is stained for no more than 30 min in 0.1% Coomassie blue in methanol: water: glacial acetic acid (5:5:1 by vol.) and then destained by diffusion in 5% (v/v) methanol, 10% (v/v) acetic acid for no longer than 60 min. The gel is then rinsed with water, placed on a transparent glass sheet over a light box, and the polypeptides of interest are cut out with a razor blade. Next the slices are trimmed so that they will fit easily into the sample wells of the second slab gel and the slices are equilibrated with 0.125M Tris-HCl (pH 6.8), 0.1% SDS by soaking each in 10 ml of this buffer for 30 min with occasional swirling. Cleveland *et al.* (5) included 1 mM EDTA in gel buffers, but one of the most useful proteases, *Staphylococcus aureus* V8 protease, reportedly requires bivalent cations for activity and so EDTA is omitted here. After soaking, slices may be used immediately or stored frozen at −20°C.

A second slab gel is prepared according to the SDS-discontinuous system, but with an exceptionally tall stacking gel (about 5 cm high) and a resolving gel 9-11 cm long. The optimal gel concentration of the resolving gel will depend on the size of the polypeptide to be digested and its peptide products but 15% acrylamide is usually a suitable concentration at least for initial studies. Concentration gradient gels may also be used to good effect (see below). The sample wells are filled with 0.125M Tris-HCl (pH 6.8), 0.1% SDS and each gel slice positioned to lie horizontally at the bottom of a preformed well using a microspatula. Spaces around each gel slice are filled by now adding this buffer containing 20% glycerol (about 10-20 μl) so that each slice is also overlayered with this. Finally, 10 μl of this buffer containing 10% glycerol, 0.001% bromophenol blue, and a known amount of protease is overlayered into each well. Some suitable proteases that can be used include *Staphylococcus aureus* V8 protease, papain, subtilisin, *Streptomyces griseus* protease, ficin, chymotrypsin, and elastase.

Electrophoresis is begun at 10 mA until the tracking dye has traversed about two thirds of the stacking gel and then the power is shut off for a set period of time. During this initial electrophoresis the protease and polypeptide substrate are co-stacked and digestion occurs. The extent of digestion depends on the type of protease, ratio of protease to polypeptide and the duration of the stacking process. Usually the power is turned off for 30-40 min and variable extents of degradation obtained by varying the ratio of protease to polypeptide (see below). After proteolysis, electrophoresis is continued at 25 mA until the tracking dye reaches the bottom of the resolving gel. The gels are stained and destained as before. If the polypeptides are radioactive, the gel is then prepared for either autoradiography or fluorography as described in Chapter 1.

Typical results are shown in *Figures 1* and *2*. *Figure 1* shows that strikingly different peptide banding patterns are observed for different proteins (in this case albumin, tubulin, and alkaline phosphatase) with a single protease and also for a single protein digested with different proteases. *Figure 2* shows a peptide map comparison of *P23* and *P23**. The former, *P23,* is the precursor of bacteriophage *T4* head protein *P23** and it contains an extra *N*-terminal amino acid sequence. As expected, a large number of peptides are common but some, derived from the additional sequence, are different.

Practical Points

(i) When polypeptides are separated on rod gels the bands should be eluted and digested before loading the sample onto the slab gel for peptide fractionation. After brief staining and destaining the bands are cut from the rod gel and either eluted electrophoretically (see Chapter 1, p. 64) or eluted by diffusion into, say, 0.05M phosphate buffer (pH 7.5), 0.1% SDS by incubation at 37°C overnight with mixing. Gel fragments are removed by centrifugation. The polypeptide is precipitated by incubation with ice-cold TCA, washed twice with ethanol-ether (1:1), and dissolved at approximately 0.5 mg/ml in 0.125 M Tris-HCl (pH 6.8), 0.5% SDS, 10% glycerol, and 0.001% bromophenol blue. The samples are heated at 100°C for 2 min, cooled, a known amount of protease added, and digestion carried out at 37°C for 30 min. SDS and 2-mercaptoethanol are then added to final concentrations of 2% and 10% respectively, proteolysis stopped by heating at 100°C for 2 min again, and 20-30 μl (10-15 μg) loaded into a sample well of a slab gel ready for peptide fractionation.

(ii) Purified proteins already in solution are also precipitated with TCA to rid the sample of buffer ions or other contaminants which could interfere with subsequent electrophoresis. Digestion then proceeds as in (i) above.

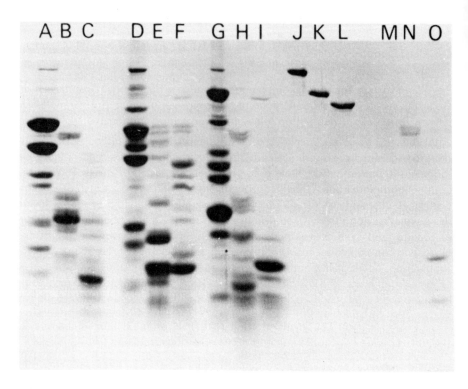

Figure 1. Peptide maps of albumin, tubulin, and alkaline phosphatase using several different proteases; proteolytic digestion allowed to occur in solution. Albumin, tubulin, and alkaline phosphatase, each at 0.67 mg/ml in sample buffer, were incubated at 37°C for 30 min with the following proteases and then 30 μl of each sample loaded onto a 20% polyacrylamide slab gel.

A to C; peptide maps of albumin, tubulin, and alkaline phosphatase after digestion with 33 μg/ml, 3.3 μg/ml, 33μg/ml, respectively, of papain (final concentrations).

D to F; peptide maps of albumin, tubulin, and alkaline phosphatase after digestion with 133 μg/ml, 67 μg/ml, 67 μg/ml, respectively, of *S. aureus* V8 protease.

G to I; peptide maps of albumin, tubulin, and alkaline phosphatase after digestion with 133 μg/ml, 67 μg/ml, 67 μg/ml, respectively, of chymotrypsin.

J to L; 2.5 μg each of undigested albumin, tubulin, and alkaline phosphatase, respectively.

M to O; papain, *S. aureus* V8 protease, and chymotrypsin, respectively, at the highest amounts used in the digestion.

(Reproduced from ref. 5 with permission.)

(iii) For peptide mapping of non-radioactive proteins, a minimum of 10 μg of each polypeptide in the initial band is required so that the peptides are easily visualised by Coomassie blue staining. Up to 20 peptides are visible using this amount of poly-peptide. For radioactive proteins the amount of material needed is usually much less and is limited only by the specific radioactivity of the protein and the sensitivity of the detection system used. The proteins may be labelled during synthesis either *in vivo* or *in vitro* using labelled amino acid precursors, usually [3H]-leucine or [35S]-methionine, and these detected by fluorography (Chapter 1). However, polypeptide mixtures may also be radioiodinated using chloramine T (6) prior to electrophoresis, or after elution of the polypeptide from the initial gel slice (2, 3). Recent methods for

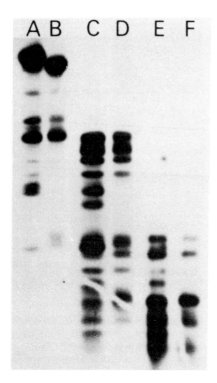

Figure 2. Peptide maps of *P23* and *P23** with *S. aureus* V8 protease; proteolytic digestion during electrophoresis. The precursor protein *P23* was cut from an SDS gel loaded with a [U-¹⁴C]-leucine-labelled envelope fraction from cells infected with a *T4* bacteriophage mutant, whilst protein *P23**, the product of *P23* processing, was cut from an SDS gel of [U-¹⁴C]-leucine-labelled phage particles. The gels were stained to visualise the protein bands and the gel slices containing *P23* and *P23** were applied to a second SDS gel (15% acrylamide) in the presence of *S. aureus* V8 protease. Digestion occurred during stacking (see Experimental Procedure). The peptides were detected by fluorography.
A, pattern of *P23*, and B, pattern of *P23**, produced by digestion with 0.005 μg of enzyme.
C, pattern of *P23*, and D, pattern of *P23**, produced by digestion with 0.025 μg of enzyme.
E, pattern of *P23*, and F, pattern of *P23**, produced by digestion with 0.5 μg of enzyme.
(Reproduced from ref.5 with permission.)

radioiodination of proteins after polyacrylamide gel electrophoresis do not even require their elution from the gel (4, 7). A detailed list of methods for labelling proteins *in vitro* is given in Appendix II.

The second method which has been used to increase the sensitivity of peptide detection is to react the polypeptide with dansyl chloride (8). The dansylated polypeptide can then be digested as usual and the peptide products detected by fluorescence in U.V. light. The sensitivity of detection is quoted as four- to eight-fold greater than Coomassie blue staining. *Figure 3* shows a typical analysis using this technique.

(iv) Although the extent of digestion can be varied by adjusting either the duration of stacking or the ratio of protease to polypeptide, it is convenient to keep the stacking period constant and to load several samples of the polypeptides to be compared, each set containing a constant amount of each polypeptide but with a different amount of

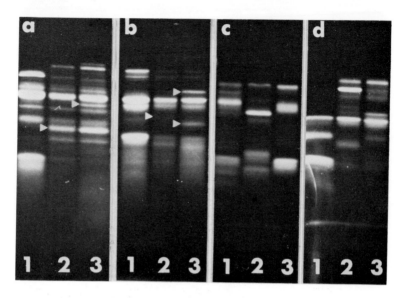

Figure 3. Peptide maps obtained using dansylated proteins.
(a) peptide maps of 1. p49, 2. p59, 3. p69, (structural proteins of densonucleosis virus) using *S. aureus* V8 protease.
(b) peptide maps of the same three viral proteins as in (a) above but using papain as protease.
(c) and (d) peptide maps of the unrelated proteins aldolase and ovalbumin, respectively, digested with 1.α-chymotrypsin, 2. *S. aureus* V8 protease, 3. papain.

The strikingly similar peptide maps of the three viral proteins reveal they have common sequences; the arrow heads indicate extra bands in the digests of proteins with higher molecular weight. The unrelated proteins, aldolase and ovalbumin, give unrelated peptide maps.
(Reproduced from ref. 8 with permission.)

the protease (e.g. *Figure 2*). This enables the investigator to select the most appropriate enzyme concentration, usually that which generates fragments with a wide range of molecular weights, leaving at most only a faint band of original protein. In many cases the peptide map observed is remarkably unchanged by up to 10-fold changes in protease concentration, but where changes do occur these serve to yield even more information on the pattern of proteolysis. Once suitable protease concentrations have been established, each set of polypeptides can be digested by a range of different proteases (e.g. *Figure 1*).

If the protease concentration required for optimal proteolysis is above the detection minimum for Coomassie blue staining then a protease band will also be evident after gel staining. To avoid confusion it is therefore always worth running one sample track containing protease only (*Figure 1*). Naturally, protease will not be detected, whatever its concentration, if radioactive polypeptides are used and detected by autoradiography or fluorography.

(v) According to the original Cleveland method, the peptide resolving gel is a uniform concentration slab gel. In practice, much sharper peptide bands are obtained if a concentration gradient gel is employed instead (*Figure 4*). A variety of gradients have been used, but linear 5-20% or 10-25% acrylamide gradients seem appropriate

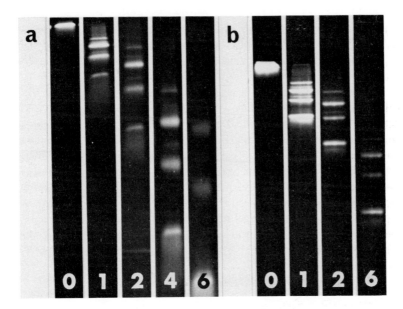

Figure 4. Increased peptide band sharpness using gradient polyacrylamide gels. Digestion products of p49 were separated on either (a) a uniform acrylamide gel (15% acrylamide, 2.66% bisacrylamide) or (b) a gradient gel (10-20% acrylamide, 5% bisacrylamide). The numbers indicate the electrophoresis time in hours after penetration of the p49 digest into the resolving gel.
(Reproduced from ref. 8 with permission.)

depending on the size of the original polypeptide and its fragments after digestion. The preparation of gradient slab gels is described in Chapter 1.

Modifications to the Basic Technique

Peptide Mapping of Heterogeneous Protein Samples

The basic method of Cleveland *et al.* (5) for comparison of two or more purified polypeptides may be adapted for an analysis of all the components of a heterogeneous mixture without purification of individual components (9). The proteins of the mixture are separated as usual in the first dimension by SDS-PAGE and then the entire gel lane containing the fractionated polypeptides is transferred to the second slab gel. All the polypeptides are then subjected to limited proteolysis in the gel and a second-dimensional electrophoresis at right angles to the first gel resolves the peptide products of each polypeptide as a series of spots below the original position of the undigested polypeptide.

The protein mixture is separated by SDS-PAGE in the first dimension using a 1.5 mm thick slab gel, with the SDS-discontinuous buffer system (p.26). At the end of electrophoresis, the gel lane containing the separated sample components is cut out and equilibrated in 50 ml of 0.125M Tris-HCl (pH 6.8), 0.1% SDS for 40 min at room temperature with gentle swirling. The second slab gel used is 1.8 mm thick (to allow easy transfer of the gel lane ready for the second electrophoresis) with a 15% acrylamide uniform concentration resolving gel, although a concentration gradient gel may give better results (see p.224). The stacking gel of the second slab gel is poured without using a sample comb to make wells; instead it is overlayered with water to obtain a flat meniscus after polymerisation.

Shortly before the transfer of the gel strip from the first electrophoresis the second slab gel is prewarmed to 50°C in an oven. A 1% agarose solution in 0.125M Tris-HCl (pH 6.8), 0.1% SDS is also prepared and kept molten at 56°C. When the gel strip equilibration is complete, the slab gel is removed from the oven, the water overlay poured off, and the top of the stacking gel quickly rinsed with some hot 1% agarose. This is replaced with fresh hot 1% agarose and the first-dimensional gel strip placed between the slab gel glass plates and quickly aligned horizontally in close contact with the stacking gel. More agarose is added if necessary to obtain a smooth top to the gel lane. Once the agarose is set, the gel is attached to the electrophoresis apparatus and the electrode reservoirs filled with reservoir buffer appropriate to the SDS-discontinuous buffer system (p. 26). The chosen protease in 1 ml of 0.125M Tris-HCl

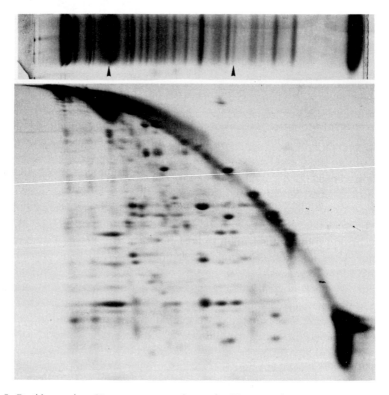

Figure 5. Peptide mapping of heterogeneous protein samples. Human erythrocyte membranes (100 μg protein) were electrophoresed on a linear 4-13% gradient slab gel and then the gel strip containing the sample was transferred to a uniform acrylamide gel (15% acrylamide) and overlayered with 1 μg of *S. aureus* V8 protease in sample buffer. The stained electrophoretogram of the same sample (50 μg) after one-dimensional electrophoresis is shown at the top of the two-dimensional gel.
(Reproduced from ref. 9 with permission.)

(pH 6.8), 0.1% SDS, 10% glycerol, 0.001% bromophenol blue is then underlayered onto the agarose gel surface. Electrophoresis to obtain co-stacking of the protease and polypeptides is carried out at 4-10 mA with interruption of the power supply for proteolysis to occur as necessary. The original authors used 0.4 μg to 25 μg of S. *aureus* V8 protease with stacking periods of between 1.5 and 7 h. After proteolysis, electrophoresis is completed at 25 mA in a few hours, or overnight at 4 mA. Gels are stained and destained as before (p.44).

Using the erythrocyte membrane as an example, about 30 bands are visible after one-dimensional SDS-PAGE and about 20 of these generated distinct peptide banding patterns (*Figure 5*). The diagonal line of spots extending from top left to bottom right of *Figure 5* is formed by undigested polypeptides in the sample. One advantage of this technique is that it allows the investigator to detect previously unsuspected sequence homologies between polypeptide bands since these give similar spot patterns (e.g. the two bands arrowed in *Figure 5*). However, the point which must always be borne in mind when using any of these methods is the assumption that the polypeptide band under study represents a single component. If this is not so, and the band comprises heterologous components which happen to coincide in mobility under the gel electrophoresis conditions used, the peptide band pattern generated will be misleading.

INTERPRETATION OF DATA

Calvert and Gratzer (10) have suggested a quantitative method for expressing the degree of similarity between proteins as determined by the Cleveland peptide mapping procedure. By taking account of the electrophoretic resolution, that part of the gel containing the peptide banding pattern is compared to N contiguous equal compartments, each of which may be empty or may contain a peptide band. In comparing two patterns, one of m and the other of n bands, distributed among the N compartments, the probability of any number of coincidences, x, is given by:—

$$P(x) = \frac{m!n!(N-m)!(N-n)!}{N!x!(m-x)!(n-x)!(N-m-n+x)!}$$

Thus, if x is sufficiently large, $P(x)$ is very small, that is, it is very improbable that such a degree of coincidence could come about by accidental matching of bands, and so the protein sequences are indeed homologous.

For example, comparison of the peptide band patterns given by the two spectrin subunits after papain digestion (10) gives:—

N = 140 (the maximum number of bands in 7 cm of gel; resolution 0.5 mm)
m = 23
n = 20
x = 11 (the number of bands with apparently identical mobility)
and hence;
$P(x)$ = 1 x 10^{-5}.

This therefore excludes the possibility that this degree of band coincidence came about by chance. A common criterion for a valid correlation is $P(x) < 0.01$.

REFERENCES

1. James,G.T. (1980) *in* Methods of Biochemical Analysis (Glick,D., ed.), J.Wiley, New York, Vol. 26, p. 165.
2. Bray,D. and Brownlee,S.M. (1973) Anal. Biochem. **55**, 213.
3. Raison,R.L. and Marchalonis,J.J. (1977) Biochemistry **16**, 2036.
4. Elder,J.H., Pickett,R.A., Hampton,J. and Lerner,R.A. (1977) J. Biol. Chem. **252**, 6510.
5. Cleveland,D.W., Fischer,S.G., Kirschner,M.W. and Laemmli,U.K. (1977) J. Biol. Chem. **252**, 1102.
6. Greenwood,F.C., Hunter,W.M. and Glover,J.S. (1963) Biochem. J. **89**, 114.
7. Christopher,A.R., Nagpal,M.L., Carroll,A.R. and Brown,J.C. (1978) Anal. Biochem. **85**, 404.
8. Tijssen,P. and Kurstak,E. (1979) Anal. Biochem. **99**, 97.
9. Bordier,C. and Crettol-Järvinen,A. (1979) J. Biol. Chem. **254**, 2565.
10. Calvert,R. and Gratzer,W.B. (1978) FEBS Lett. **86**, 247.

CHAPTER 7

Immunoelectrophoresis

RICHARD D.JURD

INTRODUCTION

Immunoelectrophoresis exploits both the differential mobility of charged proteins migrating through a gel in an electric field and the same proteins' possession of an antigenic integrity. In practice, a protein (or a mixture of proteins) is electrophoresed in an agar or agarose gel. A trough is cut in the gel parallel to the direction of the electric field, a few millimetres from the line of protein migration (*Figure 1*). After electrophoresis, antibody to the electrophoresed protein is pipetted into the trough. The antibody diffuses through the gel and forms insoluble antibody/antigen complexes with the electrophoresed protein(s). These complexes can be visualised as precipitin arcs having characteristic shapes, patterns and distances from the site of antigen application and/or the trough (*Figure 2*).

The technique can be used to:
 (i)   examine the purity of protein preparations;
 (ii) analyse the composition of protein mixtures;
 (iii) investigate the presence, in antiserum preparations, of antibodies to particular known antigens;
 (iv) check the identity or non-identity of two antigens.

Immunoelectrophoresis is particularly suitable for the characterisation of blood serum proteins, but it has numerous other applications. A further account of the relevant theoretical concepts in immunology underlying immunoelectrophoresis is given in Roitt (1); this is more fully explored in Weir (2).

Basic immunoelectrophoresis techniques have been modified and developed to yield quantitative results. Such techniques, including rocket immunoelectrophoresis and two-dimensional (crossed) immunoelectrophoresis, are described briefly.

PREPARATION OF ANTIGENS AND ANTIBODIES

Antigens

An antigen is any substance which will elicit an immune response. Common antigens include grafts of foreign tissue, pathogenic micro-organisms, and macromolecules such as lipopolysaccharides and proteins. If a protein is injected into a vertebrate animal, that animal must recognise the protein to be 'non-self' in order that antibodies may be made to it. The high degree of specificity which characterises the vertebrate immune response means that, in practice, a protein such as bovine serum albumin, for example, will be recognised as non-self by a rabbit or a rat, and antibody synthesis will be elicited. Similar antigenic determinants on similar proteins may

229

(a)

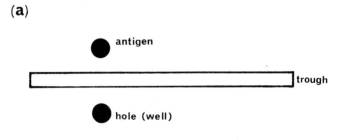

(b)

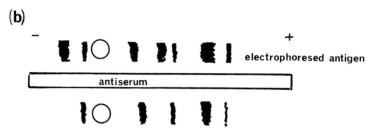

Figure 1. Principle of immunoelectrophoresis.
(a) Antigen is placed in wells cut in a gel.
(b) Antigen is electrophoresed in an electric field; antiserum is placed in a trough cut in the gel and diffuses into the gel, meeting the electrophoresed antigen.

Figure 2. Immunoelectrophoresis arcs, formed by insoluble antigen-antibody complexes.

mean that one protein may 'cross-react' with another protein. Such cross-reactivity can often be abrogated by 'absorbing' the antibody preparation with the offending, cross-reacting protein.

Antibody Production

Antibodies are usually raised in New Zealand or Dutch rabbits, although sheep and goats are frequently used when larger volumes of serum are required.[1]

Antigens to be used as immunogens should be as pure as possible since it is usually easier to prepare a pure antigen than to remove unwanted cross-reactivity in an antiserum preparation. Immunisation schedules vary widely and the ideal schedule for a

---

[1] Workers in the United Kingdom are reminded that, under the terms of the 1876 Cruelty to Animals Act, they will normally require a Home Office Licence to Experiment on Living Animals in order to immunise and bleed animals. Further advice should be sought from the experimenter's local Home Office Inspector.

given antigen will be found by experience. The author has found the following schedule to be generally reliable.

Day 0. A 3 kg rabbit is injected with 1 ml of protein antigen solution containing 0.1-5.0 mg of protein in 0.14M NaCl, buffered to pH 7.3 with 0.02M sodium phosphate ('phosphate-buffered saline') after it has been emulsified with an equal volume of complete Freund's adjuvant (Difco Laboratories Ltd.), to give 2 ml of emulsion altogether. To emulsify complete Freund's adjuvant with protein solution, the constituents are pumped in and out of a 2 ml glass hypodermic syringe fitted with a 19 gauge needle until a thick white emulsion is formed. The stability of the emulsion can be tested by dropping it onto ice-cold water: the second and third drops should float on the surface of the water and not disperse. The emulsion is injected subcutaneously into the upper arms and legs of the rabbit, 0.5 ml of emulsion at each site (*Figure 3*). It is a safe policy to immunise two rabbits in parallel, or, alternatively, in tandem, one rabbit one week after the other.

Smaller amounts of highly purified antigen (down to 50 μg of protein) may be used for the production of monospecific antibody. Even smaller amounts of antigen can be dispersed on the immunosorbent microcrystalline cellulose; 1 μg of protein has been used successfully (3). Proteins separated on polyacrylamide gels may also be used in which case the polyacrylamide acts as a supplementary adjuvant. The gel containing the protein(s) is homogenised in isotonic saline and then emulsified with complete Freund's adjuvant prior to injection.

Day 28. A second, booster injection of protein antigen is administered, as on Day 0 with the exceptions that incomplete Freund's adjuvant (Difco Laboratories Ltd.) is used instead of complete Freund's adjuvant, and that the sites of injection are subcutaneously between the shoulder blades and above the sacrum (between the hips); 1 ml of emulsion is injected at each site.

Alternatively 'alum-precipitated' protein antigen may be administered in the booster injection. The protein solution, containing up to 2 mg/ml of protein, is mixed with half its volume of 1M NaHCO$_3$ followed by one volume of 0.2M potassium aluminium sulphate solution. The protein is adsorbed onto the resulting aluminium hydroxide precipitate. The mixture is slowly stirred and left for 30 min. The precipitate is pelleted by centrifugation for 15 min at 300xg and washed three times with phosphate-buffered saline. The precipitate is suspended in 2 ml of phosphate-buffered saline and injected intramuscularly into the buttock and shoulder muscles, 0.5 ml at each site.

Day 35. The rabbit is placed on a table and is wrapped in a large cloth so that only one ear and the top of the head are exposed. The ear is gently rubbed and the fur is shaved from the posterior margin. A longitudinal nick, about 0.2 cm long, is made in the marginal ear vein. Venous blood, dripping from the nick, is collected in a 100 ml glass beaker (*Figure 4*). Approximately 25 ml of blood may be collected in this way. The flow of blood may be facilitated by working in a quiet, warm room, by reassuring the rabbit by stroking it, and by warming the ear with a reading lamp placed about 20 cm above it. If the blood flow ceases, it may be restarted by gently wiping the cut with a soft tissue. When finished, bleeding may be arrested by pressing the cut firmly for about 30 sec with a wad of cotton wool.

Day 42. A further 25 ml of blood may be obtained, as on Day 35.

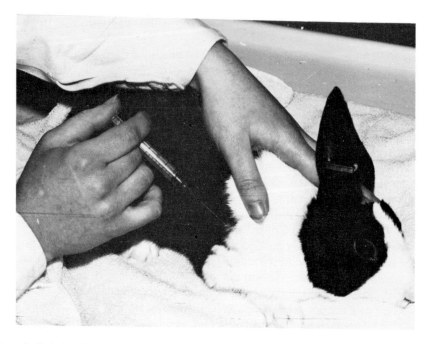

Figure 3. Technique for injecting a rabbit in the upper fore-limb. The rabbit is placed on a towel and is steadied with the left hand.

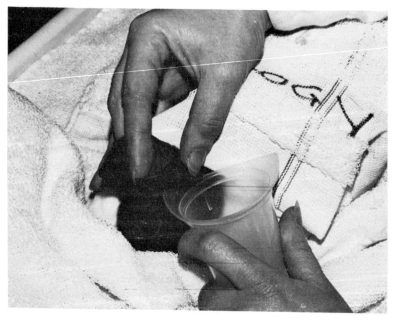

Figure 4. Bleeding a rabbit from the marginal ear vein. The rabbit is wrapped in a cloth or towel. Using the right hand, the ear is held over a small beaker.

If more antiserum is required the rabbit may be given another injection of antigen, exactly as described for Day 28, and bleeding 7 and 14 days later.

The blood clot is stirred gently to prevent it from adhering to the wall of the beaker; it is then left overnight in a refrigerator at about 2°C. Next day the straw-coloured serum is pipetted off from the somewhat contracted clot and is centrifuged for 15 min at 1,500xg. The supernatant can be stored at −10°C in 1 ml portions.

Commercially Available Antibody

Commercially prepared antisera of high titre and specificity are available for many common antigens. The catalogues of the following firms are worth inspection, although the list is by no means exhaustive:

> Gibco Biocult Ltd.
> Miles Laboratories Ltd.
> Nordic Immunological Laboratories Ltd.
> Wellcome Reagents Ltd.

Assay of Antibody Activity

Antibody activity can be tested by simple immunodiffusion. Wells are cut in the pattern shown in *Figure 5* using a No. 2 cork-borer (approx. 0.5 cm diameter) or the broad end of a Pasteur pipette in 1% agar gel (including 0.02% sodium azide) contained in a Petri-dish; the gel should be about 0.2 cm thick. Serially diluted (1:1, 1:2, 1:4, 1:8, etc.) antiserum is placed in a succession of peripheral wells. Antigen at a concentration of 1 mg/ml is placed in the central well. The plates are covered and are left in a humid environment for 48 h. The appearance of a white precipitin line between the antigen and antiserum wells indicates the presence of antibody activity in the serum. The maximum dilution of antiserum giving a visible precipitin line after 48 h gives the 'titre' of the antiserum (4).

Purification of IgG Antibody

The most useful antibodies for immunoelectrophoresis work are members of the IgG class of immunoglobulins. It is usually advantageous to prepare the IgG fraction from rabbit serum since the presence of IgM antibodies in antisera may lead to the formation of a second (IgM-antigen complex) precipitin line in the immunoelectrophoresis gel, rendering analysis of the precipitin patterns more difficult.

The serum (100 ml) is mixed with 60 ml of saturated ammonium sulphate solution made up in 0.02M Tris-HCl, 1mM EDTA (pH 8.0) for 30 min at room temperature. The precipitate is spun down at 4,000xg for 45 min, and is redissolved in 40 ml of the same buffer. Ammonium sulphate precipitation (using 24 ml of the salt solution) is repeated until the precipitate is white. The final precipitate is dissolved in 20 ml of 0.005M sodium phosphate (pH 8.0), and the solution is dialysed against this buffer.

The crude IgG solution is further purified by ion-exchange chromatography on a diethylaminoethyl (DEAE) cellulose column. Whatman DEAE cellulose (DE52) (W. & R. Balston Ltd.) is used in the ratio of 3g for every ml of protein solution and 0.5M

233

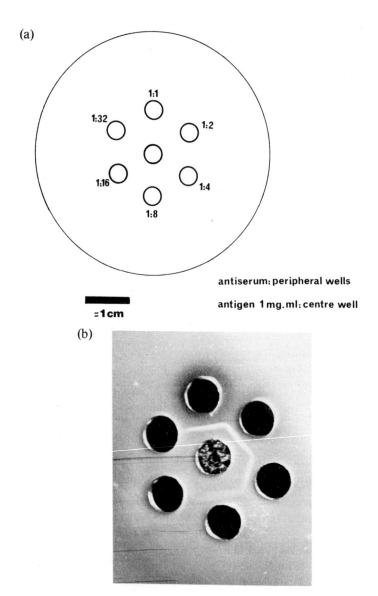

antiserum: peripheral wells

antigen 1 mg.ml: centre well

≈1cm

Figure 5. Immunodiffusion plates.
(a) Pattern of wells cut in the agar gel and dilutions of antiserum.
(b) Immunodiffusion plate. Centre well has antigen (*Limulus* haemocyanin) at a concentration of 1 mg/ml. Peripheral wells have antiserum to *Limulus* haemocyanin diluted (from top, clockwise) 1:1, 1:4, 1:16, 1:64, 1:256, 1:1024.

NaH_2PO_4 is added until the solution reaches pH 8.0. The solution is diluted with 0.005M sodium phosphate buffer (pH 8.0) so that there is about 5 ml of buffer for every gram of wet cellulose. The slurry is stirred up and allowed to settle for about 30 min to give a column 10 cm high. The supernatant is discarded, 0.5 volumes of

0.005M sodium phosphate buffer (pH 8.0) is added and the cellulose is stirred into it. The slurry is poured into a short, wide column (diameter approximately one fifth the height) with the tap open and packed by pumping 0.005M sodium phosphate (pH 8.0) through at 60 ml/h/cm^2 internal column cross-section. Typically 5 l of buffer should be used to wash a 15 cm x 3 cm column.

The dialysed protein preparation is applied to the top of the column. The starting buffer (0.005M sodium phosphate, pH 8.0), contained in the mixing chamber of a gradient mixer, is pumped through at 60 ml/h and eluent samples are monitored for protein content. The second chamber contains 0.25M sodium phosphate (pH 8.0) which is then gradually added to the starting buffer in the mixing chamber of the gradient mixer thus forming a gradient of ionic strength (the shape of which should be convex). The first protein peak eluted contains the IgG. This may be concentrated by ultrafiltration by, for example, using an Amicon apparatus in combination with a PM10 filter.

The IgG is further purified by gel chromatography on a column of Sephadex G-200 (Pharmacia Ltd.). About 20 g (dry weight) of Sephadex G-200 is swollen in 0.14M NaCl, 0.02M sodium phosphate (pH 7.3) by heating in a boiling water bath for about 6h. When the swollen gel has cooled it is degassed for 5 min by stirring it in a Buchner flask attached to a vacuum pump; this removes air bubbles which could distort the elution of the protein bands. The gel is poured down a glass rod into a vertical chromatography column (100 cm x 2.6 cm). All the gel slurry must be poured into the column at the same time. The outlet at the base of the column is opened and the gel is allowed to settle over a period of several hours. The column is packed by allowing 3 volumes of phosphate-buffered saline to flow through the column. The flow rate should be maintained at 15-25 ml/h using a peristaltic pump. Up to 7.5 ml of the IgG preparation (concentration about 2 mg/ml) may be loaded onto the column which is then run at a flow rate of between 15 and 25 ml/h, and 3 ml samples of eluate are collected in a fraction collector. Samples are assessed for protein content; the second peak to be eluted contains the IgG.

Removal of Unwanted Antibody Activity

Immunodiffusion in agar gels of antisera or IgG preparations using proteins related to the antigen may reveal unwanted cross-reactivity to shared antigenic determinants. Thus, antiserum to protein X may cross-react with proteins Y and Z because all share a given determinant. Absorption may remove this. Using the example above, anti-X can be mixed with proteins Y and Z (the exact amounts being found by experience) and left overnight at 4°C. Next morning the mixture is centrifuged at 5,000xg for 30 min to remove antibody-antigen complexes. The supernatant can be tested by immunodiffusion to see if any anti-Y or anti-Z activity remains. The minimum amount of Y and Z needed to remove unwanted antibody activity completely can be ascertained by experiment.

SIMPLE IMMUNOELECTROPHORESIS

Apparatus

Electrophoresis tank (*Figure 6*)
Power pack
Voltmeter
Well puncher – Pasteur pipette or a commercial punch (see below)
Slit former – razor-blade and ruler or a commercial device (see below)
Template for well/slit pattern
Glass microscope slides (7.5 cm x 2.5 cm)
Levelling table with spirit level
Water bath
Microsyringe (Hamilton Bonaduz AG)
Histological staining jars and slide holders
Magnetic stirrers and followers

Reagents

Purified agar (Ionagar, Oxoid Ltd., *or* Bactoagar, Difco Laboratories Ltd.)
   or alternatively agarose
Barbitone electrophoresis buffer (5)
  1 litre of buffer (pH 8.2) is prepared as below:
  5,5'-diethylbarbituric acid, sodium salt: 12 g in 800 ml water
  5,5'-diethylbarbituric acid: 4.4 g in 150 ml water

The 5,5'-diethylbarbituric acid is dissolved by heating the solution to 95°C. The two solutions are mixed, the pH is adjusted to 8.2 with 5M NaOH and 0.15 g of thiomersal is added prior to adjusting the volume to 1 litre. Note that this buffer is poisonous.

Phosphate-buffered saline solution (pH 7.3)
Staining solutions

either	amido black *or* Coomassie brilliant blue R-250	0.1 g
	5% (v/v) aqueous acetic acid	100 ml
or	azocarmine	1 g
	1M acetic acid	450 ml
	0.1M sodium acetate	450 ml

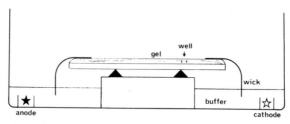

Figure 6. Cross-sectional diagram of an immunoelectrophoresis tank.

glycerol   100 ml
Destaining solution: 5% (v/v) aqueous acetic acid

Method

Either agar or agarose can be used for immunoelectrophoresis and either can be used
in this protocol. However, agarose is more expensive and it may not give better
results than the cheaper agar. About 2g of agar is heated with 98 ml of barbitone
electrophoresis buffer (5) over a boiling water bath.

Microscope slides, cleaned with ethanol, are marked with a diamond stylus to
show the anodal end and the antigen(s) and antibody used, and are placed on a levelling table or other horizontal surface.

All slides are 'pre-coated' by pouring a film of molten agar onto their surfaces,
such that the slides are just covered with agar. The agar is allowed to set and dry out.
This procedure allows the agar to act as an adhesive, preventing the overlying gel
from floating off during washing procedures.

Then 3.5 ml of molten agar are gently pipetted onto the pre-coated slides which are
left until the agar is cool and set. Holes are punched in the agar as shown in *Figure 7*
using the fine end of a Pasteur pipette; the agar plugs are sucked out. The holes may
be placed centrally, or placed off-centre, according to the protein antigen being
electrophoresed. When electrophoresing serum proteins it is usual to position the
holes nearer to the cathodal end of the slide. A trough is then cut using a scalpel or a
single-edged razor-blade; the agar strip from the trough is not scooped out until
electrophoresis is complete. Special punches to cut troughs and holes in pre-set patterns are also available commercially for those planning frequent use of this technique; for details, see below.

Samples of the protein solution (2 μl at a concentration of 1 mg/ml) in barbitone
buffer are placed in the holes (wells) using a 50 μl Hamilton precision syringe. The
slide is placed across the bridge of an electrophoresis tank (*Figure 6*) and the buffer
chambers in which the electrodes are situated are filled with barbitone buffer. Filter-
paper wicks, cut to the width of the slide, are moistened with buffer and are draped
over the ends of the agar gel in such a way that 0.5 cm of gel is covered by the wicks,
and the free ends of the wicks trail into the buffer reservoirs. Several slides may be
electrophoresed in parallel.

The tank is covered and the electrodes are connected to a power pack capable of
supplying a constant D.C. current. A current of about 10 mA per slide is applied with

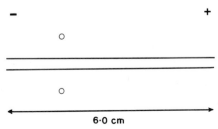

Figure 7. Pattern for cutting gels on immunoelectrophoresis slides.

a voltage of about 6 V/cm. The optimal electrophoresis time will be found by experimentation. However, for serum proteins it will be found that the addition of a very small amount of 1% bromophenol blue dye acts as a useful marker since this blue dye binds to the fast-moving albumin component allowing the migration of this protein to be monitored. Normal electrophoresis times vary between 20 min and 80 min.

After electrophoresis the slides are removed from the tank, the agar in the troughs is scooped out, and antiserum (or antibody-containing IgG preparation), approximately 0.1 ml in volume, is used to fill the troughs. The slides are then left in a humid atmosphere at room temperature for 48 h to 'develop'. They may then be examined for precipitin arcs.

Permanent preparations of immunoelectrophoresis plates may be made with ease. The unprecipitated proteins are eluted from the slide by repeated washing in stirred, one litre volumes of phosphate-buffered saline solution, each for 2 h, followed by two washes for 2 h in distilled water. The gel is dried overnight in an incubator at 37°C. After drying, the arcs are stained blue-black using amido black dye or royal blue using Coomassie brilliant blue R-250 dye. The dye plates are left for 10 min in the dye solution and the unbound dye is then removed by repeated rinses in 5% aqueous acetic acid. Alternatively, the arcs may be stained crimson using the dye azocarmine (6). The procedure is the same as for amido black except that the staining should be for 3 h, and each acetic acid rinse should be for 1 h. The dried, azocarmine-stained slides may be used directly as 'negatives' in a photographic enlarger in order to prepare photographs (*Figure 8*).

A number of commercial immunoelectrophoresis kits and accessories are available. The catalogues of Gelman Hawkesley and Shandon Southern will be of interest in this context. Millipore Ltd. manufacture a very neat, small-scale 'Immunoagaroslide' immunoelectrophoresis system, using pre-cut agarose gels mounted on cellulose acetate sheets, sold wrapped in foil pouches. Eight antigen samples can be run simultaneously on two 'Immunoagaroslides' in parallel in a specially designed electrophoresis tank (*Figure 9*).

Other methods of detection besides protein staining include (a) radioimmuno-

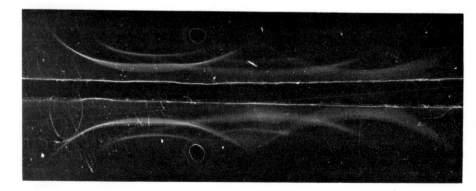

Figure 8. Immunoelectrophoresis slide, photographed by placing the slide directly in the negative carrier of a photographic enlarger. Here *Xenopus laevis* serum has been electrophoresed and reacted against anti-*Xenopus* serum antibody prepared in rabbits. Anode to right.

electrophoresis (an autoradiographical technique using radioiodinated proteins), and (b) testing for enzyme activity.

RADIOIMMUNOELECTROPHORESIS

Small amounts of antigen may be detected using this modification of the basic immunoelectrophoresis system. It is particularly useful for identifying minute quantities of components in mixtures of proteins and it involves radioiodination of the protein mixture. Detection of immunoprecipitates is then by autoradiography which is far more sensitive than the simple staining schedules described above. An example of radioimmunoelectrophoresis is a study of the emergence of new proteins (such as adult haemoglobin) during ontogenesis. A haemoglobin preparation from foetal blood is iodinated as described below, the iodinated proteins are mixed with 'cold' adult and foetal haemoglobins as 'carriers', and the preparation is electrophoresed. The electrophoresed proteins are reacted with antiserum to both adult and foetal haemoglobins and the dried, stained slide is made into an autoradiograph. The detection of silver grains over the adult haemoglobin precipitin arc would indicate the presence of adult haemoglobin in the foetal blood. The technique is much more sensitive than that which relies on simple visualisation of arcs. For another example of this technique see Jurd *et al.* (7).

Proteins (approximately 0.1 mg) are iodinated with ^{125}I using the chloramine T method of Hunter and Greenwood (8). The iodinated proteins are passed down a 20

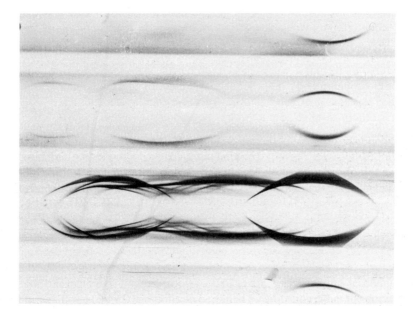

Figure 9. Millipore 'Immunoagaroslide' system slide. Four antigenic preparations are run in parallel: from top to bottom - *Xenopus laevis* bile, *Xenopus* gut fluids, *Xenopus* serum, *Xenopus* bile. The antibody in the three parallel troughs is a rabbit anti-*Xenopus* serum preparation.

cm x 2.2 cm Sephadex G-25 column, presaturated with the 'cold' protein under investigation, to remove free [^{125}I]-iodine. The radioactivity of eluent samples is measured in a gamma-radiation counter. The radioactive protein samples are pooled and concentrated by ultrafiltration. Where only small amounts of radioiodinated protein are present it is advisable to mix them with 'cold' carrier proteins. This preparation is electrophoresed and reacted with antibody.

Dried and stained immunoelectrophoresis plates are coated by dipping them at 45°C in Ilford K5 nuclear emulsion diluted with an equal volume of water. The slides are dried and left at 4°C in light-tight boxes for 10 days prior to development in Kodak D19b developer and fixation. Slides are examined under an oil film using a light microscope with a x10 eyepiece and x10, x40 or x100 objectives as appropriate (*Figure 10*).

(a)

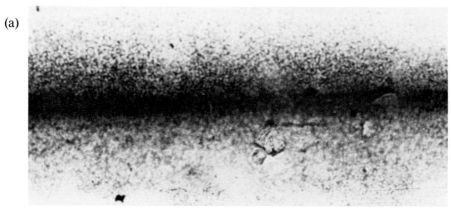

(b)

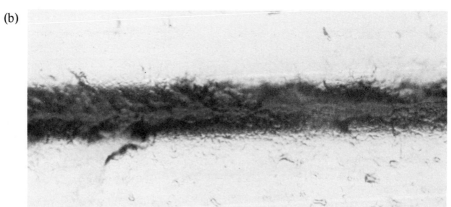

Figure 10. Radioimmunoelectrophoresis.
(a) Autoradiograph showing silver grains over a precipitin arc containing ^{125}I-labelled protein. This is ^{125}I-labelled IgM from a *Xenopus laevis* tadpole.
(b) Autoradiograph showing no silver grains over a precipitin arc formed by albumin from adult *Xenopus laevis,* used as a carrier, when reacted with rabbit anti-*Xenopus* serum antibody.
(Magnified x 20. For full details, see reference 7.)

ENZYMATIC DETECTION METHODS

Enzymic activity is another sensitive method for detecting small amounts of some antigens. Certain enzyme antigens, even when complexed with antibody on an immunoelectrophoresis plate, are capable of reacting with their substrate. In the presence of a suitable indicator, the precipitin line will be stained and thus rendered more visible. For example, xanthine dehydrogenase/anti-xanthine dehydrogenase precipitin lines will stain pink in the presence of xanthine (substrate) and 2-(p-iodo-phenyl)-3-(p-nitrophenyl)-5-phenyl tetrazolium chloride (indicator) (9). Other suitable activity stains are referenced in Appendix I of this volume.

OTHER MODIFICATIONS OF IMMUNOELECTROPHORESIS

Cross-Over Electrophoresis

This technique exploits the fact that, during electrophoresis at pH 8.2, gamma-globulins (which include most antibodies) migrate towards the cathode while most other proteins migrate towards the anode. Hence antibody and antigen migrate towards each other and form precipitin lines of antibody-antigen complexes. The technique gives quicker results than straightforward immunodiffusion since the protein migration is accelerated by the electric field.

Agar gel slides are prepared as for immunoelectrophoresis and wells are cut as shown (*Figure 11*). Antibody is placed in the well nearer the anode and antigen in that nearer the cathode. The proteins are electrophoresed and inspected within 15 min; if antibody-antigen complexes have been formed a precipitin line or arc should be visible.

'Rocket' Immunoelectrophoresis

Here immunoelectrophoresis is modified in order to quantify antigen concentrations. When an electric field is applied most protein antigens migrate anodally into an agar gel containing antibody. Antibody proteins tend to migrate towards the cathode (as in cross-over electrophoresis). When they meet antigens they will form antibody-antigen complexes. In antibody or antigen excess these complexes are soluble, but near the equivalence point precipitable complexes are formed having a 'rocket' shape. As more antigen reaches the rocket, the precipitate redissolves and migrates further towards the anode. Eventually, when no more antigen remains to migrate, the rocket becomes stationary and stable. The area of gel beneath the rocket is proportional to

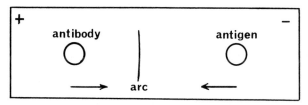

Figure 11. Principle of cross-over immunoelectrophoresis.

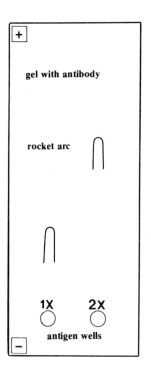

X = antigen concentration

Figure 12. Principle of rocket immunoelectrophoresis.

the antigen concentration. Avoidance of multiple lines in a rocket can be obtained by the use of monospecific IgG preparations from the rabbit antisera.

Agar (2%) in barbitone buffer (pH 8.2) is melted and cooled to 56°C. Antiserum is diluted five-fold with phosphate-buffered saline in a test-tube and warmed to 56°C. A mixture of 2 ml of warm agar and 1 ml of diluted antiserum is poured onto pre-coated microscope slides (see immunoelectrophoresis section), on a level surface and allowed to set. Two wells are punched near one end of the slide (*Figure 12*) and filled with protein antigen of known dilutions. Larger slides may be used if required, allowing the use of more wells and hence more protein samples can be electrophoresed in parallel. The proteins are electrophoresed (approximately 7.5 mA per slide, 5-10 V/cm) for about 90-150 min. Slides may be deproteinised, stained and dried as described on p.238.

The rocket arcs are now measured. Provided that electrophoresis was continued until the rockets ceased moving, the heights of the rockets are proportional to the protein antigen concentrations (*Figure 13*). The height of the rocket peak should not be less than 1 cm nor exceed 5 cm: the concentration of antibody in the gel may need adjustment to obtain this. Standard curves may be drawn (*Figure 14*) and the unknown antigen concentrations can be determined from these. Note that the standard curves do not pass through the origin.

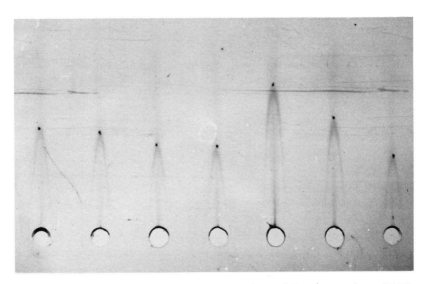

Figure 13. Rocket immunoelectrophoresis plate: here seven antigen wells have been run in parallel. The antigen is xanthine dehydrogenase from *Aspergillus nidulans:* it has been electrophoresed into a gel containing rabbit anti-xanthine dehydrogenase antibody. The three rockets on the right are of standard antigen concentrations from wild-type *A. nidulans*; the four wells on the left contained antigen preparations (of unknown concentrations) from various mutant forms of this ascomycete. (Photograph by courtesy of Dr. Heather Sealy-Lewis.)

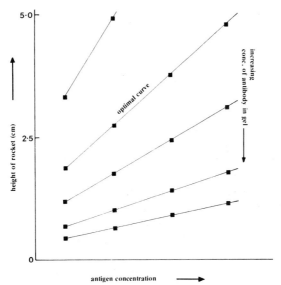

Figure 14. Standard curves for rocket immunoelectrophoresis: each curve corresponds to a given antibody concentration in the gel. Note that the curves do not pass through the origin.

Two-Dimensional Immunoelectrophoresis (Crossed Immunoelectrophoresis)

Rocket immunoelectrophoresis may be modified by the technique of two-

243

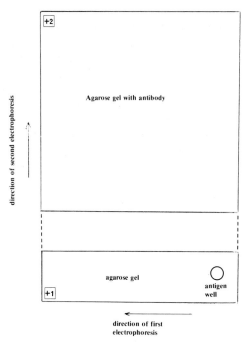

Figure 15. Principle of simple two-dimensional immunoelectrophoresis (crossed immunoelectrophoresis).

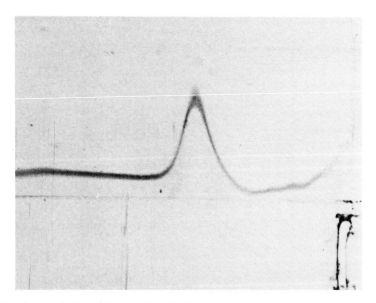

Figure 16. Photograph of precipitin arc obtained when *Aspergillus nidulans* xanthine dehydrogenase was electrophoresed, and then further electrophoresed into a gel containing rabbit anti-xanthine dehydrogenase antibody. The presence of one arc indicates the purity of the preparation; the horizontal line shows that a reference sample of xanthine dehydrogenase was incorporated into the system (see *Figure 18*). (Photograph by courtesy of Dr. Heather Sealy-Lewis.)

dimensional immunoelectrophoresis; this is particularly useful for the quantitation of mixtures of proteins and the analysis of the composition of protein mixtures. Protein antigens are separated by electrophoresis in an agarose gel and then this first electrophoretic separation is followed by a second electrophoresis, at right-angles to the direction of the first, into an antibody-containing gel (*Figure 15*).

Firstly, 3 ml of 1.5% agarose solution at 50°C are poured onto a clean microscope slide. After the agarose has set, a well is punched near one end of the slide. Protein antigen is used to fill this well, and the protein is electrophoresed for about 30-60 min (potential drop of 8 V/cm). Using a razor-blade, a 0.2 cm border is cut round the gel, squaring off its edges. The gel is slid from its slide onto the end of a 15 cm x 7.5 cm glass plate. The rest of the plate is covered with 15 ml of antibody-containing 1.5% agarose solution prepared by diluting the antibody five-fold and mixing it with an equal volume of 3% agarose solution. When the agarose has set, the plate is placed in an electrophoresis tank, broad wicks are attached and electrophoresis is performed at 1-2 V/cm for 8-16 h. The gel may then be inspected and, if necessary, dried and stained (*Figure 16*).

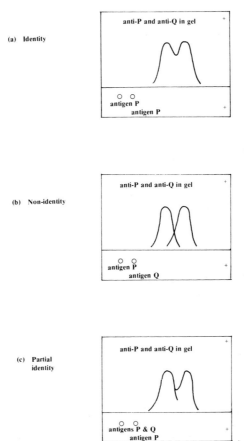

Figure 17. Principle of tandem crossed immunoelectrophoresis, showing patterns obtained when (a) identical, (b) non-identical, and (c) partially-identical antigens are electrophoresed from the two wells.

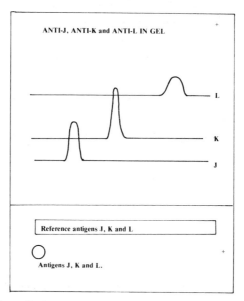

Figure 18. Principle of cross-line immunoelectrophoresis. Note that the reference antigens are added to the gel after the initial electrophoresis.

This technique is useful for giving information about the identity or non-identity of antigens. Here two or more antigens are electrophoresed in tandem (*Figure 17*). If the precipitin lines obtained after the second electrophoresis fuse there is a reaction of identity and the antigens may be said to be identical. Crossing over of the precipitin lines indicates a reaction of non-identity, while a spur shows a reaction of partial identity. In cross-line immunoelectrophoresis, reference lines for given antigens can be incorporated into the system (*Figure 18*).

For further details of these and other sophisticated modifications of the basic immunoelectrophoresis techniques described above, readers are referred to articles in Weir (2) and in Axelsen *et al.* (10). Many of these techniques allow accurate quantitation of protein antigens in mixtures of proteins.

TROUBLESHOOTING

Many of the problems associated with immunoelectrophoresis may be solved by experience and simple experimentation; for example, variation of electrophoresis times, currents and/or voltages may result in better separation and resolution of precipitin areas. A few other, common difficulties are listed below.

(i) *Antiserum of low titre.* A more rigorous immunisation schedule may need to be adopted, giving immunisations of antigen every 7 or 14 days over a 10 week period. Beware, though, of over-immunisation as excessive doses of antigen may stimulate 'high zone tolerance' (or immune paralysis) towards the antigen, may sequester all the antibody so that none is in circulation, or may so traumatise the animal that its immune response is depressed. Note that antisera also tend to lose their activity with age, and with repeated freezing and thawing.

(ii) *Antiserum cross-reacting with unwanted proteins.* A purer antigen preparation should be used for immunisation. Alternatively, the antiserum can be absorbed with the unwanted proteins.

(iii) *Proteins migrate poorly in the gel.* Vary the electrophoresis time, or the buffer pH.

(iv) *Gels float off slides during washing.* Use pre-coated slides (see immunoelectrophoresis section, p.237).

(v) *Stain taken up by parts of the gel other than precipitin arcs.* Inadequate deproteinising washes.

REFERENCES

General References

Roitt,I.M. (1977) Essential Immunology, 3rd edition, Blackwell Scientific Publications, Oxford. (An excellent introductory account of the principles of immunology).

Hudson,L. and Hay,F.C. (1976) Practical Immunology, Blackwell Scientific Publications, Oxford. (An introduction to practical immunological techniques; particularly useful for the beginner.)

Weir,D.M. (ed.) (1978) Handbook of Experimental Immunology, Vol. 1, Immunochemistry, 3rd edition, Blackwell Scientific Publications, Oxford. (Standard text on immunological methods: see Chapter 19; Immunodiffusion and Immunoelectrophoresis by O.Ouchterlony and L.Å.Nilsson.)

Axelsen,N.H., Krøll,J. and Weeks,B. (eds) (1973) A Manual of Quantitative Immunoelectrophoresis: Methods and Applications, Universitetsforlaget, Oslo. (A comprehensive account of some of the more sophisticated modifications of the basic immunoelectrophoresis principle.)

References in Text

1. Roitt,I.M. (1977) see general references.
2. Weir,D.M. (ed.) (1978) see general references.
3. Stevenson,G.T. (1974) Nature (London) **247**, 477.
4. Ouchterlony,O. (1949) Acta Pathol. Microbiol. Scand. **26**, 507.
5. Hudson,L. and Hay,F.C. (1976) see general references.
6. Uriel,J. (1964) *in* Immunoelectrophoretic Analysis (Grabar,P. and Burtin,P., eds), Elsevier,I.E.P., Amsterdam, p. 58.
7. Jurd,R.D., Luther-Davies,S.M. and Stevenson,G.T. (1975) Comp. Biochem. Physiol. **50B**, 65.

8. Hunter,W.M. and Greenwood,F.C. (1962) Nature (London) **194**, 495.

9. Scazzocchio,C., Holl,F.B. and Foguelman,A.I. (1973) Eur. J. Biochem. **36**, 428.

10. Axelsen,N.H., Krøll,J. and Weeks,B. (eds) (1973) see general references.

APPENDIX I
Bibliography of Polypeptide Detection Methods

GENERAL POLYPEPTIDE DETECTION METHODS

Staining Methods

Comparison of amido black 10B, Coomassie blue R, Fast green FCF:
Wilson, C.M. (1979) Anal. Biochem, **96**, 263.
Coomassie blue R250 used in methanol-acetic acid:
Weber, K. and Osborn, M. (1969) J. Biol. Chem. **244**, 4406.
Wilson, C.M. (1979) Anal. Biochem. **96**, 263.
Coomassie blue R250 used in isopropanol-acetic acid:
Fairbanks, G., Steck, T.L. and Wallach, D.F.H. (1971) Biochemistry **10**, 2606.
Rapid acid-based Coomassie blue stains requiring no destaining:
Blakesley, R.W. and Boezi, J.A. (1977) Anal. Biochem. **82**, 580.
Chrambach, A., Reisfeld, R.A., Wyckoff, M. and Zaccari, J. (1967) Anal. Biochem. **20**, 150.
Diezel, W., Kopperschläger, G. and Hoffman, E. (1972) Anal. Biochem. **48**, 617.
Malik, N. and Berrie, A. (1972) Anal. Biochem. **49**, 173.
Reisner, A.H., Nemes, P. and Bucholtz, C. (1975) Anal. Biochem. **64**, 509.
Staining proteins in electrofocusing gels using Fast green:
Allen, R.E., Masak, K.C. and McAllister, P.K. (1980) Anal. Biochem. **104**, 494.
High sensitivity silver stain:
Oakley, B.R., Kirsch, D.R. and Morris, N.R. (1980) Anal. Biochem. **105**, 361.
Switzer, R.C., Merril, C.R. and Shifrin, S. (1979) Anal. Biochem. **98**, 231.
Protein staining with Fe^{2+} - bathophenanthroline-sulphate:
Graham, G., Nairn, R.S. and Bates, G.W. (1978) Anal. Biochem. **88**, 434.
Tannic acid staining of native proteins to permit protein recovery:
Aoki, K., Kajiwara, S., Shinke, R. and Nishira, H. (1979) Anal. Biochem. **95**, 575.
Staining proteins prior to electrophoresis:
Bosshard, H.F. and Datyner, A. (1977) Anal. Biochem. **82**, 327.
Sun, S.M. and Hall, T.C. (1974) Anal. Biochem. **61**, 237.

Artifactual staining problems. Artifactual staining can be observed using Coomassie blue under certain conditions.
Artifactual staining when using dimethylsuberimidate for protein crosslinking:
Burke, W.F. and Reeves, H.C. (1975) Anal. Biochem. **63**, 267.
Artifactual staining with LKB Ampholines:
Otavsky, W.I. and Drysdale, J.W. (1975) Anal. Biochem. **65**, 533.

Fluorescent Dye Methods

Labelling with Fluorophore Prior to Electrophoresis

Dansyl chloride:
Schetters, H. and McLeod, B. (1979) Anal. Biochem. **98**, 329.

Stephens, R.E. (1975) Anal. Biochem. **65**, 369.

Tjissen, P. and Kurstak, E. (1979) Anal. Biochem. **99**, 97.

Fluorescamine:

Douglas, S.A., La Marca, M.E. and Mets, L.J. (1978) *in* Electrophoresis '78 (Catsimpoolas, N., ed.), Elsevier/North-Holland, Amsterdam, Vol. 2, p. 155.

Eng, P.R. and Parker, C.O. (1974) Anal. Biochem. **59**, 323.

Ragland, W.L., Benton, T.L., Pace, J.L., Beach, F.G. and Wade, A.E. (1978) *in* Electrophoresis '78 (Catsimpoolas, N., ed.), Elsevier/North-Holland, Amsterdam, Vol. 2, p. 217.

Ragland, W.L., Pace, J.L. and Kemper, D.L. (1974) Anal. Biochem. **59**, 24.

MDPF [2-methoxy-2, 4-diphenyl-3(2H)-furanone]:

Barger, B.O., White, F.C. Pace, J.L., Kemper, D.L. and Ragland, W.L. (1976) Anal. Biochem. **70**, 327.

Douglas, S.A. *et al.* (1978) see above.

Ragland, W.L. *et al.* (1978) see above.

DACM [N-(7-dimethylamino-4-methylcoumarinyl) maleimide]:

Yamamoto, K., Okamoto, Y. and Sekine, T. (1978) Anal. Biochem. **84**, 313.

o-Phthalaldehyde:

Weidekamm, E., Wallach, D.F.H. and Flückiger, R. (1973) Anal. Biochem. **54**, 102.

Labelling with Fluorophore after Electrophoresis

These methods are little used for analytical studies.

Anilinonaphthalene sulphonate:

Hartman, B.K. and Udenfriend, S. (1969) Anal. Biochem. **30**, 391.

Fluorescamine:

Jackowski, G. and Liew, C.C. (1980) Anal. Biochem. **102**, 34.

p-Hydrazinoacridine:

Carson, S.D. (1977) Anal. Biochem. **78**, 428.

Direct Detection Methods

Via protein phosphorescence:

Mardian, J.K.W. and Isenberg, I. (1978) Anal. Biochem. **91**, 1.

Detection of SDS-polypeptides by chilling:

Wallace, R.W., Yu, P.H., Dieckart, J.P. and Dieckart, J.W. (1974) Anal. Biochem. **61**, 86.

Detection of SDS-polypeptides by precipitation with K^+ ions:

Nelles, L.P. and Bamburg, J.R. (1976) Anal. Biochem. **73**, 522.

Detection of SDS-polypeptides using sodium acetate:

Higgins, R.C. and Dahmus, M.E. (1979) Anal. Biochem. **93**, 257.

Detection of SDS-polypeptides by reaction with cationic surfactant:

Takagi, T., Kubo, K. and Isemura, T. (1977) Anal. Biochem. **79**, 104.

Prelabelling with fluorescent molecules: see Fluorescent Dye Methods.

Prestaining: see Staining Methods.

DETECTION METHODS BASED ON THE USE OF RADIOISOTOPES

Radiolabelling proteins *in vitro* prior to electrophoresis: see Appendix II.
Radiolabelling proteins after gel electrophoresis:
 Christopher, A.R., Nagpal, M.L., Carroll, A.R. and Brown, J.C. (1978) Anal. Biochem. **85**, 404.
 Elder, J.H., Pickelt, R.A., Hampton, J. and Lerner, R.A. (1977) J. Biol. Chem. **252**, 6510.
Use of radiolabelled probes:
 Radiolabelled antibody or protein A - see p. 60 and 253.
 Radiolabelled lectins - see p. 252.
Indirect autoradiography (^{125}I, ^{32}P) using an X-ray intensifying screen:
 Laskey, R.A. (1980) *in* Methods in Enzymology (Grossman, L. and Moldave, K., eds), Academic press, New York, Vol. 65, p. 363.
 Laskey, R.A. and Mills, A.D. (1977) FEBS Lett. **82**, 314.
Fluorography (^{35}S, ^{14}C, ^{3}H) using PPO:
 Bonner, W.M. and Laskey, R.A. (1974) Eur. J. Biochem. **46**, 83.
 Laskey, R.A. (1980) *in* Methods in Enzymology (Grossman, L. and Moldave, K., eds), Academic Press, New York, Vol. 65, p. 363.
 Laskey, R.A. and Mills, A.D. (1975) Eur. J. Biochem. **56**, 335.
Fluorography using sodium salicylate:
 Chamberlain, J.P. (1979) Anal. Biochem. **98**, 132.
Double-label detection using X-ray film:
 Gruenstein, E.I. and Pollard, A.L. (1976) Anal. Biochem. **76**, 452.
 Kronenberg, L.H. (1979) Anal. Biochem. **93**, 189.
 McConkey, E.H. (1979) Anal. Biochem. **96**, 39.
 Walton, K.E., Styer, D. and Gruenstein, E. (1979) J. Biol. Chem. **254**, 795.

METHODS TO DETECT SPECIFIC CLASSES OF POLYPEPTIDE

Glycoproteins

Periodic Acid Schiff (PAS):
 A large number of modifications of the method have been published, some of which are listed below:
 Fairbanks, G., Steck, T.L. and Wallach, D.L.H. (1971) Biochemistry **10**, 2026.
 Glossman, H. and Neville, D.M. (1971) J. Biol. Chem. **246**, 6339.
 Matthieu, J.M. and Quarles, R.H. (1973) Anal. Biochem. **55**, 313.
 Zaccharias, R.J., Zell, T.E., Morrison, J.H. and Woodlock, J.J. (1969) Anal. Biochem. **31**, 148.
Alcian blue:
 Wardi, A.H. and Michos, G.A. (1972) Anal. Biochem. **49**, 607.
p-Hydrazino-acridine (fluorescent dye):
 Carson, S.D. (1977) Anal. Biochem. **78**, 428.
Stains-all for sialoglycoproteins:
 Green, M.R. and Pastewka, J.V. (1975) Anal. Biochem. **65**, 66.

King, L.E. and Morrison, M. (1976) Anal. Biochem. **71**, 223.
Thymol and sulphuric acid:
  Racusen, D. (1979) Anal. Biochem. **99**, 474.
Fluorescent lectins:
  Furlan, M., Perret, B.A. and Beck, E.A. (1979) Anal. Biochem. **96**, 208.
Radiolabelled lectins:
  Burridge, K. (1978) *in* Methods in Enzymology (Ginsburg, V., ed.), Academic Press, New York, Vol. 50, p. 54.
  Rostas, J.A.P., Kelly, P.T. and Cotman, C.W. (1977) Anal. Biochem. **80**, 366.
Lectins with covalently bound alcohol dehydrogenase or lactate dehydrogenase:
  Avigad, G. (1978) Anal. Biochem. **86**, 443.
Lectins with covalently bound horse-radish peroxidase:
  Wood, J.G. and Sarinana, F.O. (1975) Anal. Biochem. **69**, 320.
Crossed lectin electrophoresis:
  West, C.M. and McMahon, D. (1977) J. Cell Biol. **74**, 264.
Labelling of cell surface glycoproteins using galactose oxidase:
  Gahmberg, C.G. (1978) *in* Methods in Enzymology (Ginsburg, V., ed.), Academic Press, New York, Vol. 50, p. 204.
Labelling of glycoproteins containing terminal *N*-acetylglucosamine using galactosyl transferase:
  Wallenfels, B. (1979) Proc. Nat. Acad. Sci. U.S.A. **76**, 3223.

Phosphoproteins

Stains-all:
  Green, M.R., Pastewka, J.V. and Peacock, A.C. (1973) Anal. Biochem. **56**, 43.
Ammonium molybdate:
  Cutting, J.A. and Roth, T.A. (1973) Anal. Biochem. **54**, 386.
Acidic phosphoproteins (phosvitin):
  Hegenauer, J., Ripley, L. and Nace, G. (1977) Anal. Biochem. **78**, 308.
Radiolabelling - see p. 59.

Nucleoproteins

Ethidium bromide:
  Goodwin, G.H. (1981) *in* Gel Electrophoresis of Nucleic Acids : A Practical Approach (Rickwood, D. and Hames, B.D., eds), Information Retrieval Ltd, London.
Stains-all:
  Dahlberg, A.E., Dingman, C.W. and Peacock, A.C. (1969) J. Mol. Biol. **41**, 139.
  Goodwin, G.H. (1981): see above.
Methylene blue:
  Dahlberg *et al.* (1969): see above.
Detection of DNA-binding proteins by protein blotting:
  Bowen, B., Steinberg, J., Laemmli, U.K. and Weintraub, H. (1980) Nucl. Acid Res. **8**, 1.

Proteins with Available Thiol Groups

DACM [N-(7-dimethylamino-4-methylcoumarinyl) maleimide]:
Yamamoto, K., Okamoto, Y. and Sekine, T. (1978) Anal. Biochem. **84**, 313.
DTNB [5,5′-dithiobis (2-nitrobenzoic acid)]:
Zelazowski, A.J. (1980) Anal. Biochem. **103**, 307.

Cadmium-containing Proteins

Dipyridyl-ferrous iodide:
Zelazowski, A.J. (1980) Anal. Biochem. **103**, 307.

Collagen and Procollagen

McCormick, P.J., Chandrasekhar, S. and Millis, A.J.T. (1979) Anal. Biochem. **97**, 359.

IMMUNOLOGICAL METHODS

Incubation of the gel with radiolabelled antibody:
Burridge, K. (1978) *in* Methods in Enzymology (Ginsburg, V., ed.), Academic Press, New York, Vol. 50, p. 54.
Kasamatsu, H. and Flory, P.J. (1978) Virology **86**, 344.
Incubation of the gel with unlabelled antibody then with [^{125}I]-protein A:
Burridge, K. (1978) *in* Methods in Enzymology (Ginsburg, V., ed.), Academic Press, New York, Vol. 50, p. 54.
Adair, W.S., Jurivich, D. and Goodenough, U.W. (1978) J. Cell Biol. **79**, 281.
Bigelis, R. and Burridge, K. (1978) Biochem. Biophys. Res. Commun. **82**, 322.
Saltzgaber-Müller, J. and Schatz, G. (1978) J. Biol. Chem. **253**, 305.
Transfer of antigens from the gel to diazobenzyloxymethyl paper followed by immunological identification using [^{125}I]-protein A:
Renart, J., Reiser, J. and Stark, G.R. (1979) Proc. Nat. Acad. Sci. U.S.A. **76**, 3116.
Incubation of the gel with fluorescein-labelled antibody:
Groschel-Stewart, U., Schreiber, J., Mahlmeister, C. and Weber, K. (1976) Histochemistry **46**, 229.
Stumph, W.E., Elgin, S.C.R. and Hood, L. (1974) J. Immunol. **113**, 1752.
Incubation of the gel with antibody coupled to peroxidase followed by localisation with 3,3′-diaminobenzidine:
Olden, K. and Yamada, K.M. (1977) Anal. Biochem. **78**, 483.
Parish, R.W., Schmidlin, S. and Parish, C.R. (1978) FEBS Lett. **95**, 366.
Van Raamsdonk, W., Pool, C.W. and Heyting, C. (1977) J. Immunol. Methods **17**, 337.
Use of an agarose overlay containing antiserum to produce an immunoreplica:
Showe, M.K., Isobe, E. and Onorato, L. (1976) J. Mol. Biol. **107**, 55.

Saltzgaber-Müller, J. and Schatz, G. (1978) J. Biol. Chem. **253**, 305.
Crossed immunoelectrophoresis:
  Chua, N.H. and Blomberg, F. (1979) J. Biol. Chem. **253**, 3924.
  Converse, C.A. and Papermaster, D.S. (1975) Science **189**, 489.

ENZYME DETECTION METHODS

Three recent publications by Shaw and Prasad (1), Siciliano and Shaw (2), and Harris and Hopkinson (3a,b,c,) are particularly useful since they describe the preparation and use of reagents to detect a wide variety of enzymes. Often several different approaches to detect a particular enzyme are possible. Harris and Hopkinson (3a,b,c,) give an extensive bibliography in each case. The majority of the methods in these publications relate to starch gel but most can also be used with polyacrylamide gel. Gabriel (4) has reviewed enzyme detection methods which have been applied to polyacrylamide gel. Listed below are the enzymes which can be detected on gels, in each case with an indication of the particular compilation which can be consulted for experimental information plus other selected references. The general approaches to enzyme detection on gels, together with some cautionary notes, have been discussed earlier in this book (p. 60).

Enzyme	References
Acid phosphatase	1, 2, 3a, 4, 5.
Aconitase	1, 2, 3a, 6.
Adenine phosphoribosyl transferase (AMP pyrophosphorylase)	3a, 7, 8.
Adenosine deaminase	2, 3a, 9.
Adenosine kinase	3c, 10.
Adenylate kinase	1, 2, 3a, 11.
ADP-glycogen transferase	4, 12.
Alanine aminotransferase	2, 3a, 13.
Alcohol dehydrogenase	1, 2, 3a, 4, 14, 15.
Aldolase	1, 2, 3a, 16.
Alkaline phosphatase	1, 3a, 4, 17.
Amine oxidase	4, 18.
Amino acid oxidase	3b, 4, 19, 20.
AMP deaminase	3c, 21, 22.
Amylase	3a, 4, 23, 24.
Anthranilate synthetase	25.
Arginase	3c, 21, 26.
Arginosuccinase	3c, 21.
Aromatic amino acid decarboxylase	27.
Aromatic amino acid transaminase	1.
Arylamidases	4.
Arylsulphatases	3a, 28, 29.

Aspartate aminotransferase	1, 2, 3a, 4, 30.
Carbonic anhydrase	1, 3a, 31, 32.
Catalase	1, 2 and 3a (starch gels only), 4.
Cellobiose phosphorylase	4.
Cellulases	33.
Cholinesterase	4, 34.
Chymotrypsin	35.
Citrate synthase	3a, 36.
Creatine kinase	1, 3a, 4, 37.
3'5' Cyclic AMP phosphodiesterase	3a, 4, 38.
Cytidine deaminase	3a, 39.
Cytochrome b_5 reductase	3a, 40, 41.
Dihydrouracil dehydrogenase	43.
Dipeptidase	42. (See also peptidases.)
DNA polymerase	4, 44.
Elastase and pro-elastase	45.
Enolase	2, 3a, 46.
Esterases	1, 2, 3a, 4, 5, 47.
Folate reductase	4.
Fructose-1, 6-diphosphatase	1.
Fructosyl transferase	48.
L-Fucose dehydrogenase	49.
α-Fucosidase	3a, 4, 50.
Fumarase	1, 2, 3c, 51.
Galactokinase	3a, 4, 52.
Galactose-6-phosphate dehydrogenase	1.
Galactose-1-phosphate uridyl transferase	3a, 4, 53. (See also galactosyl transferases.)
α-Galactosidase	2, 3a, 54, 55.
β-Galactosidase	3a, 4, 56, 57, 58.
Galactosyl transferases	59.
Glucose oxidase	4.
Glucose-6-phosphate dehydrogenase	1, 2, 3a, 60.
α-Glucosidase	3a, 4, 61.
β-Glucosidase	3a, 4, 62.
β-Glucuronidase	1, 2, 3b, 63, 64.
Glutamate dehydrogenase	1, 3b, 4, 65.
Glutamate-oxaloacetate transaminase	See aspartate aminotransferase.
Glutamate-pyruvate transaminase	See alanine aminotransferase.
Glutaminase	66.
Glutathione peroxidase	3a, 67.
Glutathione reductase	1, 3a, 68.
Glutathione S transferase	69.
Glyceraldehyde-3-phosphate dehydrogenase	1, 2, 3a, 4, 70.
Glycerol kinase	4, 8.

Glycerol-3-phosphate dehydrogenase	1, 2, 3a, 4, 70.
Glycogen phosphorylase	4.
Glycolate oxidase	3c, 71.
Glycosidases	4, 72, 73.
Glyoxalase	3a, 3b, 70, 74, 75.
Guanine deaminase (guanase)	3a, 4.
Guanylate kinase	3a, 76.
Hexokinase	1, 3a, 4, 8, 77, 78.
Hexosaminidase	See ß-*N*-acetyl-hexosaminidase.
Homoserine dehydrogenase	4, 79.
Hydroxyacyl coenzyme A dehydrogenase	3b, 80.
D(-)3-Hydroxybutyrate dehydrogenase	1.
Hydroxysteroid dehydrogenase	4.
Hypoxanthine-guanine phosphoribosyl transferase	1, 3a, 4, 8, 81.
Inorganic pyrophosphatase	1, 3a, 4, 82.
Inosine triphosphatase	3a.
Invertase	83.
Isocitrate dehydrogenase	1, 2, 3a, 4, 84, 85.
Isocitrate lyase	86.
α-Ketoglutarate semialdehyde dehydrogenase	4.
α-Ketoisocaproate dehydrogenase	4.
α-Keto-β-methyl valeriate dehydrogenase	4.
Lactate dehydrogenase	1, 2, 3a, 4, 87, 88.
Leucine amino-peptidase	1, 89.
Lipase	4, 90. (See also phospholipase.)
Lipoyl dehydrogenase	4, 91.
Malate dehydrogenase	1, 2, 3a, 4, 92.
Malate synthase	93.
Malic enzyme	2, 3a, 94, 95.
Mannose-phosphate isomerase	2, 3a, 96.
α-Mannosidase	2, 3a, 73, 97.
β-*N*-acetyl-hexosaminidase	2, 3a, 98.
NADH diaphorase	See cytochrome b_5 reductase.
NAD (P) nucleosidase	3b, 99, 100.
Nitrate reductase	4, 101.
Nitrite reductase	102.
Nucleases	103, 104. (See also ribonuclease.)
Nucleoside phosphorylase	2, 3a, 105, 106.
Nucleotidase	4, 107.
Ornithine transcarbamylase	3a, 108.
Oxidases	4, 109.
PEP carboxylase	110.

Peptidases	1, 2, 3a, 111, 112. (See also dipeptidase, tripeptidase.)
Peroxidase	1, 2, 5.
Phosphatases	See relevant phosphatase.
Phosphofructokinase	3a, 4, 113, 114.
Phosphoglucoisomerase	1, 2, 3a, 115.
Phosphoglucomutase	1, 2, 3a, 3b, 4, 116.
Phosphogluconate dehydrogenase	1, 2, 3a, 4, 117.
Phosphoglycerate kinase	2, 3a, 118.
Phosphoglyceromutase	2, 3a, 119, 120.
Phosphoglycollate phosphatase	3c, 121.
Phospholipase	122.
Phosphotransferases	8.
Plasminogen activators	123.
Polynucleotide phosphorylase	4, 124.
Proteinase	123, 125, 126. (See also peptidases, subtilisin, chymotrypsin, trypsin, elastase, plasminogen activators.)
Proteinase inhibitors	127.
Protein kinase	128, 129.
Pyridoxine kinase (pyridoxal kinase)	3b, 130.
Pyrophosphatase	See inorganic pyrophosphatase.
Pyruvate carboxylase	110.
Pyruvate kinase	2, 3a, 131.
Retinol dehydrogenase	1.
Ribonuclease	1, 4, 132. (See also nucleases.)
Sorbitol dehydrogenase	1, 3a, 70.
Subtilisin	133.
Sucrose phosphorylase	4.
Sulphite oxidase	134.
Superoxide dismutase	3a, 135, 136.
Threonine deaminase	137.
Thymidine kinase	1.
Trehalase	138, 139.
Triose-phosphate isomerase	1, 2, 3a, 139, 140.
Tripeptidase	141.
Trypsin	142.
Tyrosine aminotransferase	143.
UDPG dehydrogenase	4.
UDPG pyrophosphorylase	3a, 4, 144.
UMP kinase	3a, 145.
Urease	4, 146.
Xanthine dehydrogenase	1.

References

1.   Shaw, C.R. and Prasad, R. (1970) Biochem. Genet. **4,** 297.

2.   Siciliano, M.J. and Shaw, C.R. (1976) *in* Chromatographic and Electro-phoretic Techniques (Smith, I., ed.), William Heinemann Medical Books Ltd., London, Vol. 2, p. 185.

3(a).   Harris, H. and Hopkinson, D.A. (1976) Handbook of Enzyme Electrophoresis in Human Genetics, North-Holland, Amsterdam.
  (b).   Supplement (1977).
  (c).   Supplement (1978).

4.   Gabriel, O. (1971) *in* Methods in Enzymology (Colowick, S.P. and Kaplan, N.O., eds), Academic Press, New York, Vol. 22, p. 578.

5.   Cullis, C.A. and Kolodynska, K. (1975) Biochem. Genet. **13,** 687.

6.   Slaughter, C.A., Hopkinson, D.A. and Harris, H. (1975) Ann. Hum. Genet. **39,** 193.

7.   Mowbray, S., Watson, B. and Harris, H. (1972) Ann. Hum. Genet. **36,** 153.

8.   Tischfield, J.A., Bernhard, H.P. and Ruddle, F.H. (1973) Anal. Biochem. **53,** 545.

9.   Spencer, N., Hopkinson, D.A. and Harris, H. (1968) Ann. Hum. Genet. **32,** 9.

10.   Klobutcher, L.A., Nichols, E.A., Kucherlapati, R.S. and Ruddle, F.H. (1976) *in* Birth Defects; Original Article Series, The National Foundation, March of Dimes, New York, Vol. 12, 171.

11.   Wilson, D.E., Povey, S. and Harris, H. (1976) Ann. Hum. Genet. **39,** 305.

12.   Frederick, J.F. (1968) Ann. N.Y. Acad. Sci. **151,** 413.

13.   Chen, S.H., Giblett, E.R., Anderson, J.E. and Fossum, B.L.G. (1972) Ann. Hum. Genet. **35,** 401.

14.   Sofer, W. and Ursprung, H. (1968) J. Biol. Chem. **243,** 3110.

15.   Smith, M., Hopkinson, D.A. and Harris, H. (1971) Ann. Hum. Genet. **34,** 251.

16.   Lewinski, N.D. and Dekker, E.E. (1978) Anal. Biochem. **87,** 56.

17.   Sussman, H.H., Small, P.A. and Cotlove, E. (1968) J. Biol. Chem. **243,** 160.

18.   Ma Lin, A.W. and Castell, D.O. (1975) Anal. Biochem. **69,** 637.

19.   Barker, R.F. and Hopkinson, D.A. (1977) Ann. Hum. Genet. **41,** 27.

20.   Hayes, M.B. and Wellner, D. (1969) J. Biol. Chem. **244,** 6636.

21.   Nelson, R.L., Povey, S., Hopkinson, D.A. and Harris, H. (1977) Biochem. Genet. **15,** 1023.

22.   Anderson, J.E., Teng, Y.S. and Liblett, E.R. (1975) *in* Birth Defects; Original Article Series, The National Foundation, March of Dimes, New

York, Vol. 11, p. 295.

23. Heller, H. and Kulka, R.G. (1968) Biochim. Biophys. Acta **165**, 393.

24. Boettcher, D. and De La Lande, F.A. (1969) Anal. Biochem. **28**, 510.

25. Grove, T.H. and Levy, H.R. (1975) Anal. Biochem. **65**, 458.

26. Farron, F. (1973) Anal. Biochem. **53**, 264.

27. Landon, M. (1977) Anal. Biochem. **77**, 293.

28. Shapira, E., De Gregorio, R.R., Matalon, R. and Nadler, H.R. (1975) Biochem. Biophys. Res. Commun. **62**, 448.

29. Chang, P.L., Ballantyne, S.R. and Davidson, R.G. (1979) Anal. Biochem. **97**, 36.

30. Davidson, R.G., Cortner, J.A., Rattazi, M.C., Ruddle, F.H. and Lubs, H.A. (1970) Science **169**, 391.

31. Hopkinson, D.A., Coppock, J.S., Mühlemann, M.F. and Edwards, Y.H. (1974) Ann. Hum. Genet. **38**, 155.

32. Drescher, D.G. (1978) Anal. Biochem. **90**, 349.

33. Erickson, K.E. and Petterson, B. (1973) Anal. Biochem. **56**, 618.

34. Harris, H., Hopkinson, D.A. and Robson, E.B. (1962) Nature (London) **196**, 1296.

35. Gertler, A., Trencer, Y. and Tinman, G. (1973) Anal. Biochem. **54**, 270.

36. Craig, I. (1973) Biochem. Genet. **9**, 351.

37. Yue, R.H., Jacobs, H. K., Okabe, K., Keutel, H.J. and Kuby, S.A. (1968) Biochemistry **7**, 4291.

38. Monn, E. and Christiansen, R.O. (1971) Science **173**, 540.

39. Teng, Y.S., Anderson, J.E. and Giblett, E.R. (1975) Am. J. Hum. Genet. **27**, 492.

40. Williams, L. and Hopkinson, D.A. (1975) Hum. Hered. **25**, 567.

41. Kaplan, J.C. and Beutler, E. (1967) Biochem. Biophys. Res. Commun. **29**, 605.

42. Suguira, M., Ito, Y., Hirano, K. and Sawaki, S. (1977) Anal. Biochem. **81**, 481.

43. Hallock, R.O. and Yamada, E.W. (1973) Anal. Biochem. **56**, 84.

44. Jovin, T.M., Englund, P.T. and Bertsch, L.L. (1969) J. Biol. Chem. **244**, 2996.

45. Dijkhof, J. and Poort, C. (1977) Anal. Biochem. **83**, 315.

46. Sharma, H.K. and Rothstein, M. (1979) Anal. Biochem. **98**, 226.

47. Coates, P.M., Mestriner, M.A. and Hopkinson, D.A. (1975) Ann. Hum. Genet. **39**, 1.

48. Russell, R.R.B. (1979) Anal. Biochem. **97**, 173.

49. Schachter, H., Sarney, J., McGuire, E.J. and Roseman, S. (1969) J. Biol. Chem. **244**, 4785.

50. Turner, B.M., Beratis, N.G., Turner, V.S. and Hirschhorn, K. (1974) Clin. Chim. Acta **57**, 29.

51. Edwards, Y.H. and Hopkinson, D.A. (1978) Ann. Hum. Genet. **42**, 303.

52. Nicholls, E.A., Elsevier, S.M. and Ruddle, F.H. (1974) Cytogenet. Cell Genet. **13**, 275.

53. Ng, W.G., Bergren, W.R., Fields, M. and Donnell, G.N. (1969) Biochem. Biophys. Res. Commun. **37**, 354.

54. Beutler, E. and Kuhl, W. (1972) J. Biol. Chem. **247**, 7195.

55. Beutler, E., Guinto, E. and Kuhl, W. (1973) Am. J. Hum. Genet. **25**, 42.

56. Norden, A.G.W. and O'Brien, J.S. (1975) Proc. Nat. Acad. Sci. U.S.A. **72**, 240.

57. Alpers, D.H. (1969) J. Biol. Chem. **244**, 1238.

58. Alpers, D.H., Steers, E., Shifrin, S. and Tomkins, G. (1968) Ann. N.Y. Acad. Sci. **151**, 545.

59. Pierce, M., Cummings, R.D. and Roth, S. (1980) Anal. Biochem. **102**, 441.

60. Criss, W.E. and McKerns, K.W. (1968) Biochemistry **7**, 125.

61. Swallow, D.M., Corney, G., Harris, H. and Hirschhorn, R. (1975) Ann. Hum. Genet. **38**, 391.

62. Beutler, E., Kuhl, W., Trinidad, F., Teplitz, R. and Nadler, H. (1971) Am. J. Hum. Genet. **23**, 62.

63. Fondo, E.Y. and Bartalos, M. (1969) Biochem. Genet. **3**, 591.

64. Franke, U. (1976) Am. J. Hum. Genet. **28**, 357.

65. Nelson, R.L., Povey, M.S., Hopkinson, D.A. and Harris, H. (1977) Biochem. Genet. **15**, 87.

66. Davis, J.N. and Prusiner, S. (1973) Anal. Biochem. **54**, 272.

67. Beutler, E. and West, C. (1974) Am. J. Hum. Genet. **26**, 255.

68. Kaplan, J.C. and Beutler, E. (1968) Nature (London) **217**, 256.

69. Board, P.G. (1980) Anal. Biochem. **105**, 147.

70. Charlesworth, D. (1972) Ann. Hum. Genet. **35**, 477.

71. Duley, J. and Holmes, R.S. (1974) Genetics **76**, 93.

72. Price, R.G. and Dance, N. (1967) Biochem. J. **105**, 877.

73. Gabriel, O. and Wang, S.F. (1969) Anal. Biochem. **27**, 545.

74. Kompf, J., Bissbort, S., Gussman, S. and Ritter, H. (1975) Humangenetik **27**, 141.

75. Parr, C.W., Bagster, I.A. and Welch, S.G. (1977) Biochem. Genet. **15**, 109.

76. Jamil, T., Fisher, R.A. and Harris, H. (1976) Hum. Hered. **25**, 402.

77. Katzen, H.M. and Schimke, R.T. (1965) Proc. Nat. Acad. Sci. U.S.A. **54**, 1218.

78. Rogers, P.A., Fisher, R.A. and Harris, H. (1975) Biochem. Genet. **13**, 857.

79. Ogilvie, J.W., Sightler, J.H. and Clark, R.B. (1969) Biochemistry **8**, 3557.

80. Craig, I., Tolley, E. and Borrow, M. (1976) *in* Birth Defects; Original Article Series, The National Foundation, March of Dimes, New York, Vol. 12, 114.

81. Vasquez, B. and Bieber, A.L. (1978) Anal. Biochem. **84**, 504.

82. Fischer, R.A., Turner, B.M., Dorkin, H.L. and Harris, H. (1974) Ann. Hum. Genet. **37**, 341.

83. Babczinski, P. (1980) Anal. Biochem. **105**, 328.

84. Henderson, N.S. (1968) Ann. N.Y. Acad. Sci. **151**, 429.

85. Turner, B.M., Fisher, R.A. and Harris, H. (1974) Ann. Hum. Genet. **37**, 455.

86. Reeves, H.C. and Volk, M.J. (1972) Anal. Biochem. **48**, 437.

87. Allen, J.M. (1961) Ann. N.Y. Acad. Sci. **94**, 937.

88. Werthamer, S., Freiberg, A. and Amaral, L. (1973) Clin. Chim. Acta **45**, 5.

89. Strongin, A.Y., Azarenkova, N.M., Vaganova, T.I., Levin, A.D. and Stepanov, V.M. (1976) Anal. Biochem. **74**, 597.

90. Nachlase, M.M., Morris, B., Rosenblatt, D. and Seligman, A.M. (1960) J. Biophys. Biochem. Cytol. **7**, 261.

91. Millard, S.A., Kubose, A. and Gal, E.M. (1969) J. Biol. Chem. **244**, 2511.

92. Allen, S.L. (1968) Ann. N.Y. Acad. Sci. **151**, 190.

93. Volk, M.J., Trelease, R.N. and Reeves, H.C. (1974) Anal. Biochem. **58**, 315.

94. Cohen, P.T.W. and Omenn, G.S. (1972) Biochem. Genet. **7**, 303.

95. Povey, S., Wilson, D.E., Harris, H., Gormley, I.P., Perry, P. and Buckton, K.E. (1975) Ann. Hum. Genet. **39**, 203.

96. Nichols, E.A., Chapman, V.M. and Ruddle, F.H. (1973) Biochem. Genet. **8**, 47.

97. Poenaru, L. and Dreyfus, J.C. (1973) Biochim. Biophys. Acta **303**, 171.

98. Okada, S., Veath, M.L., Lerov, J. and O'Brien, J.S. (1971) Am. J. Hum. Genet. **23**, 55.

99. Flechner, L., Hirschhorn, S. and Bekierkunst, A. (1968) Life Sci. **7**, 1327.

100. Ravazzolo, R., Bruzzone, G., Garrè, C. and Ajmar, F. (1976) Biochem. Genet. **14**, 877.

101.   Ingle, J. (1968) Biochem. J. **108,** 715.

102.   Hucklesby, D.P. and Hageman, R.H. (1973) Anal. Biochem. **56,** 591.

103.   Lacks, S.A., Springhorn, S.S. and Rosenthal, A.L. (1979) Anal. Biochem. **100,** 357.

104.   Rosenthal, A.L. and Lacks, S.A. (1977) Anal. Biochem. **80,** 76.

105.   Spencer, N., Hopkinson, D.A. and Harris, H. (1968) Ann. Hum. Genet. **32,** 9.

106.   Edwards, Y.H., Hopkinson, D.A. and Harris, H. (1971) Ann. Hum. Genet. **34,** 395.

107.   Dvorak, H.F. and Heppel, L.A. (1968) J. Biol. Chem. **243,** 2647.

108.   Baron, D.N. and Buttery, J.E. (1972) J. Clin. Pathol. **25,** 415.

109.   Feinstein, R.N. and Lindahl, R. (1973) Anal. Biochem. **56,** 353.

110.   Scrutton, M.C. and Fatebene, F. (1975) Anal. Biochem. **69,** 247.

111.   Lewis, W.H.P. and Harris, H. (1967) Nature (London) **215,** 315.

112.   Rapley, S., Lewis, W.H.P. and Harris, H. (1971) Ann. Hum. Genet. **34,** 307.

113.   Brock, D.J.H. (1969) Biochem. J. **113,** 235.

114.   Niessner, N. and Beutler, E. (1974) Biochem. Med. **9,** 73.

115.   De Lorenzo, R.J. and Ruddle, F.H. (1969) Biochem. Genet. **3,** 151.

116.   Kühn, P., Schmidtmann, U. and Spielmann, W. (1977) Hum. Genet. **35,** 219.

117.   Fildes, R.A. and Parr, C.W. (1963) Nature (London) **200,** 890.

118.   Beutler, E. (1969) Biochem. Genet. **3,** 189.

119.   Omenn, G.S. and Cheung, S.C.Y. (1974) Am. J. Hum. Genet. **26,** 393.

120.   Chen, S.H., Anderson, J., Giblett, E.R. and Lewis, M. (1974) Am. J. Hum. Genet. **26,** 73.

121.   Barker, R.F. and Hopkinson, D.A. (1978) Ann. Hum. Genet. **42,** 1.

122.   Shier, W.T. and Troffer, J.T. (1978) Anal. Biochem. **87,** 604.

123.   Heussen, E. and Dowdle, E.B. (1980) Anal. Biochem. **102,** 196.

124.   Klee, C.B. (1969) J. Biol. Chem. **244,** 2558.

125.   Andary, T.J. and Dabich, D. (1974) Anal. Biochem. **57,** 457.

126.   North, M.J. and Harwood, J.M. (1979) Biochim. Biophys. Acta **566,** 222.

127.   Filho, J.X. and De Azevedo Moreira, R. (1978) Anal. Biochem. **84,** 296.

128.   Hirsch, A. and Rosen, M. (1974) Anal. Biochem. **60,** 389.

129.   Gagelman, M., Pyerin, W., Kübler, D. and Kinzel, V. (1979) Anal. Biochem. **93,** 52.

130. Chern, C.J. and Beutler, E. (1976) Ann. Hum. Genet. **28,** 9.

131. Imamura, K. and Tanaka, T. (1972) J. Biochem. **71,** 1043.

132. Biswas, S. and Hollander, V.P. (1969) J. Biol. Chem. **244,** 4185.

133. Lyublinskaya, L.A., Belyaev, S.V., Strongin, A. Ya, Matyash, L.F. and Stepanov, V.M. (1974) Anal. Biochem. **62,** 371.

134. Cohen, H.J. (1973) Anal. Biochem. **53,** 208.

135. Beauchamp, C. and Fridovitch, I. (1971) Anal. Biochem. **44,** 276.

136. Beckman, G., Lundgren, A., Tarnvik, A. (1973) Hum. Hered. **23,** 338.

137. Hatfield, G.W. and Umbarger, H.E. (1970) J. Biol. Chem. **245,** 1736.

138. Killick, K.A. and Wang, L.W. (1980) Anal. Biochem. **106,** 367.

139. Kaplan, J.C., Teeple, L., Shore, N. and Beutler, E. (1968) Biochem. Biophys. Res. Commun. **31,** 768.

140. Scopes, R.K. (1964) Nature (London) **201,** 924.

141. Suguira, M., Ho, Y., Hirano, K. and Sawaki, S. (1977) Anal. Biochem. **81,** 481.

142. Gertler, A., Tencer, Y. and Tinman, G. (1973) Anal. Biochem. **54,** 270.

143. Walker, D.G. and Khan, H.H. (1968) Biochem. J. **108,** 169.

144. Manrow, R.F. and Dottin, R.P. (1980) Proc. Nat. Acad. Sci. U.S.A. **77,** 730.

145. Giblett, E.R., Anderson, J.A., Chen, S.H., Teng, Y.S. and Cohen, F. (1974) Am. J. Hum. Genet. **26,** 627.

146. Shaik-M, M.B., Guy, A.L., Pancholy, S.K. (1980) Anal. Biochem. **103,** 140.

APPENDIX II
Reagents for the Isotopic Labelling of Proteins

A wide variety of reagents is commercially available for the isotopic labelling of proteins and these are summarised below. Following the summary, the reagents are listed in alphabetical order, with a description of their properties and selected references to demonstrate their use.

SUMMARY

Reacting groups	Labelling reagent
—NH$_2$ Free amino groups (*N*-terminal or lysine residues)	Acetic anhydride Bolton and Hunter reagent Dansyl chloride Ethyl acetimidate 1-Fluoro-2,4-dinitrobenzene Formaldehyde Isethionyl acetimidate Maleic anhydride Methyl 3,5-diiodohydroxybenzimidate Phenyl isothiocyanate Sodium borohydride/potassium borohydride Succinic anhydride *N*-Succinimidyl propionate
—SH Thiol groups (cysteine residues)	Acetic anhydride Bromoacetic acid Chloroacetic acid *p*-Chloromercuribenzenesulphonic acid *p*-Chloromercuribenzoic acid Dansyl chloride *N*-Ethylmaleimide Iodoacetamide Iodoacetic acid
⬡—OH Phenolic hydroxyl groups (tyrosine residues)	Acetic anhydride Dansyl chloride Iodine

	Dansyl chloride Iodine
Imidazole groups (histidine residues)	
—CH$_2$OH Aliphatic hydroxyl groups (serine/threonine residues)	Acetic anhydride DFP

REAGENTS

Acetic anhydride (^{3}H or ^{14}C)

Acetic anhydride is a non-specific acylating agent that may be used to label serine, lysine, threonine, tyrosine or cysteine and *N*-terminal residues.

1. Avivi,P., Simpson,S.A., Tait,J.F. and Whitehead,J.K. (1954) Proceeding 2nd Radioisotope Conference, Oxford; Butterworths, London; Vol. 1, p. 313.
2. O'Leary,M.H. and Westheimer,F.H. (1968) Biochemistry **7**, 913.
3. Ostrowski,K., Barnard,E.A., Sawicki,W., Chorzelski,T., Langner,A. and Mikulski,A. (1970) J.Histochem. Cytochem. **18**, 490.
4. Heinegård,D.K. and Hascall,V.C. (1979) J. Biol. Chem. **254**, 921.
5. Whitehead,J.K. (1958) Biochem. J. **68**, 662.
6. Barnard,E.A., Wieckowski,J. and Chiu,T.H. (1971) Nature (London) **234**, 207.
7. Brems,D.N. and Rilling, Hans,C. (1979) Biochemistry **18**, 860.
8. Gersten,D.M. and Goldstein,L.S. (1979) Int. J. Appl. Radiat. Isot. **30**, 469.

Bolton and Hunter reagent (^{125}I)

This reagent is specific for free amino groups and has been used extensively for the iodination of proteins under mild conditions.

1. Kågedal,B. and Källberg,M. (1977) Clin. Chem. **23**, 1694.
2. Roberts,R., Sobel,B.E. and Parker,C.W. (1978) Clin. Chim. Acta **83**, 141.
3. Culvenor,J.G. and Evans,W.H. (1977) Biochem. J. **168**, 475.
4. Pinder,J.C., Phethean,J. and Gratzer,W.B. (1978) FEBS Lett. **92**, 278.

Bromoacetic acid (^{14}C)

The haloacetic acids, bromo-, chloro- and iodoacetic acid, are the most widely used protein alkylating reagents. They react primarily with cysteine residues. The reactivity depends on the halide, and decreases in the order iodoacetic acid > bromoacetic acid > chloroacetic acid.

Bromoacetic acid is a fairly mild alkylating reagent but it is not very selective, reacting with most thiol groups except those in very protected positions.

1. Glick,D.M., Goren,H.J. and Barnard,E.A. (1967) Biochem. J. **102**, 7c.
2. Fanger,M.W., Hettinger,T.P. and Harbury,H.A. (1967) Biochemistry **6**, 713.

Chloroacetic acid (^{14}C)

Chloroacetic acid is less reactive than bromoacetic acid. It is therefore used for very selective alkylations since it will only react with the most reactive thiol groups in the protein.

1. Gerwin,B.I. (1967) J. Biol. Chem. **242**, 451.

p-Chloromercuribenzenesulphonic acid (^{203}Hg)

Organic mercurials react rapidly and specifically with thiol groups at about pH 5. The mercurials absorb strongly in the ultraviolet and the reaction may be used quantitatively.

1. Velick,S.F. (1953) J. Biol. Chem. **203**, 563.

p-Chloromercuribenzoic acid (^{203}Hg)

This reagent is used in a similar way to *p*-chloromercuribenzenesulphonic acid for labelling thiol residues with ^{203}Hg.

1. Waterman,M.R. (1974) Biochim. Biophys. Acta **371**, 159.
2. Boyer,P.D. (1954) J. Am. Chem. Soc. **76**, 4331.
3. Bucci,E. and Fronticelli,C. (1965) J. Biol. Chem. **240**, PC551.
4. Guha,A., Englard,S. and Listowsky,I. (1968) J. Biol. Chem. **243**, 609.
5. Erwin,V.G. and Pedersen,P.L. (1968) Anal. Biochem. **25**, 477.

Dansyl chloride (5-dimethylamino-1-naphthalene sulphonyl chloride) (^{3}H or ^{14}C)

Dansyl chloride reacts with amino, thiol, imidazole and phenolic hydroxyl groups. In general, the reaction with aliphatic hydroxyl groups is very slow.

This reagent is used widely to detect very small weights of protein by means of the intense fluorescence of the sulphonamide formed when it reacts with the terminal amino group of proteins or peptides. These sulphonamides are stable in hot acid and the assay method using dansyl chloride has nearly one hundred-fold greater sensitivity than that using 1-fluoro-2,4-dinitrobenzene.

1. Chen,R.F. (1968) Anal. Biochem. **25**, 412.
2. Gray,W.R. and Hartley,B.S. (1963) Biochem. J. **89**, 59P.
3. Schultz,R.M. and Wassarman,P.M. (1977) Anal. Biochem. **77**, 25.
4. Venn,R.F., Basford,J.M. and Curtis,C.G. (1978) Anal. Biochem. **87**, 278.
5. Airhart,J., Kelley,J., Brayden,J.E. and Low,R.B. (1979) Anal. Biochem. **96**, 45.

DFP (Di-isopropyl phosphorofluoridate) (^{3}H)

DFP is a specific reagent for serine residues. It is a pseudosubstrate reacting with active-site serine residues in many proteases and esterases. The electrophilic phosphorus reacts with the hydroxyl group forming a stable modified enzyme.

1. Darzynkiewicz,Z. and Barnard,E.A. (1967) Nature (London) **213**, 1198.
2. Budd,G.C., Darzynkiewicz,Z. and Barnard,E.A. (1967) Nature (London) **213**, 1202.
3. Ostrowski,K., Barnard,E.A., Darzynkiewicz,Z. and Rymaszewska,D. (1964) Exp. Cell Res. **36**, 43.
4. Rogers,A.W., Darzynkiewicz,Z., Salpeter,M.M., Ostrowski,K. and Barnard,E.A. (1969) J. Cell Biol. **41**, 665.
5. Fischer,E.P. and Thompson,K.S. (1979) J. Biol. Chem. **254**, 50.

Ethyl acetimidate (^{14}C)

This reagent reacts specifically with free amino groups under relatively mild conditions. It penetrates cells without impairing membrane function, and labels the protein under physiological conditions.

Ethyl acetimidate is rapidly hydrolysed by water, but the products do not interfere with the main reaction.

269

1. Hunter,M.J. and Ludwig,M.L. (1962) J. Am. Chem. Soc. **84**, 3491.
2. Whiteley,N.M. and Berg,H.C. (1974) J. Mol. Biol. **87**, 541.

N-Ethylmaleimide (^{14}C)

This reagent reacts specifically with more exposed thiol groups and may be used over a wide temperature range at neutral pH. Although *N*-ethylmaleimide may be used quantitatively to determine thiol groups, it has also been widely used to study the effects on enzyme activity of substitution of active-site thiol groups.

1. Sekine,T., Barnett,L.M. and Kielley,W.W. (1962) J. Biol. Chem. **237**, 2769.
2. Lai,Tzen-son (1971) J. Chin. Chem. Soc. (Taipei) **18** (3), 145.
3. Barns,R.J. and Keech,D.B. (1968) Biochim. Biophys. Acta **159**, 514.
4. Yamada,S. and Ikemoto,N. (1978) J. Biol. Chem. **253**, 6801.
5. Kielley,W.W. and Barnett,L.M. (1961) Biochim. Biophys. Acta **51**, 589.
6. Riggs,A. (1961) J. Biol. Chem. **236**, 1948.
7. Gadasi,H., Maruta,H., Collins,J.H. and Korn,E.D. (1979) J. Biol. Chem. **254**, 3631.

1-Fluoro-2,4-dinitrobenzene (^{3}H or ^{14}C)

This reagent is widely used under mildly alkaline conditions to identify terminal amino acids. The *N*-terminal residue can be separated after hydrolysis of the modified peptide and identified by comparison with standards.

At strongly alkaline pH, 1-fluoro-2,4-dinitrobenzene also reacts with phenolic, thiol and imidazole groups, but these dinitrophenyl groups may be displaced by treatment of the modified protein at pH 8 with 2-mercaptoethanol.

1. Whitehead,J.K. (1961) Biochem. J. **80**, 35P.
2. Schultz,R.M., Bleil,J.D. and Wassarman,P.M. (1978) Anal. Biochem. **91**, 354.
3. Travis,J. and McElroy,W.D. (1966) Biochemistry **5**, 2170.
4. Gerber,G.B. and Remy-Defraigne,J. (1965) Anal. Biochem. **11**, 386.

Formaldehyde (^{14}C)

Formaldehyde is used primarily with a reducing agent such as sodium cyanoborohydride in reductive methylation of free amino groups. However, because of its high reactivity and water solubility it may also be used to bring about crosslinking by reaction with thiol and amino groups.

1. Dottavio-Martin,D. and Ravel,J.M. (1978) Anal. Biochem. **87**, 562.
2. Rice,R.H. and Means,G.E. (1971) J. Biol. Chem. **246**, 831.
3. Nelles,L.P. and Bamburg,J.R. (1979) Anal. Biochem. **94**, 150.
4. Peterson,D.T., Merrick,W.C. and Safer,B. (1979) J. Biol. Chem. **254**, 2509.
5. Tolleshaug,H., Berg,T., Frölich,W. and Norum,K.R. (1979) Biochim. Biophys. Acta **585**, 71.
6. MacKeen,L.A., DiPeri,C. and Schwartz,I. (1979) FEBS Lett. **101**, 387.

Iodine (^{125}I)

Usually [^{125}I]-iodine is used for labelling proteins because it has a longer half-life than ^{131}I (60 days compared to 8 days for ^{131}I) and is safer to use since the γ-rays are much less penetrating. Iodination is carried out using sodium iodide in an oxidising environment which favours the formation of the cation $^{+}$I. The formation of these ions can be catalysed either chemically or enzymatically although the choice of conditions is critical if protein activity is to be preserved (1). Tyrosine is the amino acid most commonly modified though in more alkaline conditions histidine is also iodinated.

An alternative very mild procedure for labelling with ^{125}I uses the Bolton and Hunter reagent which is specific for amino groups (p. 266).

1. Bolton,A.E. (1977) Radioiodination techniques (Review 18), Radiochemical Centre, Amersham, England.
2. Samols,E. and Williams,H.S. (1961) Nature (London) **190**, 1211.

271

3. Greenwood,F.C., Hunter,W.M. and Glover,J.S. (1963) Biochem. J. **89**, 114.
4. Redshaw,M.R. and Lynch,S.S. (1974) J. Endocrinol. **60**, 527.
5. Thorell,J.I. and Johansson,B.G. (1971) Biochim. Biophys. Acta **251**, 299.

Iodoacetamide (^{14}C)

Iodoacetamide reacts with cysteine residues. This carboxyamidation reaction is often used to convert every cysteine residue within a protein to a derivative that is stable to acid hydrolysis except in the presence of oxygen.

1. Truitt,C.D., Hermodson,M.A. and Zalkin,H. (1978) J. Biol. Chem. **253**, 8470.
2. Inagami,T. (1965) J. Biol. Chem. **240**, PC 3453.
3. Inagami,T. and Hatano,H. (1969) J. Biol. Chem. **244**, 1176.
4. Heinrikson,R.L. (1966) J. Biol. Chem. **241**, 1393.
5. Anderson,J.M. (1979) J. Biol. Chem. **254**, 939.
6. Toste,A.P. and Cooke,R. (1979) Anal. Biochem. **95**, 317.
7. Nusgens,B. and Lapiere, Ch.M. (1979) Anal. Biochem. **95**, 406.
8. Kröger,M., Sternbach,H. and Cramer,F. (1979) Eur. J. Biochem. **95**, 341.

Iodoacetic acid (^{3}H or ^{14}C)

Iodoacetic acid is the most reactive of the haloacetic acids (see bromoacetic acid and chloroacetic acid). It reacts primarily with cysteine residues but is not very selective. With decreasing pH, the reactivity increases and iodoacetic acid will also react with methionine, histidine, lysine, aspartate and glutamate residues.

1. Takahashi,K., Stein,W.H. and Moore,S. (1967) J. Biol. Chem. **242**, 4682.
2. Baldwin,G.S., Waley,S.G. and Abraham,E.P. (1979) Biochem. J. **179**, 459.
3. Harris,I., Meriwether,B.P. and Harting Park, J. (1963) Nature (London) **198**, 154.
4. Colman,R.F. (1968) J. Biol. Chem. **243**, 2454.
5. Price,P.A., Moore,S., and Stein,W.H. (1969) J. Biol. Chem. **244**, 924.
6. Li,T.K. and Vallee,B.L. (1965) Biochemistry **4**, 1195.
7. Crestfield,A.M., Stein,W.H. and Moore,S. (1963) J. Biol. Chem. **238**, 2413.
8. Neumann,R.P., Moore,S. and Stein,W.H. (1962) Biochemistry **1**, 68.
9. Weinryb,I. (1968) Arch. Biochem. Biophys. **124**, 285.

10. Baldwin,G.S., Waley,S.G. and Abraham,E.P. (1979) Biochem. J. **179**, 459.
11. Wiman,K., Trägardh,L., Rask,L. and Peterson,P.A. (1979) Eur. J. Biochem. **95**, 265.
12. Holmgren,A. (1979) J. Biol. Chem. **254**, 3664.
13. Anderson,P.J. (1979) Biochem. J. **179**, 425.

Isethionyl acetimidate (^{14}C)

Unlike ethyl acetimidate (see earlier), isethionyl acetimidate is unable to penetrate intact cells and may therefore be used to label the proteins on the outer surfaces of the membranes. It reacts specifically with free amino groups under relatively mild conditions.

For references to the use of this material, see ethyl acetimidate.

Maleic anhydride (^{14}C)

This reagent may be used for the reversible alkylation of amino groups. Maleyl proteins tend to be water soluble and stable at neutral pH but are rapidly hydrolysed on acidification. This hydrolysis is more rapid than that of the corresponding succinyl derivatives (see later).

The much slower maleylation of thiol groups is not reversed by acidification.

1. Butler,P.J.G., Harris,J.I., Hartley,B.S. and Leberman,R. (1969) Biochem. J. **112**, 679.

Methyl 3,5-diiodohydroxybenzimidate (^{125}I)

This reagent is used to label lysine residues and terminal amino groups. It is a milder reagent than the Bolton and Hunter reagent, and has the advantage of preserving the charge on the protein.

1. Wood,F.T., Wu,M.M. and Gerhart,J.C. (1975) Anal. Biochem. **69**, 339.
2. Ulevitch,R.J. (1978) Immunochemistry **15**, 157.
3. Morgan,J.L., Holladay,C.R. and Spooner,B.S. (1978) FEBS Lett. **93**, 141.
4. Morgan,J.L., Holladay,C.R. and Spooner,B.S. (1978) Proc. Nat. Acad. Sci. U.S.A. **75**, 1414.
5. Miller,N.E., Weinstein,D.B. and Steinberg,D. (1978) J. Lipid Res. **19**, 644.
6. Subramani,S., Bothwell,M.A., Gibbons,I., Yang,Y.R. and Schachman,H.K. (1977) Proc. Nat. Acad. Sci. U.S.A. **74**, 3777.

Phenyl isothiocyanate (^{14}C or ^{35}S)

This reagent reacts preferentially with terminal amino groups and is used to sequence peptides by step-wise degradation (Edman degradation). Each phenylthiohydantoin formed can be identified by comparison with standards.

1. Callewaert,G.L. and Vernon,C.A. (1968) Biochem. J. **107**, 728.
2. Laver,W.G. (1961) Biochim. Biophys. Acta **53**, 469.

3. Laver,W.G. (1961) Virology **14**, 499.
4. Geising,W. and Hornle,S. (1973) *in* Peptides (Proceedings of the 11th Peptide Symposium, 1971), published 1973, p. 146 (in German).
5. Levy,N.L. and Dawson,J.R. (1976) J. Immunol. **116**, 1526.

Potassium borohydride (^{3}H)

This reagent may be used as an alternative to sodium borohydride (see below) in reductive methylation of free amino groups. It has the advantage of being water soluble and stable in aqueous solution for short periods, unlike sodium borohydride which is hydrolysed almost instantaneously.

$$
\begin{array}{c}
\text{O} \\
\parallel \\
-\text{C}-\text{NH}-\underset{\overset{\mid}{\underset{\overset{\mid}{\text{NH}}}{\text{C}=\text{O}}}}{\text{CH}}-(\text{CH}_2)_4-\text{NH}_2 \;+\; \text{HCHO} \longrightarrow
\begin{array}{c}
\text{O} \\
\parallel \\
-\text{C}-\text{NH}-\underset{\overset{\mid}{\underset{\overset{\mid}{\text{NH}}}{\text{C}=\text{O}}}}{\text{CH}}-(\text{CH}_2)_4-\text{N}=\text{CH}_2
\end{array}
\end{array}
$$

$$\downarrow \text{KBH}_4$$

$$
\begin{array}{c}
\text{O} \\
\parallel \\
-\text{C}-\text{NH}-\underset{\overset{\mid}{\underset{\overset{\mid}{\text{NH}}}{\text{C}=\text{O}}}}{\text{CH}}-(\text{CH}_2)_4-\text{NH}-\text{CH}_3
\end{array}
$$

1. Kumarasamy,R. and Symons,R.H. (1979) Anal. Biochem. **95**, 359.

Sodium borohydride (^{3}H)

This reagent is used together with formaldehyde (see earlier) in the reductive methylation of free amino groups. The formaldehyde reacts with the amino group to form a Schiff's base which is then reduced with the borohydride. Using tritiated sodium borohydride the N-[^{3}H]-methyl derivative is produced.

$$
\begin{array}{c}
\text{O} \\
\parallel \\
-\text{C}-\text{NH}-\underset{\overset{\mid}{\underset{\overset{\mid}{\text{NH}}}{\text{C}=\text{O}}}}{\text{CH}}-(\text{CH}_2)_4-\text{NH}_2 \;+\; \text{HCHO} \longrightarrow
\begin{array}{c}
\text{O} \\
\parallel \\
-\text{C}-\text{NH}-\underset{\overset{\mid}{\underset{\overset{\mid}{\text{NH}}}{\text{C}=\text{O}}}}{\text{CH}}-(\text{CH}_2)_4-\text{N}=\text{CH}_2
\end{array}
\end{array}
$$

$$\downarrow \text{NaBH}_4$$

$$
\begin{array}{c}
\text{O} \\
\parallel \\
-\text{C}-\text{NH}-\underset{\overset{\mid}{\underset{\overset{\mid}{\text{NH}}}{\text{C}=\text{O}}}}{\text{CH}}-(\text{CH}_2)_4-\text{NH}-\text{CH}_3
\end{array}
$$

1. Biocca,S., Calissano,P., Barra,D. and Fasella,P.M. (1978) Anal. Biochem. **87**, 334.
2. De La Llosa,P., Marche,P., Morgat,J.L. and De La Llossa-Hermier,M.P. (1974) FEBS Lett. **45**, 162.
3. Means,G.E. and Feeney,R.E. (1968) Biochemistry **7**, 2192.
4. Moore,G. and Crichton,R.R. (1973) FEBS Lett. **37**, 74.

5. Chansel,D., Sraer,J., Morgat,J.L., Hesch,R.D. and Ardaillou,R. (1977) FEBS Lett. **78**, 237.
6. Ascoli,M. and Puett,D. (1974) Biochim. Biophys. Acta **371**, 203.
7. Keul,V., Kaeppeli,F., Ghosh,C., Krebs,T., Robinson,J.A. and Retey,J. (1979) J. Biol. Chem. **254**, 843.

Succinic anhydride (¹⁴C)

Succinylation of free amino groups is carried out in mildly alkaline solution using conditions similar to those for acetylations with acetic anhydride. However, whereas acetylation of the cationic group yields an electrically neutral product, succinylation yields an anionic product. For this reason, succinylated proteins may be more soluble than the acetylated ones.

Although succinylation is reversible under acid conditions, the reaction is less facile than hydrolysis of the corresponding maleylated proteins (see maleic anhydride).

1. Habeeb,A.F.S.A., Cassidy,H.G. and Singer,S.J. (1958) Biochim. Biophys. Acta **29**, 587.
2. Chu,F.S., Crary,E. and Bergdoll,M.S. (1969) Biochemistry **8**, 2890.
3. Frist,R.H., Bendet,I.J., Smith,K.M. and Lauffer,M.A. (1965) Virology **26**, 558.

N-Succinimidyl propionate (³H)

This reagent is specific for free amino groups and reacts in an analogous manner to the Bolton and Hunter reagent (see earlier). It has the advantage of being a smaller molecule than the Bolton and Hunter reagent and hence causes less alteration to the protein structure.

No publication describing the use of this compound has so far appeared in the literature. However the reagent is used at The Radiochemical Centre (Amersham) to prepare α-bungarotoxin, *N-[propionyl-^{3}H]*propionated. The labelled product has been found to have a biological activity comparable with that of the native protein.

ACKNOWLEDGEMENT

We wish to express our thanks to The Radiochemical Centre (Amersham, Bucks, England) for allowing us to base this bibliography on their Technical Bulletin 79/6.

APPENDIX III
Molecular Weights and Isoelectric Points of Selected Marker Proteins

The list given below is not intended to be comprehensive but rather a selection of marker proteins which are defined in terms of the number of subunits, subunit molecular weight (in the *absence* of thiol reagent)[1], and isoelectric points, and most of which are commercially available in an essentially pure form. A more comprehensive list is available from the original sources:

Molecular weights; Handbook of Biochemistry and Molecular Biology, (3rd Edition; 1976), Proteins Vol. II (Fasman,G.D., ed.), CRC Press, Cleveland, p. 326. Isoelectric points; Malamud,D. and Drysdale,J.W. (1978) Anal. Biochem. **86**, 620.

An additional list of protein molecular weights and isoelectric points has been published by Righetti,P.G. and Caravaggio,T. (1976) J. Chromatogr. **127**, 1.

Any standard proteins which are found to consist of more than one species should be used with great caution since the protein desired may be only a minor species of the mixture.

---

[1]Molecular weights of suitable polypeptides for use in SDS-PAGE in the *presence* of thiol reagents are given in *Table 6* of Chapter 1.

Appendix III

Protein	Species	Tissue	Number of Subunits	Subunit Mol. Wt.	Isoelectric Point
Adenine phospho-ribosyltransferase	Human	Erythrocyte	3	11,000	4.8
Nerve growth factor	Mouse	Salivary gland	2	13,259	9.3
Ribonuclease	Bovine	Pancreas	1	13,700	7.8
Haemoglobin	Rabbit	Erythrocyte	4	16,000	7.0
Micrococcal nuclease	*S. aureus*	-	1	16,800	9.6
β-Lactoglobulin	Bovine	Serum	2	17,500	5.2
Ceramide trihexosidase	Human	Plasma	4	22,000	3.0
Adenylate kinase	Rat	Liver (cytosol)	3	23,000	7.5
Trypsinogen	Cow	Pancreas	1	24,500	9.3
Chymotrypsinogen A	Bovine	Pancreas	1	25,700	9.2
Triosephosphate isomerase	Rabbit	Muscle	2	26,500	6.8
Galactokinase	Human	Erythrocyte	2	27,000	5.7
Arginase	Human	Liver	4	30,000	9.2
Deoxyribonuclease I	Cow	Pancreas	1	31,000	4.8
Uricase	Pig	Liver	4	32,000	6.3
Glycerol-3-phosphate dehydrogenase	Rabbit	Kidney	2	34,000	6.4
Malate dehydrogenase	Pig	Heart	2	35,000	5.1
Alcohol dehydrogenase	Yeast	-	4	35,000	5.4
Deoxyribonuclease II	Pig	Spleen	1	38,000	10.2
Aldolase	Yeast	-	2	40,000	5.2
Pepsinogen	Pig	Stomach	1	41,000	3.7
Hexokinase	Yeast	-	2	51,000	5.3
Lipoxidase	Soybean	-	2	54,000	5.7
Catalase	Cow	Liver	4	57,500	5.4
Alkaline phosphatase	Calf	Intestine	2	69,000	4.4
Acetylcholinesterase	*Electrophorus*	-	4	70,000	4.5
Glyceraldehyde-3-phosphate dehydrogenase	Rabbit	Muscle	2	72,000	8.5
β-Glucuronidase	Rat	Liver	4	75,000	6.0
Lysine decarboxylase	*E. coli*	-	10	80,000	4.6
Glycogen synthetase	Pig	Kidney	4	92,000	4.8
Phosphoenolpyruvate carboxylase	*E. coli*	-	4	99,600	5.0
Phosphoenolpyruvate carboxylase	Spinach	Leaf	2	130,000	4.9
Urease	Jack bean	-	2	240,000	4.9

APPENDIX IV
Suppliers of Specialist Items for Electrophoresis

Many of the larger companies have subsidiaries in other countries whilst most of the smaller companies market their products through agents. The name of a local supplier is most easily obtained by writing to the relevant address listed here, which is usually the head office.

W & R Balston Ltd.; Springfield Mill, Maidstone, Kent, England.

BDH Chemicals Ltd.; Poole BH12 4NN, Dorset, England.

Bio-Rad Laboratories; 2200 Wright Avenue, Richmond, Calif. 94804, U.S.A.

Boehringer Mannheim GmBH Biochemica; P.O. Box 310120, 6800 Mannheim 31, West Germany.

Buchler Scientific Instruments; Fort Lee, N.J. 07024, U.S.A.

Calbiochem-Behring; La Jolla, Calif. 92037, U.S.A.

Difco Laboratories Ltd.; P.O. Box 14B, Central Avenue, East Moseley, Surrey, England.

Eastman Kodak Co.; 343 State Street, Rochester, N.Y. 14650, U.S.A.

Fisher Scientific; Pittsburgh, Pa. 15219, U.S.A.

Gelman Sciences Incorporated;
600 South Wagner Road, Ann Arbor, Mich. 48106, U.S.A.
12 Peter Road, Lancing, Sussex, England.

Gibco Biocult Ltd.; 3 Washington Road, Sandyford Industrial Estate, Paisley, Renfrewshire PA3 4EP, Scotland.

Hamilton Bonaduz AG; CH 7402 Bonaduz, Switzerland.

Hamilton Co.; Reno, Nev., U.S.A.

Hanimex (UK); Faraday Road, Dorcan, Swindon SN3 5HW, Wiltshire, England.

Hoefer Scientific Instruments; 520 Bryant Street, San Francisco, Calif. 94107, U.S.A.

Ilford Ltd.; Christopher Martin Road, Basildon SS14 3ET, Essex, England.

Ingold Instrumentation; 113 Hartwell Avenue, Lexington, Mass. 02173, U.S.A.

ISCO; P.O. Box 5347, Lincoln, Neb. 68505, U.S.A.

Joyce-Loebl Ltd.; 48 Princes Way, Team Valley, Gateshead on Tyne NE11 OUJ, England.

LKB Produkta AB; S-16125, Bromma, Sweden.

Marine Colloids Division, FMC Corporation; P.O. Box 308, Rockland, Mass. 04841, U.S.A.

Microchemical Specialities; Berkley, Calif. 94707, U.S.A.

Miles Research Products Division, Miles Laboratories Ltd.;
Stoke Poges, Slough SL2 4LY, England.
Elkhart, Ind. 46514, U.S.A.

Millipore Corporation;
Bedford, Mass. 01730, U.S.A.
Millipore House, Abbey Road, London NW10 7SP, England.

MRA Corporation; Clearwater, Fla. 33515, U.S.A.

National Technical Information Service (NTIS); Springfield, Va. 22161, U.S.A.

New England Nuclear Research Products;
  NEN Chemicals GmBH, Postfach 401240, 6072 Dreieich, West Germany.
  NEN Corporation, Boston, Mass., U.S.A.
Nordic Immunological Laboratories; P.O. Box 54, Maidenhead, Berkshire, England.
Oxoid Ltd.; Wade Road, Basingstoke RG24 OPW, Hants, England.
Pharmacia Fine Chemicals AB; P.O. Box 175, S-75104, Uppsala 1, Sweden.
Polysciences; Paul Valley Industrial Park, Warrington, Pa. 18976, U.S.A.
Radiochemical Centre; Amersham, Buckinghamshire, England.
Raven Scientific Ltd.; Sturmer End, Haverhill, Suffolk, England.
RND Optical Systems Inc.; 2466 Bowersox Road, New Windsor, Md. 21776, U.S.A.
Savant Instruments; Hicksville, N.Y. 11801, U.S.A.
Serva Feinbiochemica GmBH & Co.; D-6900, Heidelberg 1, P.O. Box 105260, West Germany.
Shandon Southern; 93-96 Chadwick Road, Astmoor Industrial Estate, Runcorn, Cheshire, England.
Sigma Chemical Co.; 3500 Dekalb Street, St. Louis, Mo. 63118, U.S.A.
Small Parts Inc.; 6901 N.E., 3rd Avenue, Miami, Fla. 33138, U.S.A.
Thomas A.H.; Philadelphia, Pa., U.S.A.
Union Carbide Ltd.;
  Stamford, Conn., U.S.A.
  Peter House, Oxford Street, Manchester 1, England.
Wellcome Reagents, The Wellcome Foundation Ltd.; 303 Hithergreen Lane, London SE13 6TL, England.

U.V. scanning of polyacrylamide gels,
44

Valence determination, 131-132

Y_o, 13-14, 126-128

Zone electrophoresis, 1
Zone-sharpening, 7-11, 39

Forthcoming companion volume

Gel Electrophoresis of Nucleic Acids:
A Practical Approach

Edited by D. Rickwood and B. Hames

The introduction of polyacrylamide gel electrophoresis, originally for the size separation of RNA, marked the beginning of an increasingly significant role for electrophoresis. Currently gel electrophoresis is a key method in the field of nucleic acid research. Gel electrophoresis is important not only for separating different size molecules of DNA and RNA but also for the separation of nucleoproteins as well as for the sequencing of DNA. This book covers all aspects of the gel electrophoresis of nucleic acids with the emphasis on the practical aspects of these procedures.

200 pp (est).
0 904 147 24 X (soft) £7.50/U.S.$18.00

Also available ......

Centrifugation: A Practical Approach
Edited by D. Rickwood

"Centrifugation is a fundamental laboratory technique, and this book provides an excellent general guide for practical experiments in both established and new areas . . . All aspects of centrifugation are covered and the book will appeal to a wide range of readers, from beginners to research workers experienced in the art . . . This book is highly recommended and should become the reference text on modern centrifugation techniques."
Separation News

"As there is a paucity of publications covering the same field at an elementary level, this book should find wide acceptance, especially as the price of the paperback should encourage younger workers to have a personal copy."
Nature

224 pp. 40 illus., 1978
0 904 147 10 X (hard) £8.00/U.S.$16.00
0 904 147 11 X (soft) £4.00/U.S.$9.00

 IRL Press Ltd

1 Abbey Street, Eynsham, Oxford, OX8 1JJ, England
1911 Jefferson Davis Highway, Arlington, VA 22202, USA

Research journals from IRL Press

The EMBO Journal
Monthly – starting 1982
Executive Editor: J. Tooze

An authoritative new journal to be published on behalf of the European Molecular Biology Organization. *The EMBO Journal* aims to provide rapid publication for reports of significant, original research in molecular biology and related areas. A free 16-page booklet containing full details of the journal is available on request.

Current Eye Research
Monthly
Executive Editors: C.A. Paterson and N.A. Delamere

The first rapid publication journal for the international eye research community. Covers all aspects of eye research including clinical research, anatomy, physiology, biophysics, biochemistry, pharmacology, developmental biology, microbiology and immunology.

Carcinogenesis
Monthly
Executive Editors: A. Dipple and R.C. Garner

A major multi-disciplinary journal designed to bring together all the varied aspects of research which will ultimately lead to the prevention of cancer in man. The journal publishes full papers and short communications which warrant prompt publication (within 9 weeks of acceptance) in the areas of carcinogenesis; mutagenesis, factors modifying these processes such as DNA repair, genetics and nutrition, metabolism of carcinogens; the mechanism of action of carcinogens and promoting agents; epidemiological studies; and the formation, detection, identification and quantification of environmental carcinogens.

Nucleic Acids Research
24 issues p.a.
Executive Editors: R.T. Walker, D. Söll, M. Deutscher, T. Platt and A.M. Weiner

The leading international journal for the rapid publication of research on the physical, chemical, biochemical and biological aspects of nucleic acids.

Journal of Plankton Research
Quarterly
Executive Editor: D.H. Cushing

Deals with both zoo- and phytoplankton in marine, freshwater and brackish environments.

Chemical Senses
Quarterly
Executive Editors: H.R. Moskowitz, E.P. Koster and J. Boeckh

An international journal covering taste, smell and all aspects of chemoreception.

Send for more information and a free sample copy of the journal which covers your field of interest.

 Published by IRL Press Ltd.